T0074581

HOW WE AGE

How We Age

THE SCIENCE OF LONGEVITY

COLEEN T. MURPHY

PRINCETON UNIVERSITY PRESS

PRINCETON & OXFORD

Published by Princeton University Press
41 William Street, Princeton, New Jersey 08540
99 Banbury Road, Oxford OX2 6JX

press.princeton.edu

All Rights Reserved
ISBN: 978-0-691-18263-6
ISBN (e-book): 978-0-691-25033-5

British Library Cataloging-in-Publication Data is available

Editorial: Alison Kalett and Hallie Schaeffer
Production Editorial: Jenny Wolkowicki
Cover design: Karl Spurzem
Production: Danielle Amatucci
Publicity: Matthew Taylor and Kate Farquhar-Thomson

Jacket image: Dmitry Markov152 / Shutterstock

This book has been composed in Arno Pro

10 9 8 7 6 5 4 3 2 1

This book is dedicated to my parents

CONTENTS

ACKNOWLEDGMENTS

THIS BOOK TOOK me several years to write and is largely based on information I gathered while teaching my class, "Molecular Mechanisms of Longevity: The Genetics, Genomics, and Cell Biology of Aging," at Princeton University. Together, my students and I learned a great deal about the aging and longevity field, far beyond my own lab's research. I thank all of the students of the class over the years for their enthusiasm and their contributions. This class changed the way I think about the field.

Speaking of the aging field, I thank you all for doing such interesting work and being excellent colleagues who love to share their work. I have tried my best to cover as many relevant papers as I can in these chapters (a recent combined reference count was something close to a thousand citations) but I am sure I have missed something, and I apologize for that. I'd like to thank two anonymous reviewers and Nicholas Stroustrup for their suggestions. The aging and longevity field moves fast; every week there is another amazing discovery, and I'm sure that the week after I submit my final draft, there will be a paper that I wish I had time to include—but I have to stop somewhere!

I also want to thank the members of my lab. Not only did they do much of the work that I describe in detail in various sections of the book, but they also made comments on early versions of the manuscript (in particular, Dr. Rachel Kaletsky and Dr. Titas Sengupta). But beyond that, they make coming to work fun—every day is the chance to discover something that no one else has seen before, and I love to share that moment with my trainees, who are truly colleagues, as well. I can't imagine a better job than getting to try to solve these puzzles together, especially since what we find might help someone's life someday.

My family—my husband Zemer, son Benjamin, and daughter Lena— might be even more excited than I am when this book is finalized. They deserve a medal for every time they've had to hear me explain that I need to work on the book again—they have been very patient! But they ALSO read drafts (my daughter started reading it when she was ten, and then designed several beautiful book cover suggestions), engaged in lively discussions, and gave me constructive criticism and many suggestions, which I deeply appreciate. Zemer, a bacterial cell biologist himself, was a particularly good critic, and he raised excellent points that I hope I have sufficiently addressed.

Finally, I would like to thank my parents, Dr. John J. and Constance K. Murphy, for their support. They encouraged my interest in STEM at a young age, even fighting with the school board many times to make sure I got a better math education than our rural school district wanted to offer, and pressuring them to offer options that weren't given to girls back then. Because of my parents, it never occurred to me that there was something I could not do, and I hope I convey that message to other women and girls as well. Moreover, my parents are excellent role models in every possible way, demonstrating the importance of education and hard work. Even now in retirement, they are role models for healthy aging, as they are busy every day on their farm, prepping for weekend farmer's markets, arranging flowers for weddings, teaching ballroom dancing classes, volunteering for Meals-on-Wheels, teaching Master Gardening classes, and just continuing to be interesting individuals every day. I thank them for a lifetime of love and support.

AD	Alzheimer's disease
AGE	advanced glycation end product
AICAR	5-aminoimidazole-4-carboxamide ribonucleoside
ALS	amyotrophic lateral sclerosis
AMH	anti-Mullerian hormone
AMHR	anti-Mullerian hormone reeptor
AMP	adenosine monophosphate
AMPK	adenosine monophosphate kinase
APP	amyloid precursor protein
ART	artificial reproductive technology
ASD	autism spectrum disorder
ATP	adenosine triphosphate
BCAAS	branched chain amino acids
BLSA	Baltimore Longitudinal Study of Aging
BMI	body mass index
CALERIE	Comprehensive Assessment of Long-Term Effects of Reducing Calorie Intake study
CDC	Centers for Disease Control
CHD	coronary heart disease
circRNA	circular RNA
CJD	Creutzfeldt-Jakob disease
CLS	chronological lifespan
CMA	chaperone-mediated autophagy

CR	caloric restriction
CRS	Caloric Restriction Society
CSF	cerebrospinal fluid
CVD	cardiovascular disease
DAE	DAF-16 associated element
DAMPS	mitochondrial-derived damage-associated molecular patterns
DBE	DAF-16 binding element
DHEA	dehydroepiandrosterone
DNA	deoxyribonucleic acid
DNAM	DNA methylation
DNP	dinitrophenol
DR	dietary restriction
ER	endoplasmic reticulum
ERV	endogenous retrovirus
ESCS	embryonic stem cells
ETC	electron transport chain
FDA	Food and Drug Administration
FGF21	fibroblast growth factor 21
FMD	fasting-mimicking diet
FMT	fecal microbial transplant
GHR	growth hormone receptor
GWAS	genome-wide association studies
HA	hyaluronic acid
HDL	high density lipoprotein
HLA	human leukocyte antigens
HMM	hidden Markov models
HMW	high molecular weight
HSCS	hematopoietic stem cells
HSPS	heat shock proteins

IBD	inflammatory bowel disease
IF	intermittent fasting
IGF-1	insulin-like growth factor 1
iGWAS	informed GWAS
IIS	insulin/IGF signaling
IPSCS	induced pluripotent stem cells
IRS	IGF-1 receptor substrate
ISR	integrated stress response
ITP	Intervention Testing Program
IVF	in vitro fertilization
LLFS	Long Life Family Study
LQ	longevity quotient
LS	Laron syndrome
MDPS	mitochondrial-derived peptides
MEP	mother enrichment program
MHC	major histocompatibility complex
MHL	major histocompatibility locus
MHT	menopausal hormone therapy
miRNA	mitochondrial RNA
MP	male pheromone
mRNA	messenger RNA
MTDNA	mitochondrial DNA
mTOR	mammalian/mechanistic TOR
NAD	nicotinamide adenine dinucleotide
NDGA	nordihydroguaiaretic acid
NFTS	neurofibrillary tangles
NIA	National Institute on Aging
NIH	National Institutes of Health
NLS	nuclear localization signal
NMD	nonsense-mediated decay

NR	nicotinamide riboside
NSCS	neural stem cells
OSNS	olfactory sensory neurons
OTU	operational taxonomic unit
OXPHOS	oxidative phosphorylation
PCR	polymerase chain reaction
PTSD	post-traumatic stress disorder
RAGE	receptor for AGEs
rDNA	ribosomal DNA
RLS	replicative lifespan
RNA	ribonucleic acid
RNAi	RNA interference
ROS	reactive oxygen species
rRNA	ribosomal RNA
SASP	senescence-associated secretory phenotype
SCFA	short-chain fatty acid
SCPAB	Simons Foundation's Collaboration on Plasticity in the Aging Brain
SLAM	Study of Longitudinal Aging in Mice
SNPS	single nucleotide polymorphisms
SOD	superoxide dismutase
SPPB	Short Physical Performance Battery
sRAGE	soluble receptor for advanced glycation end product
TAME	Targeting Aging with Metformin trial
TE	transposable element
TEI	transgenerational epigenetic inheritance
TERT	telomerase reverse transcriptase
TGF-BETA	transforming growth factor beta
TOR	target of rapamycin
TRE	time-restricted eating

TRF	time-restricted feeding
UCSF	University of California, San Francisco
UPR	unfolded protein response
UPS	ubiquitin proteasome system
UW	University of Wisconsin, Madison
VNO	vomeronasal organ
WGS	whole-genome sequencing

HOW WE AGE

Introduction

What doesn't fit is often what is getting at something exciting!

—DR. EVELYN WITKIN, AMERICAN GENETICIST
WHO TURNED 100 ON MARCH 9, 2021

IN THE LATE 1990S, I was a graduate student in the lab of Jim Spudich, in the Department of Biochemistry at Stanford University. I studied how the motor protein myosin—the molecular motor that powers our muscles and makes our hearts pump—works, by swapping parts from myosins of "slow" and "fast" organisms, and then testing how those swaps affected its activity. I loved that protein; understanding how a sequence of amino acids arranged the right way could take energy and turn it into movement by swinging its "lever arm" a small distance was one of the most interesting questions I could imagine at the time. But when I explained my research to people at parties who asked me, "What do you do?" they would nod and politely smile, then ask when I would graduate. That would be the end of the discussion.

That all changed a few months later after I heard a fantastic talk by Dr. Cynthia Kenyon, a professor from the University of California, San Francisco (UCSF). Cynthia is a lively, engaging speaker and she told the audience about her lab's work on aging and longevity in a small worm, the nematode *Caenorhabditis elegans*. Her lab had found that changing a *single gene* could double the lifespan of these animals,[1] and she showed movies of the mutant worms crawling around at an age when normal worms were already decrepit and dying. This was an "Aha!" moment that made it clear that she wasn't talking about extending the end of life, but rather the youthful, healthy part of life, an outcome that we would all like to experience. That gene, called *daf-2*, turned out to encode an insulin/IGF-1 receptor, meaning it could matter for people, too, since our bodies also have insulin. After hearing her talk, I knew what I wanted to do: find out how those mutant worms were so healthy. Soon after, I asked Cynthia if I could come to her lab for my postdoctoral research,[2] and she agreed. At that point, when people asked me what I was going to do, there was a noticeable difference. It turns out that almost everyone is interested in aging research, and everyone has an opinion about it. It quickly became obvious that one's likelihood of supporting the idea of aging research is generally correlated with one's age, and I got several exhortations to "work faster!"

I decided to write this book after developing a class at Princeton, "Molecular Mechanisms of Longevity: The Genetics, Genomics, and Cell Biology of Aging," to teach students about my research field. While preparing for that class, I realized that we (the royal We, being researchers in the field of aging and longevity) have made many molecular insights in the past two decades that would be good to convey to the general public. While the popular science market for longevity books is saturated—no one needs another celebrity's viewpoint on aging or another diet book, and several excellent introductory books already exist—at least a few people might want to have a molecular explanation of the exciting work that has been done in this arena. As I will explain, we have found out a LOT in the past two decades about how longevity is regulated, which can give us clues about how we might slow aging. We now have a better grasp of the genetic pathways and cellular processes that communicate from one cell to another how to tune the rate of aging, and we also better understand the reasons that longevity is regulated at all. These insights have then led to ideas about how to slow age-related decline, and we have some good candidates for those medicines now. Some of this excitement has recently been turned into serious biotech development, with many companies focused on longevity and aging springing up in the past few years.

I have been lucky enough to be right in the middle of things since 2000, since new genes that control longevity had just been revealed. The millennium was a real turning point: after bacteria and yeast, *C. elegans* was the first multicellular organism whose genome was sequenced, and *Drosophila* quickly followed. Those large-scale projects were a direct benefit of the approaches developed for the Human Genome Project and allowed biologists to carry out experiments that had not been previously possible on a genome-wide scale. RNA interference (RNAi), a mechanism that causes the messenger RNA (mRNA) of a gene of interest to be degraded, was first described in detail by Craig Mello and Andrew Fire in *C. elegans* in 1998,[3] and it was quickly employed by the worm field to test *all of the genes in the genome* for every characteristic of interest—including aging—through new tools to easily knock down gene expression levels.[4] This ability to rapidly test many genes in worms quickly led to an explosion of functional genomics (that is, testing of all genes in a genome for a particular activity), and the field has been expanding in many directions ever since.

I got into the aging field because I was fascinated with the question of how longevity and aging are controlled genetically and biochemically. The tools that were newly available at the time, genomic expression microarrays and RNAi, allowed a previously unimaginable ability to probe long-lived mutants (that is, animals with changes to their DNA that affect a gene) and to learn what was going on inside them. The existence of complete genome sequences for all of these organisms also ushered in new genomic approaches, such as DNA microarrays and later next-generation sequencing, allowing the analysis of every gene simultaneously and giving us unprecedented insights into the inner workings of cells as they age. The amount of data available to researchers has been exploding ever since. Genetic and genomic methods have led the way in longevity research, and large-scale studies of metabolism have added to our understanding. Meanwhile, new molecular tools, particularly the gene-editing tool CRISPR and stem-cell approaches, offer the exciting possibility that we might even modify ourselves to achieve better health.[5]

Because of the nature of the question—understanding how aging works—the field is extremely broad. One can attack the aging question from many different viewpoints: demography, population genetics, evolution, model-system genetics, molecular biology, cell biology, nutrition science, and pharmacology. All of these perspectives are helpful in understanding how aging works and whether we can slow it down. While I will tell you about my lab's work (and I'll try not to *only* talk about our work), I will also explain the latest work throughout the field. It's a fast-moving field, with new discoveries all the time, and inevitably a few things will be missed, but I'll try to give you a good understanding of not only what we know but *how* we know it—the work that was done to figure things out.

What you will not find in this book are descriptions of what I or other scientists eat, or weigh, or how often we exercise—all information that has somehow become the norm for pop-sci books and articles on aging and the researchers who work on aging. As a scientist, I can't stand reading this information—those are all "*n* of 1" experiments whose results we don't yet know, so I won't report them—it's just bad science. Additionally, I've noticed an odd cult-of-personality air about some aging books, and those cults usually leave out the contributions of female scientists. And I'm not a longevity evangelist; I'm not trying to sell you something, no supplements or drugs or diet plans. I just want to tell you what we know about aging and how we came to these conclusions.

Finally, I won't be using the popular phrase, "..., *at least in worms and flies*," which seems to pepper most books on aging. I am an unapologetic model-system advocate, for one simple reason: almost everything we know at the molecular level about the underlying mechanisms controlling (regulating) longevity is because of the work that was done first in invertebrate model systems, and then *tested later* in higher organisms (mammals like mice), a fact that is often overlooked and underreported. Beyond that, the tools that allow us to do the work, all the way up through human cells, have been identified, characterized, and tested in these simpler model systems before being adapted for use in mammals. (The most powerful yet may be CRISPR, which was first discovered in bacteria.) Without model systems, our understanding of longevity regulation would be very poor indeed. For that reason, I won't just be talking about studies of humans with some verification in mice, but I'll try to describe how we really learned about the molecular goings-on inside all of our cells, which relies on studies in small invertebrate systems. For the Sarah Palins of the world, who do not acknowledge the contributions of fundamental ("basic") research to medicine,* this will be a shock, but for the rest of you I hope it will give a fairer insight into how scientists actually learn how things work, and how we might apply what we've learned to help people live better, longer—as Palin would say, *I kid you not.*

In this book, I hope to let you know what we've discovered about longevity in recent years. But before diving into the science, I'll discuss *why* we should study aging—it's not always immediately obvious, but understanding aging could help our whole society in the long run, even economically (chapter 1)—longevity is not just for billionaires. There are many evolutionary theories about *why* we age (chapter 2), but molecular techniques are now helping us better understand this question and adjust our theories accordingly. In

---

* "You've heard about some of these pet projects, they really don't make a whole lot of sense and sometimes these dollars go to projects that have little or nothing to do with the public good. Things like fruit fly research in Paris, France. I kid you not" (Sarah Palin quoted in Adam Rutherford, "Palin and the Fruit Fly," *Guardian*, October 27, 2008, https://www.theguardian.com/commentisfree/2008/oct/27/sarahpalin-genetics-fruit-flies).

chapter 3, we'll start to see how modern genetic and genomic techniques can reveal the secrets of centenarians' long lifespans; but to experimentally test them we need to use model organisms—that is, well-studied animals we can grow in the lab and genetically manipulate so that we can test hypotheses (chapter 4). Of course, in order to study aging, we have to establish some definitions of what it means, and how we can measure these changes with age (chapter 5). In later chapters, I'll describe what we currently know about longevity pathways (chapters 6–10) and interventions in detail, so that you'll recognize the molecules that are being targeted for clinical treatment (chapter 17). Reproduction and mating are intimately linked with longevity, as I'll describe in chapters 11 and 12. What we can sense can also influence how long we live (chapter 13), while aging can affect what we can sense and our cognitive function (chapter 14). Some of the newest thoughts in the field concern how we might inherit factors from our ancestors that affect aging (chapter 15), and that what we eat and the microbes that inhabit our gut might also influence aging (chapter 16). Finally, I'll discuss the current state of longevity biotech, and how we might go about finding treatments for age-related decline (chapter 17).

We are right in the middle of the business of understanding the processes that regulate aging, and it is an exciting time because we are still in that era of discovery. I don't want to imply that we know all of the answers at this time. Instead, what I hope to convey is what we do know and, more importantly, *how* we know it, and what we might be able to do with that wealth of data. With this information at our disposal, we should all be able to make wise decisions about how to manage our own longevity.

1

Ethics and Economics
of Longevity

IS IT RIGHT TO STUDY AGING?

It's paradoxical that the idea of living a long life appeals to everyone, but the idea of getting old doesn't appeal to anyone.

—ANDY ROONEY

THE SEARCH FOR immortality cuts across all times and cultures. Ancient texts emphasize the long life of leaders that span many generations: the Old Testament says Methuselah lived 969 years, Hindu lore puts Bhishma's age at 128, the *Epic of Gilgamesh* estimates the Mesopotamian king's reign lasted for 126 years, Attila the Hun reportedly lived to be 124, and Alexander the Great was rumored to have found a river of Eden. (While Ponce de León has been linked with the search for the fountain of youth when he landed in Florida, the bigger motive seems to have been the search for gold, cheap labor, and land to provide to the Spanish crown.) Even today, in cultures without official birth records, there are many alleged supercentenarians whose age is likely quite exaggerated. (In fact, I'm pretty sure that the old woman we met in a Masai village is not 105, but who am I to say?) Nonetheless, the interest in finding the secret to long life has continued to fuel our imaginations. This search has also stoked our fears, as Oscar Wilde illustrated in *The Picture of Dorian Gray*. Similarly, the 589 years Gollum achieved by keeping his precious ring is hardly an advertisement for healthy extension of lifespan. We must ask ourselves: Is the search for longevity wise?

Today we have extreme diets, questionable treatments and medicines, cry-onics, and multiple Silicon Valley–funded efforts to identify new anti-aging treatments. This search for an aging "cure" has only accelerated as scientific advancements make the possibility of altering our fate ever more alluring—but this possibility also increases our need to distinguish quackery from legiti-mate potential aging treatments. Before we dive into explaining how we know anything about the mechanisms that may control our longevity, we should discuss the pros and cons of such research.

What We Die of Depends on Who We Are

The twentieth century was witness to perhaps the greatest increase in life ex-pectancy ever, from 48 to 74 years in men and 51 to 80 years in women in the United States.[1] Although we generally think of old age when we hear the term "life expectancy," in fact most of this drastic increase has been achieved through improved living conditions and medicines that have decreased infant and childhood mortality, not improvements in aging treatments. In general, the rise in life expectancy is due to the elimination of factors that previously cut young, healthy lives short. In fact, one of the biggest dips in life expec-tancy in recent history was caused by the 1918 influenza virus, which affected largely young people.*

Public health efforts that increase early-life and midlife survival, such as general sanitation (handwashing, clean water, toilets) and the availability of

---

* How the SARS-CoV-2/Covid-19 pandemic will affect life expectancy will not be known for a few years, but the first wave cut life expectancy for African American males up to two years; how subsequent variants will impact life expectancy remains to be seen. Older people (>65 years) were most affected early in the pandemic, but the demographics of mortality have shifted as vaccines have become available: vaccination in the >65-year-old group greatly ex-ceeded that in younger groups, just as the Delta variant arose. As a result, there was a shift toward younger deaths in later months, as total deaths exceeded those due to the 1918 flu pandemic. In June 2022, David Leonhardt suggested that, due to get-out-the-vaccine outreach efforts in the Black and Hispanic communities and antivaccination and antimasking attitudes in white (particularly Republican) communities, there was a narrowing of the racial gap, and perhaps a flip in mortality, with whites dying at a higher rate in mid-2022 (David Leonhardt, 2022, "Covid and Race," *New York Times, The Morning,* June 9, 2022, https://www.nytimes.com /2022/06/09/briefing/covid-race-deaths-america.html). However, whether this study was properly age adjusted has been questioned, and it might be an example of Simpson's paradox. In fact, Black Americans may not live long enough to even make it into those top aging brackets, because of many racial inequalities in society and medical care, irrespective of Covid.

antibiotics, have decreased overall mortality. Early-life medical interventions, most notably vaccinations against childhood diseases, have greatly reduced childhood mortality. Together, these changes in general cleanliness, public health interventions, and vaccinations have increased substantially our chances of living long. Consequently, these increases in lifespan are accompanied by a rise in age-related diseases, decreasing quality of life with age. Only recently have we increased lifespan through better treatment of late-life diseases, such as cardiovascular disease and cancer.

We must also acknowledge that while life expectancy has steadily increased, it has not always increased equally. For some populations, these improvements have not yet arrived; this is true not only in developing countries, but also for historically underserved populations in the United States.[2] One particularly egregious exception to this general increase in life expectancy is for Black men; the epidemiologist Sherman James has described the "John Henry" effect, in which an individual's success in coping with difficult conditions is accompanied by a lifetime of stress, leading to an acceleration of aging-related morbidities (e.g., heart disease, diabetes, and hypertension) and early death.[3] This discrepancy has been severely exacerbated by Covid-19, with minority populations (Native Americans, Hispanics, and African Americans) disproportionately affected early in the pandemic,[4] and Black men in particular dying at higher rates, leading to a subsequent decrease of 2.5 years in life expectancy.

For other demographics, life expectancy may have already peaked and may now be on the decline. The economist and Nobel Prize winner Sir Angus Deaton and economist Anne Case, both at Princeton, showed in 2015 that alcohol and drug use among white, middle-aged Americans has led to increased rates of overdose and suicide ("deaths of despair"), causing life expectancy to plateau and decrease in recent years. The Centers for Disease Control (CDC) reported in 2015 that, for the first time in 22 years, life expectancy of Americans had declined—and was led by declines in the life expectancy of Americans under 65. Similarly, Eileen Crimmins (University of Southern California) has found that life expectancy correlates with income in the United States;[5] in particular, states that are largely Democratic have longer lifespans, which correlates with higher incomes, less inequality, and better health care, while largely Republican states suffer more from obesity and metabolic disease, which decrease lifespan. In 2017, an analysis of voting patterns and health revealed that measures of poor health (diabetes, heavy alcohol consumption, obesity, lack of physical activity) predicted 43% of Trump's vote.[6]

Together, these findings highlight the fact that we already know how to increase average lifespan: through better preventative health care, vaccines against childhood diseases, reduced inequality, decreasing rates of smoking and obesity, reduced stress, and other lifestyle changes that can maximize the current potential lifespan . . . but these changes require a political will that extends beyond the actions of individuals, socioeconomic investment in addition to basic equity in food and job security. These investments include public health efforts to promote universal vaccination, education about behavioral risk factors, and efforts to stem the opioid epidemic. In the latter case, there is finally some good news: efforts to track every opioid pill and limit prescriptions are finally reducing opioid-related deaths. The United States could have a longer average lifespan if we were to invest more in efforts to stem inequalities.

Women live, on average, longer than men, but one danger they face that men are not exposed to is childbirth.* Pregnancy and childbirth are the biggest health risks a woman can take, and this fact is reflected in "maternal mortality," defined as deaths related to pregnancy and occurring within 42 days after the end of pregnancy. Much of this risk depends on where you live and, at least in part, what your ethnicity is. Overall, maternal mortality has declined significantly since the 1950s, and in some countries, it is almost rare. For example, maternal mortality rates have dropped to only 3 deaths per 100,000 births in Finland, Greece, Iceland, and Poland, but in parts of Africa rates are in the high hundreds per 100,000, with Sierra Leone at a staggering 1360 deaths per 100,000 births.[7] But maternal mortality is not an issue just in the developing world: the United States has an embarrassingly high maternal mortality rate[8]—the same as Qatar, and worse than all of Europe and most other developed countries. In fact, the rate has increased in recent years, from 7.2 deaths per 100,000 in 1987 to 18 per 100,000 in 2014.

Risk by age is a U-shaped curve, with lowest risk in the mid-20s. Worldwide, the highest-risk pregnancy group is girls aged 10–14, which is coupled with other medical issues in countries considered "fragile states" (by the World Health Organization)—but risk for this age group may rise in the United States as more child victims of rape are refused abortion access in some states after the Supreme Court's disastrous 2022 Dodd decision overturning *Roe v.*

---

* Of course this applies to any pregnant person, but most demographic studies have not yet included this information.

Wade (the right to abortion). Of course, other countries have recognized that legal abortion is part of health care, and have expanded access—including Ireland, Gibraltar, and San Marino in Europe, and Mexico, Argentina, Uruguay, Columbia, and Chile in the Americas—and have found that there is a corresponding drop in maternal mortality.[9] In developed countries, aging is a major factor in maternal mortality, as rates are highest in women aged 35 and over.[10] "Geriatric pregnancy" (the unpleasant term I learned when I had my first kid at age 38) contributes significantly to the increasing rates of maternal mortality. A study of women in the District of Columbia from 2008 to 2014 showed that the 40+ age-group had the highest rate of maternal mortality, at 142 per 100,000 births.

A large part of the poor statistics in the United States is due to the shockingly high rate of maternal mortality for Black women, which is four times the rate of death of white women and is often accompanied by preeclampsia during pregnancy. This issue eclipses socioeconomic factors—even tennis player Serena Williams, one of the world's top female athletes, reported being ignored by her doctors and almost dying soon after childbirth; five-time Olympic sprinter Allyson Felix also suffered from preeclampsia and had to deliver her child at just 32 weeks. Infant mortality rates and deaths of Black children after surgery parallel these trends (and in general men fare better than women after surgery). While part of this may be due to differences in some risk factors (high blood pressure, diabetes, obesity, age), and part due to differences in prenatal health care, at least some of this disparity may be due to the dismissal of Black women's health concerns by medical professionals, differences in the amount of time that doctors spend with Black patients, and quality of health care in hospitals that serve mostly Black and brown patients. Some states, such as California, have improved their rates of maternal mortality through more careful postpartum monitoring, but such measures, as well as universal prenatal care, still need to be applied uniformly across the country to improve these rates in the United States.

In short, life expectancy is still not entirely an issue of getting older people to live longer but also survival through childhood and adolescence, and, for women, making it safely through reproduction as well.*

---

* This is being written during the SARS-CoV2 pandemic, so the statistics are not yet known, but early reports have suggested higher fatality rates for older people and particularly men in China. Whether that will be true in other countries and across all ages is not yet known.

Is There a Life Expectancy Plateau?

Given the fact that we already have some clues how to live longer, it would be fair to ask: When it comes to living long, have we already hit the wall? In 2016, the biologist Jan Vijg (Albert Einstein College of Medicine) ignited a fiery debate, because he and his colleagues noted that as more and more people live to be centenarians, few of them live beyond 115 years, concluding that 115 years must be the maximum human lifespan.[11] However, as several demographers quickly pointed out, the 115-year estimate hinges on a single data point; that is, attempting to draw a line through data that include the exceptionally long lifespan of the famed French "supercentenarian" Jeanne Calment, who lived a record-breaking 122 years, skews the slope.[12] While mostly proving that controversy rather than sound science is likely to get one published sometimes, the matter highlighted how interested many people are in knowing how long we, as a species, could live. And soon thereafter, demographer James Vaupel and his colleagues suggested that mortality—that is, the odds of dying at any time—actually *decreases* by age 105.[13] If true, this would suggest that once you've made it that far, it's hard to predict when you might die. With this information in mind, it's not clear what the maximum will end up being.

One very important note that seems to get lost in these discussions of "maximum lifespan" is that current centenarians have not benefitted from any of the knowledge that we are now obtaining about the molecular underpinnings of longevity—so that "only 115" estimate, right or wrong, is *without* the benefit of aging therapeutics. The development of drugs to treat age-related decline is a major goal in the study of the molecular genetics of aging. The longevity and aging research field aims to find ways to extend healthy lifespan, which should result in an increase beyond what current demographics suggest is the maximum lifespan. Even if we have hit the maximum that we can reach with the currently available tools and medicines, the hope of the field is to move this needle, or at least that of healthy lifespan, in a positive direction.

Is It Ethical to Study Aging?

There is a tremendous interest in understanding how we age and how we might slow this process. When I started researching aging, most people I met who heard about my work were very excited and could immediately understand the point. From time to time, however, I would encounter someone who was repulsed by the idea of trying to understand the basis of aging, suggesting that it is unethical to manipulate aging rates. This debate is reignited—at least on Twitter—every time there is an announcement of another Silicon Valley billionaire-founded longevity company, and sometimes scientists in other fields suggest that funding aging research is a waste of money.

On the one hand, the critics do have a point: at its extreme, one could imagine a society where the wealthy never age because they have access to expensive drugs that allow them to stave off death, while the less well-off languish and die early. But let's be honest: to some degree, this is already a problem in the United States because of the unequal distribution of medical care and health insurance, which allows the rich a much better chance of surviving almost every disease and, as a result, living longer, as Crimmins's work highlights. This is a socioeconomic problem that we need to solve for reasons independent of any potential longevity extension therapies.

Another knock against aging research is the potential for overpopulation; but this seems unlikely, as there would be no accompanying increase in reproduction, because the affected population would all be post-reproductive (unless, of course, the aim were to actually extend the reproductive span [see chapter 8]—but this, too, would be unlikely to cause a huge spike in population). If we were to cure everyone of cancer, heart disease, and diabetes, we would also see a rise in population—but it would seem outlandish to most of us to suggest that we should stop working on any of these diseases.

And finally, there are economic issues associated with supporting an ever-older population, such as an increase in the number of retired workers. However, the goal of most serious aging research is to extend *healthy lifespan*.[14] Jay Olshansky and colleagues made the argument that healthy-lifespan extension should save money, not cost us more, in *Aging: The Longevity Dividend*.[15] Olshansky argues that by maintaining a healthy life longer, we will decrease, not increase, the economic impact of aging. In that same book, economist Dana Goldstein notes that changes to our current social security structure might need to be made to accommodate such demographic shifts,[16] so the economic concerns will need to be addressed. Nevertheless, an amplification of current disparities in health-care access should be addressed, but this is not a scientific question—we already know how to solve these problems, it just requires the political will to do so.

We Already Know How to Live Longer

When we think about how to extend lifespan, we should acknowledge that we already know ways to live to our full potential. Most of these have to do with a healthy lifestyle: eating a healthy diet, getting enough exercise and sleep, preventing and addressing medical problems, and avoiding excessive stress. In this regard, we already know one simple trick to increase one's lifespan: be rich. We can see these disparities reflected in actions that are often considered "lifestyle" (and somehow moral) choices when in fact they are the result of economic circumstances. For example, diet: we all know that we should eat more fruits and vegetables to stay healthy, but those are often expensive and often not available in the "grocery deserts" of inner cities. Moreover, rich people don't have to alter what they eat based on their monthly paycheck. Exercise is another luxury: rich people might humblebrag about how many hours they work per week, or their latest marathon or spin class, but they are not pulling a second shift to make sure their kids are fed. They might complain on Facebook or Twitter about being stressed, but probably don't worry about whether they will experience violence in their own neighborhood. They are unlikely to have to deal with the daily stress of having their legitimacy questioned at work or at school, particularly by authorities like police. And, finally, rich people get frequent preventative health care, can get treatment for their health problems outside of an emergency room, and don't go broke choosing whether to get treated or not—all of these disparities in health care have long-term consequences for lifespan. (The only thing that some rich people are doing wrong very recently is ignoring the benefits of vaccination, but their selfish reliance on herd immunity means they are unlikely to suffer any consequences.)

All of these factors—diet, exercise, sleep, stress, and proper health care—are major determinants of lifespan beyond whatever genetic hand we are dealt. In addition to these lifestyle factors, there are what we call "gene × environment" effects, in which your genetic background affects how your biology deals with a particular environmental stress, which may in turn have longevity effects. In the lab, using our model organisms, we usually replicate what might be considered the best-case scenarios, so that layers of stress induced by extreme socioeconomic differences are not usually what we test. But we should acknowledge that the best way to live at least one's maximum potential is to be rich and privileged enough to be able to have access to the best longevity-affecting lifestyle choices.

Extension of Healthy Life Is the Goal: Compression of Morbidity

We are often asked, "Won't making people live longer just stretch out the amount of time they are sick?" Maybe, but only if we do it wrong. In fact, one might argue that how we are doing it now *is* the wrong way: we are battling diseases one by one, giving drugs for one disease after another, rather than treating the source of age-related disease, aging itself. And because we don't yet understand how to prevent neurodegeneration, in some cases we are increasing body healthiness without maintaining cognitive function, leading to a potentially catastrophic increase in the number of people suffering from Alzheimer's disease and other age-related dementias. The failure of pharmaceutical companies to address this problem thus far (see any of the recent pharma companies' surrender in yet another Alzheimer's drug clinical trial) means that, at least for now, only academic labs can solve this puzzle.

One of the biggest misconceptions about the study of aging is that we are merely trying to stretch out the end of life, which, understandably, is not very attractive to most people. In fact, when I give talks about my research, if I don't show a video of the extremely long-lived little worms I study, most people assume that we are simply extending the end stage of their lives—but in the video they can see that the long-lived worms are youthful and healthy, not decrepit. These long-lived mutant worms have extended the healthy part of their lives. That's what we would like to do for humans, to increase the "healthspan" of people. A major goal of aging research is what Jim Fries termed the "compression of morbidity."[17] The idea is that one would maximize the healthy part of lifespan and then quickly decline at the end of life, compressing the "frail" part of life as much as possible. I think most people would agree that this would be a laudable goal for aging research, since we all acknowledge that increased lifespan in the absence of increased health is not beneficial. In fact, there has been a recent shift in the field from the study of "maximum lifespan" to developing new ways to measure what we call "healthspan"—different aspects of *quality of life* that change with age. Studying healthspan will likely help us uncover ways to both extend longevity and to compress morbidity and frailty.[18] Ultimately, in our study of aging, we must change the definition of a long life to include the concept of a high quality of life. The realistic outcomes from this work will extend healthy life, not just for the superrich or longevity obsessed, but also for the average population.

So how will we do it? By understanding how aging and longevity are regulated, and then finding ways to engage these mechanisms to activate them in humans. Luckily, most of this book will focus on mechanisms that not only extend lifespan, but also improve health.

2

Why Do We Age?

WE ACCEPT AGING AS A natural part of life, but why do we age in the first place? Are we obligated to age? In this chapter, I describe past and current theories of aging. Aging at its most basic level can be thought of as a loss of homeostasis, an inability to repair damage at a sufficient rate for the organism to remain unchanged with time. One simple model is that during development and early adulthood, organisms use energy to repair damage in order to continue reproducing; after reproduction has ceased, however, there may be little to no evolutionary pressure to prevent damage-induced decline, so less energy is expended on repair—which leads to unrepaired damage, which we then see as "aging." The need to be able to respond to changing environmental conditions led to the development of slowed reproductive rate, and along with it, longevity-extending mechanisms to allow delayed reproduction in certain conditions. Therefore, aging itself can be viewed not as a selected trait but rather a by-product of the cessation of organismal maintenance after reproduction. Framing aging as this loss of maintenance accompanying the loss of reproduction that might be prevented or slowed is useful in our evaluation of approaches to address aging.

―――――

I had the opportunity to visit Cuba back in the late 1990s, and one of the most striking sights in Havana was the omnipresence of old, brightly colored American cars. These 1950s relics were frozen in automotive time—at least superficially—because of a block on foreign imports imposed by Fidel Castro after the 1959 revolution. The amazing thing was that so many were still on the road, kept up through repair and replacement of precious parts over the years. Judging from the state of these vehicles, such maintenance must be a time- and effort-consuming task.

It is an overused analogy in the aging field that our bodies are like those old cars, disintegrating with time. The analogy falls apart once you realize we put energy into the repair of our cells, and we can adjust the rate at which we decline through genetic pathways. Our cells are constantly repairing and replacing damaged parts, using energy to make those repairs. With enough technology, we may learn to do the same thing, by kick-starting the body's natural mechanisms, by using drugs to induce repair, or perhaps through cellular replacement. In fact, some of the most exciting recent aging work has suggested that we might be able to slow or even reverse some aspects of aging. For example, worm oocytes "rejuvenate" all their proteins right before they are fertilized,[1] so that a new egg is always fresh, no matter how long it has been sitting around in the mother's uterus. One hope of the aging field is that in the future we may be able to engineer in new parts or cells to replace the old ones—resulting in a functional body but possibly from mix-and-match parts, much like those old American cars in Cuba have under their shiny hoods.

For many people, one of the most interesting questions about aging is *why does it happen in the first place?* First, a disclaimer: science is really good at answering *how* things work, but the *why* can be a bit trickier—we have to infer the reason why by reconstructing the past from what we can observe experimentally in the present, so this inference is naturally limited. For that reason, there have been a lot of theories about aging that have come and gone over the years as we've gathered more data. (For a very in-depth discussion, see Steven Austad's 1996 book *Why We Age.*)* While most of this book will focus much more on what we now know about the molecular mechanics of aging and longevity regulation than on evolutionary theories of aging, a cursory review of these theories might be helpful when thinking about how such mechanisms may have evolved.

Instead of asking why we age, it might be useful to flip the question on its head and instead ask *why wouldn't we?* Like all things in the universe, animals are subject to the second law of thermodynamics: the total entropy of an isolated system can never decrease over time. Much like keeping my house clean, maintaining a pristine, functioning cell and genome takes work; without this constant work, entropy (aka my family) takes over, and the next thing you know, everything's a mess and no one knows where anything is. Cells take in energy and

* New York: John Wiley.

use it to repair and clean up damage, but they are always fighting to maintain order, which is energy-consuming. The fact that caloric restriction extends lifespan seems counterintuitive when phrased this way, but cells can also adjust their activity (especially protein production) rates, which creates less damage and thus requires less energy to repair—much like how clean my house would look if the rest of my family were away and I were the only one there for a week.

One of the most intuitive and oldest theories of aging is a simple "wear and tear" model. This is how one could think of an aging car—the more it gets used and the older it is, the more worn down it will get, and the more it will need repairs. Walter Cannon, a noted physiologist who used his findings to save the lives of World War I soldiers suffering from shock (by treating their acidic blood with sodium bicarbonate), first developed the concept of "homeostasis" to describe his observation that certain systems are in place that help keep the body balanced and functioning. In the simplest terms, aging is what happens when cells aren't able to repair damage fast enough. This loss of "cellular homeostasis" results in the accumulation of damage and eventual cellular dysfunction. At the molecular level, this damage includes things like oxidatively damaged proteins and DNA mutations.

Unlike a car, however, we don't depreciate the minute we drive off the lot, or "start aging the minute we are born," as is oft repeated. I've been to a lot of aging talks, and one popular opening slide always bugs me: the picture shows the head of a child, then an adult, then a middle-aged and an older adult. The photo implies that aging is a continuous downward trajectory from birth, ignoring what we all know about biology: in fact, the cells and tissues of juveniles are remarkably healthy, repairing and replacing themselves to maintain function. We have several distinct life stages, and while the date that we "start to age" might be difficult to define, we don't think of teens being more "aged" than elementary school kids, for good reason. Genetic programs are in place to direct development from the juvenile state to reproductive adulthood. You develop to adulthood, and at some time *after* that you start to age. (You might have noticed this yourself, if you've hit 40 or so.) While there is evidence of damage accumulation, these problems don't seem to arise at the same rate in different organisms, or even in the same organism at different ages. Therefore, something more complex must be going on to regulate rates of aging.

The mechanisms that allow juvenile organisms to *replace* their cells might be different than those that allow *repair* of existing cells, and being able to harness each of those appropriately would be powerful. In fact, there are a few organisms that appear to be immortal. Many seemingly immortal animals constantly replace their own cells. Certain marine animals, flatworms like planaria, and some trees could be considered to experience "negligible senescence" because their cells are constantly being replaced. These organisms can replace their cells through the activity of stem cells, the type of "mother" cell that can give rise to all other cell types upon division. Planaria can be split into many parts and generate whole animals from these bits, because its stem cells are sprinkled throughout their bodies. When stressed, adult Japanese "immortal jellyfish" (*Turritopsis dohrnii*) can—like Benjamin Button, Max Tivoli, and Merlin—essentially reverse their development and revert to a juvenile stage (polyp), through a process called "transdifferentiation." The case can be made that these animals are not truly immortal, but instead maintain a constant functional state through the birth of new cells to replace old ones. Make no mistake: the maintenance of stem cells is critical for our health as well, as we need them to maintain our immune system, our muscles, our skin, and even our brains. Figuring out how to keep stem cells working—instead of petering out with age—is a major effort in the aging field. While we will never be immortal, we can try to understand and then harness the tricks that seemingly immortal animals use to maintain themselves.

———

One of the first to study aging seriously was the Russian scientist Elie Metchnikoff, who in addition to his Nobel Prize–winning (1908) discoveries of macrophages and other groundbreaking work in immunology and microbiology, coined the terms "gerontology" and "thanatology" (the study of death, not the aptly named Marvel character). His observations suggested the roles of the immune system and microbiome (see chapter 16) in aging, and its slowing through probiotics. Metchnikoff is also credited with ideas similar to hormesis ("My general conclusion from these facts is that it is logical to lay down the principle that the higher elements of our body could be strengthened by subjecting them to the action of small doses of the appropriate cytotoxic serums"),*[2] and at least one origin of the theory of "programmed aging," which on its surface is similar to August Weismann's idea that aging is preprogrammed—but the two differed on whether natural death could be evolutionarily beneficial. In fact, Metchnikoff was decidedly in the anti-aging

* Hormesis is the concept that nonlethal stresses will induce a response that will ultimately strengthen the organism.

camp, recognizing that many factors, including poor diet, disease, and infection, may cut short a person's maximum potential lifespan. For these reasons he might be considered a founding father of gerontology and "longevity science," with a focus on practical treatments to slow aging.

Raymond Pearl's "rate-of-living" theory (1928) hinged on the idea that resources are always finite and limiting—thus the faster an animal metabolizes energy, the shorter it will live. Such an idea might fit with the general notion that mice live shorter than elephants, but not with the fact that there are very long-lived species (e.g., bats and naked mole rats) that are the same size as mice. That is, bats and naked mole rats have extremely large "longevity quotients,"[3] because their lifespans are much longer than what you would predict from their size. It also doesn't fit with the fact that small, yippy hyperactive dogs generally live longer than large dogs.

Another early theory of aging, Peter Medawar's "mutation accumulation" theory (1952), is bit like the car model, invoking the concept of stochastic damage, but after reproduction. At first, this seems reasonable; most middle-aged people can relate to the idea that things just start to fall apart at some point. The mutation accumulation theory suggested that after reproduction, problems that arise with age, particularly those that are caused by random mutation, cannot be selected against, and therefore are likely to be deleterious and cause aging. He used the analogy of test tubes breaking over time: even though the tubes themselves are not visibly aging, with enough time, they will—like your wedding china over the years, or the cars in Cuba—break and be lost in a random manner.

George Williams's antagonistic pleiotropy theory (1957) incorporated the distinction between early and late life phases.[4] Williams suggested that some genes may confer a benefit to an organism early in life—such as fast development and a large number of progeny—but the activity of that same gene may be deleterious late in life, leading to age-induced decline. Genetic programs that link nutrient levels to aging rates could be considered to be in this class. (I will note a few examples of these models as we progress through the book.) The antagonistic pleiotropy theory also better fits with the observation that animals with shorter lifespans often mature faster and reproduce more.

Williams, notably, was of the opinion that simple "fixes" of aging—and, thus, single causes or single genes affecting longevity—are unlikely: "*This conclusion banishes the 'fountain of youth' to the limbo of scientific impossibilities.*"[5] Of course, as I'll describe below, the much later discovery of single-gene mutations that double lifespan in worms and later in flies refuted William's assertion that lifespan extension was an impossible goal.*[6] The widespread notion that a single gene could not alter lifespan was so pooh-poohed that it likely slowed progress in aging research for years, and limited acceptance of experimental results beyond that (and people still often repeat the idea that single genes can't alter lifespan despite almost 30 years of data refuting this theory). Of course, as we will see in later chapters, because that single gene is a "master regulator" of a whole cascade of downstream genes, it really controls hundreds of other genes simultaneously—and none of those individual genes by itself has a huge effect, which does fit with Williams's theory. But single genes can indeed be critical regulators of lifespan. Theories that do not acknowledge this fact are now obsolete, and we should make this point better known.

Sean Curran, Gary Ruvkun, and others set out to test one aspect of the theory of antagonistic pleiotropy using an elegant method of post-developmental gene knockdown using RNA interference.[7] They found that many highly conserved, essential genes (that is, the animal could not develop normally without them) actually *extended* lifespan when knocked down only in adulthood. Why might this be? A proteomics analysis showed that these essential genes are overrepresented in the set of age-related aggregation proteins,[8] suggesting that these proteins may become less useful with age and act as aggregation seeds—that is, the beginning of the sticky blobs that may grow and become damaging to cells—becoming deleterious with age. Of course, not every essential gene will extend lifespan when knocked down, and not every gene that exhibits antagonistic pleiotropy is essential, but testing genes late in life helps divorce their effects from developmental roles.

An alternative to the idea of aging being caused by stochastic damage is Mikhail Blagosklonny's notion of a "quasi-program of aging."[9] This "developmental hyperfunction quasi-program" is a mechanistic extension of antagonistic pleiotropy, sort of antagonistic pleiotropy on steroids (and to be honest, I'm not sure there is a great difference from antagonistic pleiotropy). The hyperfunction theory suggests that there are catastrophic results for "leaving on" systems used in development when they are no longer needed, leading to damage that ultimately kills the animal. Blagosklonny uses the example of a bathtub: one can imagine filling a tub with water, but after it is full, keeping the faucet running could be disastrous ("a 'quasi-program' to

---

* It should be noted that these mutants do slow development, if severe versions ("alleles") are used, which does fit with the antagonistic pleiotropy theory. But they can also be knocked down after development and still extend lifespan.

flood your apartment"). He suggests that hyperfunction of developmental processes in humans results in such pathologies as diabetes, obesity, tumors, hypertension, and so on—basically all the features of age-related disease. One could imagine that as reproduction ends, the imperfect shutting off of some reproductive functions might lead to the deleterious effects that Blagosklonny describes.

This theory was linked to previously observed effects of age-associated accumulation of developmental proteins. In two separate studies of C. elegans age-related morphological decline, Delia Garigan and Cynthia Kenyon (2002) and Laura Herndon and Monica Driscoll (also 2002) reported that lipid-binding proteins called vitellogenins that normally are carried into developing oocytes can accumulate in the rest of the animal's body with age, particularly in shorter-lived animals.[10] This suggested that the post-reproductive accumulation and aggregation of these lipoproteins in non-reproductive tissues after the lipids were no longer needed or properly packaged into developing eggs could be deleterious. In fact, I found in my postdoctoral research with Cynthia Kenyon, and others have since confirmed, that long-lived mutants suppress vitellogenin production, and the reduction of those lipid-carrying proteins does extend lifespan.[11] Later, it was suggested that this post-reproductive mis-accumulation of vitellogenins in aging worms is an example of Blagosklonny's "hyperfunction quasi-program," with run-on developmental processes that become detrimental with age.[12]

———

One of the oldest and most popular theories of aging—*so popular it just won't die!*—is the idea that damage induced by reactive oxygen species (ROS) is the primary driver of aging. In fact, this is likely the theory that you have read most often in any popular media about aging. Denham Harman, one of the early pioneers of molecular aging research, first proposed the "free radical" theory of aging in 1956, and for decades it remained the primary theory associated with aging. Linus Pauling, the famous Caltech chemist of double Nobel Prize fame, championed the corollary of this notion, that antioxidants should be the ultimate anti-aging medicine. (Reportedly, Pauling downed megadoses of vitamin C like candy, and it's hard to argue whether he would have lived longer with or without it, considering he made it to the age of 93, and Harman to 98! But don't do it yourself, as excess vitamin C just gets excreted, and megadosing of vitamin C can cause kidney stones.)

The ROS theory was so well accepted that it seemed to be unquestioned and, frankly, unavoidable when I joined the aging field in the early 2000s. At that time, the prevailing model in the field was that superoxide dismutase (SOD) and catalase, two enzymes that work together to eliminate oxygen radicals, are *solely* responsible for the longevity of long-lived animals and of those with mutations in insulin signaling, a major longevity pathway—and in fact, those genes are upregulated in the long-lived mutants.* One popular idea was that the long-lived *C. elegans* insulin/IGF-1 receptor mutant *daf-2* owed its longevity entirely to the upregulation of the superoxide dismutase *sod-3*;† this gene was indeed shown by Honda and Honda to be highly upregulated in *daf-2* mutants.[13] Superoxide dismutases work together with enzymes called catalases to rid the cell of dangerous superoxide radicals. But my microarray gene-expression experiments on the *daf-2* longevity mutant (see chapter 6) showed that hundreds of genes, not just SOD and catalase, are upregulated in those mutants, suggesting that the superoxide dismutase/catalase pair alone was unlikely to account for all of the mutant's lifespan extension. Even more importantly, by knocking down the individual gene functions, we found that each of these upregulated genes—including superoxide dismutase and catalase—is responsible for only a small fraction (5%–10%) of the mutant's long lifespan. Adding up the activities of *all* of the genes, not just antioxidants, is necessary to achieve *daf-2*'s remarkably long lifespan. *Thus, battling ROS cannot be the only thing that keeps the insulin longevity mutants healthy.* (Frustratingly, a few lay articles and reviews about our work described it as having found that antioxidant genes extend lifespan, just because they were among the genes on the upregulated list, while ignoring the dozens of other genes that had the same effect on lifespan—so missing the whole point, actually.) In fact, in 2012, Jeremy van Raamsdonk and Siegfried Hekimi showed that worms with all five of their superoxide dismutase genes knocked out (*sod-12345*) are more sensitive to stress, but have normal lifespans under unstressed conditions.[14] So while ROS may be part of the problem, they are unlikely to be the whole cause of aging.

---

* In fact, a *Nature* paper that claimed that loss of *sod-3* completely abolished the long lifespan of the *C. elegans* insulin/IGF-1 receptor mutant *daf-2* was retracted when it was found that the "double mutant" contained neither *daf-2* nor *sod-3* mutations, so it was a wild-type strain with wild-type lifespan! Awkward.

† *C. elegans* mutants and genes are usually named for their characteristics (phenotypes) and italicized. "daf" means the mutant is defective for proper formation of dauers, a diapause state. *daf* mutants either go into dauer when they shouldn't, like *daf-2*, or don't go into dauer when they should, like *daf-16*.

Nevertheless, there continues to be some support for Harman's ROS theory. For example, in 2010, Steve Austad and the students in the Aging course at the Marine Biology Labs at Woods Hole carried out a comparative study of two clam species from similar environments but with different lifespans, the extremely long-lived *Arctica islandica* (with a maximum lifespan estimated to be somewhere around 400–500 years!) and the related but shorter-lived (100 years) *Mercenaria mercenaria*. Their results suggested that, in general, the long-lived species is better able to deal with oxidative stress.[15] However, there were tissue-specific exceptions to this rule, and the basal antioxidant activities were not different, meaning that free radical elimination cannot be the sole source of *Arctica islandica*'s extreme longevity. This study, like many, provides some support for the ROS theory, but is far from concluding that oxidative stress protection is the *only* mechanism that allows long-lived organisms to survive. Rochelle Buffenstein, whose lab studies the exceptional longevity of naked mole rats (see more in chapter 4), laid it out pretty clearly in a paper she titled "Antioxidants Do Not Explain the Disparate Longevity between Mice and the Longest-Living Rodent, the Naked Mole-Rat" (2005).[16]

The ROS theory has been really pounded into the ground, as more recent studies suggest not only that antioxidants cannot be the sole source of longevity but that in some contexts they may actually be deleterious. In fact, expression of antioxidant protective genes, like superoxide dismutase, was suppressed in mice treated with vitamin C for 18 months, which may be why they were not longer lived.[17] Antioxidants—even the ones that Linus Pauling swore by—can actually be detrimental when applied under slightly stressful conditions that normally extend lifespan; that is, when given antioxidants, the animals who endured stressful conditions actually survived *less* well. Even more concerning, when healthy young men exercised for four weeks while taking (or not taking) vitamins C and E, the exercise was beneficial only for the participants who did *not* consume the vitamins![18] Not exactly a ringing endorsement for the antioxidant industry. Together, these results suggest that some oxidative damage might be a good thing, or at least necessary in order to elicit other beneficial effects. That is likely because the oxidative damage can in fact act as a signal that indicates the need for broader repair.

This brings me to the somewhat Nietzschean concept of *hormesis*—those conditions that are stressful but don't end up killing the animal can help it live longer in the end. For example, both Tom Johnson's (2002) and Gordon Lithgow's (2006) labs demonstrated that the repeated application of nonlethal stresses, such as oxidative stressors or pulses of heat early in life, increased the lifespan of *C. elegans*.[19] It turns out that small stresses do in fact increase ROS, but instead of causing damage as the ROS theory suggests, they seem to increase the animal's stress *responses*, which in turn increase lifespan. This was originally thought to be through direct activation of stress responses, but more recent data suggest that there are more complicated signaling pathways between subcellular components, particularly mitochondria, and between cells that induce the hormetic stress response and communicate this signal. Michael Ristow's lab (ETH Zurich) showed that the application of antioxidants blunts this signal, and as a result, the animals never turn on the stress responses that give them a boost, and instead die *earlier*.

Thus, despite its vexing pervasiveness in both the scientific and popular literature, and its definite contribution to some aspects of aging, ROS damage is unlikely to be the *sole cause* of aging. Moreover, gobbling up antioxidants is unlikely to save us, and might even prevent us from coping with some stressful conditions. But ROS and its responses do play a role in longevity regulation that we need to understand.

————

Other theories link reproduction and longevity. In 1979, Thomas Kirkwood proposed the "disposable soma" theory, which is the idea that resources (that is, energy) can be allocated either to maintain the germline *or* to the soma (the rest of the body), but not both.[20] This is an attractive theory, because it invokes a trade-off between long life and reproduction, which fits our general intuition that *you can't have everything*. The idea of resource allocation and trade-off is invoked with regularity, and in general, the fact that many long-lived organisms have a small number of large progeny, rather than a large number of small progeny, and that many longevity conditions reduce reproductive output meet the general expectations of this model.

One might predict, based on the disposable soma theory, that sterility should induce long lifespan—and there was some anecdotal evidence supporting such a view. For example, the average lifespan of Korean eunuchs was reported to be about 14 years longer than that of men with similar or better socioeconomic status,[21] and the incidence of centenarians was increased 130-fold in the castrated population (although these claims were also debated).[22] Superficially, these results might suggest that sterility will lead to long life. However, the authors also suggested a second possibility, that reducing the production of male hormones might extend lifespan, and in chapter 9 I'll talk a bit more about this idea, which seems closer to being correct.

A detailed study by *C. elegans* geneticist Cynthia Kenyon and student Honor Hsin directly addressed the *does sterility increase lifespan?* question. Hsin had started working in Kenyon's lab at UCSF as a preteen homeschooled student, and with a set of elegant experiments they abolished the simple notion that sterility itself is the whole answer, just using a tiny laser beam and worms.

To understand what they did, I first need to explain a little bit about the early development of *C. elegans*. In the first larval stage, called L1, there is not much of a germline or gonad yet; in fact, the precursor for both consists of only four cells, named Z1, Z2, Z3, and Z4. (Yes, that's all of them!) The beauty of *C. elegans* is that every cell develops the same way every time (called stereotypical development). Each of *C. elegans*'s cells is identifiable and has been given a name, and owing to detailed lineage tracing studies, we know what each cell will become as it develops, which is part of why the worm became such a powerful genetic model system. The two inner cells, Z2 and Z3, will divide and become the worm's proliferating germline cells, while the outer cells, Z1 and Z4, are the precursors to all of the non-germline, support cells, called the "somatic gonad."

Kenyon and Hsin took advantage of this simplicity to address the sterility question. By using a laser beam to fry ("ablate," or kill) the precursor cells that would normally build the germline (Z2 and Z3), they were able to create worms that still had a somatic gonad, but lacked all germline cells. This germline ablation obviously made the worms sterile, and greatly extended their lifespan. This would seem to support the theory that resources were shifted from the germline to the soma, exactly as the disposable soma theory would predict. However, Kenyon and Hsin did another, paradigm-shifting experiment: in a separate group of worms, they *also* killed the somatic gonad precursor cells (Z1 and Z4)—that is, they destroyed *both* the germline cells and the support tissue around those proliferating cells, so these worms never developed any reproductive system at all. These worms lacking both a germline and the surrounding somatic tissues were also sterile—and thus, used no resources for reproduction—but were *not* long-lived (figure 1).[23]

That is, Hsin and Kenyon had uncoupled sterility from longevity. Therefore, the longevity of the germline-less animals could *not* be due to simple resource allocation but rather had to be due to the loss of a signal that must come from a healthy, intact germline that normally promotes aging. In fact, Hsin and Kenyon's data suggest that there must be two antagonistic signals, one *pro-aging*

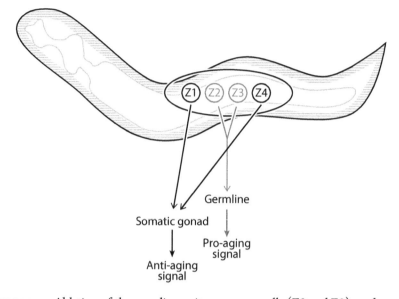

FIGURE 1. Ablation of the germline or its precursor cells (Z2 and Z3) renders *C. elegans* sterile and long-lived, while ablation of the somatic gonad + germline (all four precursor cells) renders the animals sterile with a normal lifespan, suggesting that there are antagonistic pro- and anti-aging signals emanating from the germline and gonad, respectively. These results also argue against a direct mechanism for the disposable soma hypothesis.

signal from the germline and a different *pro-longevity* signal from the somatic gonad. Together, the balance of these signals tells the rest of the soma (body) how long the animal should live, based on the status of both the germline and gonad. That is, when the germline is absent, only the pro-longevity signal functions, resulting in a long-lived animal; when both are absent, the lifespan is normal. (And my lab later showed that mating shifts the pro-aging signal to high.)[24] Therefore, some aspects of the disposable soma theory may be correct, but it is not as direct as originally proposed: rather than directly allocating resources on the fly, animals have programs and signals that control longevity, most likely in *anticipation* of future reproductive needs. Such a model would involve signaling hormones in the control of lifespan, rather than the direct loss of reproduction and reallocation of resources.

It is worth noting that there are no known longevity mutants or conditions that completely lack effects on reproduction, suggesting that the two are definitely linked—but of course correlation does not imply causality. The connection between reproduction and longevity may not be as direct as the disposable soma theory would imply, as there are mutants that live very long (sometimes up to 10 times as long) and yet have very small decreases in reproduction, certainly not proportional to their extensions of lifespan.* More importantly, Andrew Dillin and Cynthia Kenyon showed in 2002 that the reduction of expression of the worm's DAF-2 insulin/IGF-1 receptor *only in adulthood*—which avoided any effects on development or reproduction—still doubled lifespan.[25] That is, for *daf-2* mutants to live longer, it is not necessary for them to produce fewer progeny. This work showing the decoupling of reproduction from longevity is hugely important in its refutation of the disposable soma theory, but is not as well known in the field as it should be. Recent work by Collin Ewald's lab supported this conclusion; these researchers showed that degradation of the DAF-2 receptor protein itself well past the reproductive period extended lifespan,[26] further supporting a disconnect between longevity and reproduction that Dillin and Kenyon had previously shown. And recently, my own lab found that individual worms that have many progeny and/or reproduce longer also live longer[27]—again suggesting that the disposable soma theory may not hold for individuals, but rather that the ability to reproduce longer may be linked with long lifespan, as Perls and colleagues showed for women who were able to have children later.[28] These results mean that it is not absolutely necessary to reduce reproductive output or to alter signaling during the reproductive period in order to increase lifespan—which is good news for most of us!

————

Are animals "programmed to die"?[29] We've all heard of the precipitous demise of Pacific salmon, who swim hundreds of miles upstream to spawn, and then die. The fish experience immunosuppression due to high corticosteroid levels, which in turn causes rapid organ deterioration and subsequent death. These "semelparous" animals reproduce only once (the "big bang") and then die right after they've finished reproducing, which seems like a pretty clear case of programmed death. Semelparous species rely on high numbers of progeny

---

* Work done on hermaphroditic *C. elegans* can be a bit misleading if the researchers don't mate the hermaphrodites with males. That's because hermaphroditic progeny number is limited by the number of self-sperm the animals make before they switch over to making oocytes, but they still have excess oocytes that can be used if mated with a male, which would reveal total progeny number. Since self-sperm progeny number doesn't actually measure the total number of possible progeny, self-progeny data shouldn't be used to interpret a relationship between brood size and longevity, although such reports have often been used to support the disposable soma theory.

with little to no parental care (a reproductive strategy known as "r selection"), although the female salmon do enhance the fitness of their progeny by protecting the nest site as long as they can before they die. In the case of salmon, aging—or at least death—certainly appears to be programmed.

A controversial theory that has come into and out of favor (and back in?) is that of "group selection." The idea here is that aging and subsequent death of an individual may confer a benefit to the whole herd—for example, by getting out of the way and freeing up more resources for the young; older, diseased animals are more likely to be caught by predators, which would be beneficial to younger animals.[30] This is an altruistic view of aging, a sort of *Logan's Run* approach to population maintenance.* Group selection may in fact be beneficial not because of the food spared but because it may act as a mechanism to accelerate the rate of evolution—for example, by propagating genes that could benefit the group rather than the individual.[31] In this case, the evolutionary benefit may be something less direct than shifting resources from individuals of different ages, but would still confer a benefit to the young if aged individuals are eliminated.

Perhaps because I am not an evolutionary biologist, I find this theory hardest to either refute or test; it may be true, but it is also harder to imagine how it would be useful in most cases. For example, I have often had people tell me with certainty that the old worms must age and die so the young ones have more to eat. That sounds great until you do the experiment and realize that most of the food is eaten by developing juveniles anyway—after reproduction, old worms hardly eat at all, so getting rid of them doesn't seem likely to make much of a dent in the food supply, whether they die on day 15 or day 50. But of course, our experiments are done in the lab, with an excess of food; it is possible that the situation would be different in the wild, with fluctuating levels of nutrient availability. (Maybe this would work if the young worms would cannibalize the old dead worms, or if bacteria flourished on the dead and provided nutrients for young worms, so again, slightly indirectly.) Perhaps such a phenomenon would be more critical for species that are always on the verge of food limitation, and, more importantly, where aged adults continue to consume a significant fraction of the available food source, so their elimination

---

* For the young folks: in this dystopia, everyone was terminated at the age of 30 to maintain sufficient resources (Michael Anderson's film *Logan's Run* [Metro-Goldwyn-Mayer, 1976] was based on the 1967 graphic novel of the same name by William F. Nolan and George Clayton Johnson [Bluewater Comics]).

would in fact benefit younger animals. Under such circumstances, there might be an advantage to speeding development and maximizing progeny production in good times.

Predation and extrinsic factors are important to consider as well. The classic example of predation affecting longevity is that of mice and bats, two similarly sized mammals with a tenfold difference in longevity (2–3 vs. 30 years). Previously it was assumed that the key difference in lifespan between these animals might be due to bats' ability to hibernate, but Steve Austad's group has eliminated this difference as the source of exceptional longevity, both because some long-lived bats do not hibernate and because the protective effect that allows hibernating bats to survive longer might simply be the lack of predation during hibernation.[32] After surveying 19 different hibernating mammalian species, Turbill and colleagues also concluded that hibernation itself does not cause long lifespan, even though these bats have higher maximum lifespans than other species their size.[33] Instead, hibernation is associated with lower predation (a lower risk of being eaten) and slower life histories. Of course, one of the other major differences between these two animals is the bat's ability to fly, so it is theorized that the bat's long lifespan can be attributed to its ability to escape predators. The effect of escaping predation may reveal the ability of a species to live long in the absence of pressure to reproduce fast, and that in turn may affect aging. Of course, evolutionarily this might not be as simple as moving from a short-lived/many progeny/no wings animal to a long-lived/few progeny/with wings animal, but could have a large influence on lifespan.

Lifting the threat of predation may have greatly affected the lifespans of other animals, including fish and some cephalopods. Deep-sea octopuses apparently have no predators,[34] and may live up to two decades, while shallow-water octopuses, having lost their protective shells, have relatively short lifespans (two years), despite their impressive bag of protective tricks to mimic and camouflage themselves. Reznick and colleagues (2006) tested the effect of predation on different phases of life history (pre-reproductive, reproductive, and post-reproductive) by examining guppies (*Poecilia reticulata*) from pools with different predation levels in Trinidad.[35] By moving guppies from high- to low-predation environments, they found—counter to previous predictions—that the shifted guppies' mortality rates were lower both initially and throughout their lives, with longer median lifespans and higher fecundity (reproduction). They also found that all of the guppies had an extended post-reproductive lifespan. Together, this study showed that only reproductive lifespan, which directly affects individual fitness, was selected in response to extrinsic mortality factors such as predation. (Further studies on wild strains of animals [e.g., worms and killifish] may also address additional theories on longevity effects from ecological conditions and the evolution of longevity as a trait.)

There is also a limit to the escape from predation. As Sean Carroll explains in *The Serengeti Rules*, some animals have evolved to increase in size and, in doing so, simply become too large for a predator to attack.[36] This sounds like a great strategy, but the cost is that because they are so large, they then become much more susceptible to changes in the food supply. The ability to adapt to differences in nutrient availability by altering reproductive development and duration might become especially important when resources are limited.

Animals who never lived beyond their reproductive spans because they perished at the teeth of predators would never have experienced aging. So the real question is not "Why do we age?" but rather "Why do we live so long beyond our reproductive spans?" That is, evolutionarily speaking, a relatively new problem. Human life expectancy has only recently risen, and only a few animals seem to live a long time beyond their reproductive spans. In humans this has been attributed to the positive effect of grandmothers on the fitness of their grandprogeny, the grandmother hypothesis.[37] The idea here is that grandmothers who are no longer reproductive themselves, but help with the raising and support of their grandprogeny, may increase the evolutionary fitness of those grandprogeny, and so there is some selective pressure for women's longer lifespan. Several studies of the contributions of human grandmothers lend support to this hypothesis, as do studies of large mammals. On the other hand, men live only a few years less than women, and there are also animals with negligible progeny care that exhibit the same life histories, with longer post-reproductive lifespans. Together, this poses the question of how much of a role the grandmother hypothesis truly plays in the regulation of lifespan across all species.

A few years ago, I became interested in understanding factors that correlate with the length of post-reproductive lifespan, so I recruited an undergraduate to do some fact-finding. For his senior thesis work, George Maliha found that within mammals, post-reproductive lifespan is proportional to gestation time, which may result in larger offspring. For non-mammals, however, we found that post-reproductive life span is best correlated with the ratio of the progeny's size to the mother's size.[38] That is, *the bigger the baby, the longer the mother's survival after reproduction ends.* Conversely, animals that produce tiny progeny seem to be able to reproduce even when aged and relatively infirm and die soon after, thus leading to shorter post-reproductive life spans.

For example, some species of sea urchins can live for many decades, even centuries.[39] These spiny, tube-footed balls can be relatively large, and they reproduce by releasing clouds of microscopic eggs and sperm that meet each other outside the parent. They can reproduce almost until they die, and thus have a very short—almost nonexistent—post-reproductive lifespan. By contrast, mothers who give birth to progeny that are relatively large have a long post-reproductive lifespan. This made sense to us, as it seems intuitive that a mother has to be in better shape if the progeny is extremely large, and thus might have "leftover health" that would lead to a longer lifespan. In fact, human babies have evolved to have relatively giant heads, compared with their mothers; the evolution of large heads correlates with the development of big brains but the cost is that mothers die in childbirth at a non-negligible rate. The fact that the post-reproductive lifespan scales with progeny-to-mother size ratio across many species means that we do not need to invoke the grandmother hypothesis to explain long post-reproductive lifespans. That's good, because we see that C. elegans, the hermaphroditic nematode, lives a long time after it has produced all of its progeny—and the worms don't babysit their own kids, much less their grandprogeny. This of course does not mean that the grandmother hypothesis is altogether wrong, but it might be a modification on top of a larger biological effect that already extended post-reproductive lifespan for other reasons.

The demographer Annette Baudisch (Max Planck Institute for Demographic Research) has modeled the effects of prolonging life after reproduction on mutation fixation, countering the notion that post-reproductive lifespan plays no role in lifespan determination. Instead, her group's modeling suggests that a beneficial mutation is more likely to be fixed in a population with longer post-reproductive lifespan.[40] That is, both reproductive lifespan and timing of death, not just reproduction, can evolve in response to selective pressures—a new twist on the role that post-reproductive lifespan may play in the evolution of aging.

––––––

I gave an "at home" talk early in my career at Princeton, focusing primarily on my postdoctoral work on gene expression of long-lived animals (see chapter 6) and my lab's new project using worms to study cognitive decline with age (chapter 16). In the Q&A session afterwards, an evolutionary biologist raised his hand and told me that all my work was wrong (I'm paraphrasing here), because there can be no selection for long life. I was taken aback, partly because longevity wasn't even the focus of my talk that day, and partly because I had shown the audience in the introduction that insulin signaling mutants *do* live longer, so it didn't seem like a matter of debate. It soon became clear that the main reason he had come to my talk was to tell me that all of the results of the field for the prior decade must be incorrect. I pointed out that there was no denying the fact that there are single-gene mutants that can greatly extend lifespan, regardless of any theory about whether they *should* exist or not. More-over, these genes and their effects on lifespan are conserved evolutionarily, so this was not just a weird worm phenomenon. It was quite obviously aggravat-ing to him that all of the discoveries of single-gene longevity mutants in worms and flies didn't fit with Williams's simple theory.

But it's always useful to reexamine our theories when our data don't seem to match them. *Regulated* programmed theories of aging seem to fit our obser-vations best. That is, an animal might need to respond to changing conditions (e.g., during a famine) and adjust its life history rates (growth, reproduction, longevity) accordingly. The evolutionary biologist was absolutely correct that nothing can be selected post-reproductively, so *longevity itself* is not a trait that could evolve directly through selection. What he may have been reacting to (rather than my talk) was the popular press depiction that there is a set "program" whose purpose is to extend lifespan—and he would probably be surprised to hear that I would agree with him. As we will learn later in this book, the genes that we have found that extend lifespan by and large play a role in regulating other functions that *can* be selected for, including the ability to respond to nutrient levels to adjust *developmental* and *reproductive* rates. Both insulin signaling mutants and dietary restriction conditions extend lifespan and reproduction, seemingly flying in the face of the disposable soma theory and the fact that longevity should not be selected, but consistent with the no-tion that reproductive rates and timing should be adjusted to match nutrient intake. The pre- and post-reproduction distinction is also important when considering the impact of the extrinsic mortality rate (mostly affected by pre-dation) upon aging, such as in the case of the guppies in different predation environments.

All of these facts make more sense when reframed as adjustments an animal makes to allow *reproduction* as a response to available nutrient conditions—the animal adjusts its reproductive rates to optimally use nutrients when they are there, and slow things down when there is no or little food. These concepts can all be reconciled by one simple tweak to the theory: we should view aging and longevity not as the *main goal* of an animal but as a *side-effect* of post-reproductive survival, and, importantly, one that can be modulated. Reproduction, not longevity, is the selected trait, and the factors or signals that modulate longevity do so primarily to optimize reproduction. Instead of talking about programs that lock an animal into one decision or another to control longevity, the important regulators allow *plasticity*—that is, the ability to flexibly respond to differences in conditions in order to reproduce.

Optimization does not always mean "have progeny as fast as possible"—if there are not enough nutrients, then it makes no sense to reproduce right away, since all your progeny will then be doomed to die of starvation; instead, reproduction should be slowed down. If an animal does slow down reproduction, then it needs to be able to do two things for this to be worth the effort: (1) maintain the quality of its germline until reproduction is restarted so that its progeny will be healthy, and (2) maintain somatic tissues so that when reproduction is restarted, the mother doesn't die before she has successfully produced her progeny. The latter, somatic quality maintenance, is essentially what we measure with lifespan. Therefore, linking the two quality systems makes sense for the animal, even though the only "selection" is for reproduction.

Another point to consider is the fact that this is not a closed system; as long as there are nutrients, energy can be used to maintain and repair, at least as long as those maintenance and repair systems are functional. During reproduction, there may be a signal to continue maintaining and repairing tissues, both germline and somatic; after reproduction is done, the system really doesn't need to maintain either the germline or the soma—but the animals that have the highest somatic quality at the end of reproduction will live the longest post-reproductively, as our multispecies post-reproductive–lifespan analysis suggested.

How could such adjustability develop, evolutionarily speaking? We have already mentioned hibernation, but you could imagine more subtle shifts in metabolism in response to restricted nutrient availability that might slow down how "fast" an animal works. An animal that can slow down its metabolism in hard times might win out over generations, even if it doesn't develop the fastest or produce the most progeny, if conditions are always nonoptimal. But if conditions are stable and food is abundant, then the animal that develops to maturity fastest and produces the most progeny rapidly will overtake the population. A fluctuating environment would favor animals who can best respond to all kinds of conditions. From this set of guidelines, it's easy to see how mechanisms that convey information about nutrient levels to tune metabolic rates could have been selected, since the animals that have such systems could make the right choices all the time. Some animals, like those who hibernate or go into spore-like "diapause" states, have taken this concept to the extreme, but we share with these animals the same metabolic regulatory systems. Eusocial insects, like ants and bees, who have designated "castes" to carry out different tasks within a colony, are another example of extremes in this metabolic adjustment: queens and workers in the same colony come from the same genetic background, but experience extremely different lifespans (years vs. weeks) because of their metabolic programming and compartmentalization of functions. Adjusting the rates of life history traits allows animals to best survive in their particular environment.

Thus, my annoying colleague was both right and wrong, and it's worth re-examining what we know about longevity maintenance and adjusting our theories to account for real data. There is some promising work in mice suggesting that late-life interventions, particularly in females, might have health benefits.[41] For humans, the billion-dollar question will be whether we actually *can* affect longevity post-reproductively.

———

So if worms can live twice or even three times as long as they usually do simply by tweaking the function of an insulin/IGF-1 receptor, why don't we find "mutant" people who live 250 years? In fact, people are living pretty long now; as life expectancy has increased, so has the maximum lifespan. While there is some debate about what that maximum will be (and a recent estimated maximum of 115 has been vigorously refuted—see chapter 1), given the observation that mortality rates seem to level off after a certain age, it seems unlikely that any human will reach the age of 250 any time soon, in the absence of major interventions. Why might that be? What's keeping us in this range?

One possibility is the notion of restricted tolerance. We can survive to adulthood only by making it through development in a reasonable time; the possible mutations that might double our lifespans would also likely slow our development so severely that we would never reach adulthood.* Even in *C. elegans*, strong mutations that almost completely block insulin/insulin-like

* In fact, there are medical reports of patients who have such mutations, causing development to halt altogether around the age of two.

growth factor 1 (IGF-1) signaling cause the worms to stop developing in their first larval stage, or to get stuck in the "dauer" diapause state—such a mutant would never win out in the wild, of course. In humans, Laron syndrome, which causes a defect in the IGF-1 signaling pathway,[42] leads to dwarfism and other disorders, although the patients rarely die of cancer and other common age-related diseases. It's clear that we cannot greatly affect the insulin/IGF-1 pathway, particularly during development, without having unintended negative effects beyond lifespan.

Temperature may be another factor in restricted tolerance. Unlike invertebrates and cold-blooded animals, which must deal with whatever conditions they are exposed to, we maintain our own internal temperature. This warm-bloodedness may severely restrict our metabolic plasticity, limiting the range over which we can dial our metabolism up and down. The ability of our bodies to generate heat through brown fat is highly regulated and restricted to very early development, possibly to avoid this energetically costly process. Exactly for this reason, there is excitement about shifting white fat into brown fat in adults for weight loss—for example, through the use of mitochondrial uncouplers. Such chemicals, like 2,4-dinitrophenol (DNP) and "fen-phen" as over-the-counter diet pills, have a sordid history, and are banned now because they are so toxic. This is perhaps not surprising, given the fact that DNP has alarming effects: it was noticed that many workers in French munitions factories using DNP to make explosives lost weight, sweated excessively, had high body temperatures (up to 107°!), and often died. Runaway thermogenesis will kill us, as will hypothermia, so we can utilize only thermal longevity mechanisms that act within a restricted range.

Finally, we must acknowledge that even within model systems, there can be confusion between what is naturally observed and the mutants we work with in our labs. In the wild, even *C. elegans* may have a smaller range of lifespans, like people. These long-lived mutants thrive in our labs, but for the reasons I've already discussed they might never be found in nature: out in the wild, they cannot compete with "wild-type" (normal) worms, who can deal better with a fluctuating environment, while the longevity mutants are locked into one state. That is, the lack of ability to change state in response to a changing environment is a disadvantage in the wild. All of this does *not* mean that the findings for longevity mutants have no relevance for humans, but rather that these kinds of extreme mutants hit the same genes that would need to be modified in adulthood, after development and reproduction are done, in order to elicit the same kinds of longevity effects in humans. The answer might be in slight variations to these genes, such as those that are found in the single nucleotide polymorphisms (SNPs) that are uncovered in genome-wide association studies (GWAS)—and, in fact, genome studies of centenarians has suggested exactly this conclusion, that tiny changes to functioning insulin/IGF-1 signaling pathway genes might confer a longevity benefit to individuals.[43]

———

The "Why do we age?" question could be answered if we better understood, at the molecular level, this shift from development and then maintenance to the loss of homeostasis with age. While Dobzhansky correctly observed that "nothing in biology makes sense except in the light of evolution," we can better understand the path that evolution might have taken if we can identify the shared molecular and signaling pathways that regulate longevity in multiple animals.[44] In fact, a molecular understanding of longevity mechanisms has already ruled out some of the oldest, most popular evolutionary theories of aging; our theories have to be adjusted to account for new data, not the other way around. It seems that in order to understand why we age, we must first understand *how we age*, and that might lead us to understand *how we can slow it down*. To answer those questions, we need to understand what is going on at the molecular level—like those Cuban cars, to find out what's going on under the hood.

3

Studying the Genetics of Human Longevity

CENTENARIANS AND WHAT
WE CAN LEARN FROM THEM

No single subject is more obscured by vanity, deceit, falsehood and deliberate fraud than the extremes of human longevity.

—*GUINNESS BOOK OF WORLD RECORDS*, 1984 EDITION

EVERY YEAR OR SO, someone takes on the deadliest job title in the world: the Oldest Living Person. At the time of this writing, Kane Tanaka of Japan holds that distinction, at 119 years.* Ms. Tanaka is a member of a growing group of exceptionally long-lived humans, or *supercentenarians*, who have passed the age of 110. Of course, there is a lot of uncertainty in maximum lifespan, because in many parts of the world, a century ago, birth records were not kept for many people. This has led to almost legendary lifespan claims in some cultures. One example is an Indonesian man named Sodimedjo, or Mbah Gotho, who was rumored to be 146 at his death in 2017, but of course there is no credible evidence to support this claim. But there are now plenty of examples of people (mostly women) with credible birth records who have lived beyond 110 years.

---

* During the writing of this book, several of the world's Oldest Living Persons died, including Ms. Tanaka, and were replaced by new, living record holders.

The most famous of the supercentenarians is of course Jeanne Calment, the French woman who lived to the age of 122, even outliving the lawyer who paid her monthly for 30 years in hopes of occupying her apartment after her death. As a side note, during the course of this book's writing, Nikolai Zak, a Russian mathematician, suggested that Calment's extreme age might in fact be the result of an identity-switching scam carried out by her daughter Yvonne to avoid paying estate taxes upon Jeanne's death (which was documented as Yvonne's death) in 1934.[1] According to Zak, their scam was intended to fool not demographers but rather tax collectors. Nevertheless, Yvonne—who was 23 years younger than her mother—lived so long (either 99 years, or 122 if she were actually Jeanne) that Yvonne/Jeanne became a French national treasure as the "world's oldest person"—or, alternatively, the truth had become too awkward to reveal. Zak presents some fairly convincing photographic evidence, historical context, and circumstantial evidence to support his theory, but also includes a bit of sketchy reasoning, such as the juicy tidbits that "Jeanne" and her "son-in-law" got along famously while raising Yvonne's son (which is framed by Zak as some of the most incriminating evidence). The case is still being debated (others have argued strenuously that Zak is wrong and Calment did live to 122) but it should be noted that even if Jeanne Calment loses her status as the longest-lived human, Kane Tanaka will take her place as the longest-living person, with nothing to suggest that her amazing 119 years was achieved through deception.

There is an ongoing debate about how long humans can live. This debate raged at least in part because of the publication of a well-publicized *Nature* paper suggesting that 115 years is the maximum lifespan that humans can achieve.[2] As I noted previously, this analysis can consider only current conditions—that is, lifespan without any longevity-specific intervention of any kind, simply the medicine that is currently available—it does <u>not</u> imply that we can never live any longer if appropriate treatments were developed. In any case, the authors suggested that we have already reached the maximum human lifespan, using Calment's age of 122 at her death as an outlier peak that affected the rest of the slope, leading to essentially a plateau around 115.

Vijg's "115 year maximum" publication ignited a firestorm of tweets and then several rebuttal letters.[3] Some of these pointed out that the study used Jeanne Calment's single data point of 122 in 1997 and an odd "portioning" approach to treat the data as two groups rather than one, and this approach had skewed the regressions, leading the authors to conclude that maximum lifespan has leveled out at 115. In my favorite of these rebuttals, one group illustrated this analytical problem by using the example of long-jump performance.[4] By using the same portioning approach—that is, treating the maximum long-jump distances each year after Mike Powell's world-record-setting jump in 1991 separately from those before, instead of properly including all data since 1960 as a whole—they showed that this approach would lead to the erroneous conclusion that long-jump performance has declined with time. Read that again: using the same analytical approach, one would conclude that since 1991, humans have been getting *worse* at long jump, because of Mike Powell's exceptional performance. Of course, this is a ludicrous conclusion. The reality is that instead there has been a statistically significant *increase* in performance since 1960, and Bob Beamon's and Powell's performances are outliers, like Jeanne Calment's exceptional lifespan. Additionally, the increasing number of very old people in countries like Japan suggests that the maximum human lifespan is not currently known and likely has not yet been reached. Later, Jim Vaupel and colleagues also somewhat controversially suggested that mortality—that is, the likelihood of dying—levels out around the age of 105, rather than continuing to increase with age, counter to expectation.[5] What this means is that once one reaches 105, the risk of dying no longer increases with time. Thus, it is likely that as the population of centenarians increases, the likelihood of more and more people making it past 115 will also increase.

So what's the secret to supercentenarian longevity? It's hard to know what practical advice there is to be gleaned from these super-agers, particularly since their own experience is by definition an "*n* of 1"—that is, one experience is not statistically significant, and they can know only what they have done themselves, so it's not a controlled experiment by any means. Although many supercentenarians attribute their longevity to specific foods or drink, their descriptions of their diets and health regimens appear to offer no consistent themes—except, perhaps, the surprisingly frequent mention of alcohol and "avoiding men," which might just be practical, given that most supercentenarians are women who have long outlived their peers. In fact, the 10 longest-lived women are all longer-lived than their male counterparts—the #10 longest-lived woman, Misao Okawa, lived over 117 years, while the longest-lived man so far, Jiroemon Kimura, died at 116.

One important point that may seem obvious but needed to be tested is whether long-lived individuals are healthy. This question is at the heart of aging research, since there is often the assumption that extending lifespan will just extend the frail part of life, when we would instead like to extend healthy lifespan. One could imagine that there might be some very frail people who just stay alive a long time in that state and make it to 100. Or, instead, centenarians could be, on average, more like our long-lived *daf-2* mutant worms, who stay healthy longer, and that is how they live so long. Remarkably, centenarians for the most part seem like *daf-2* mutants: they live longer by staying healthy longer. When individuals with exceptional longevity from the Longevity Genes Project and New England Centenarian Study were examined for the onset of age-related diseases (cancer, cardiovascular disease, diabetes, hypertension, etc.) it was found that they really did stave off these diseases longer and stay healthy longer than the reference participants did.[6] Their morbidity had indeed been compressed, as Fries had suggested.[7] Future centenarians seem to stave off age-related diseases longer than do their normal-lived counterparts.

Despite their avoidance of age-related diseases, healthy lifestyles seem to be the exception rather than the rule among supercentenarians. Parisian Robert Marchand retired from competitive cycling only at age 106 (he kept riding until age 108) and lived three more years, but Marchand's athleticism seems to be an outlier, as most centenarians have *not* attributed their own longevity to exercise or healthy living. In fact, many seem to have lived extraordinarily long *in spite of* unhealthy behaviors, such as smoking, drinking, and odd diets. The famous example is usually Jeanne Calment, who smoked most of her life (but perhaps we should adjust this to say that one might live only to 99?). Alcohol, ice cream, and bacon are frequently mentioned in response to the question "What's the key to your long life?" Jack Reynolds (106), who made a habit of setting a new Guinness World Record every year on his birthday, asserted that two shots of whiskey a day is the magic bullet, while Agnes Fenton (110) favored Miller High Life and Johnny Walker Blue. Theresa Rowley (104) credited Diet Coke for her longevity but acknowledges that her father's long life (102) might have more to do with it, particularly since Diet Coke wasn't around in his time. In fact, if there is a theme at all, it might be that if you are destined to live exceptionally long, it might not matter what your lifestyle is—hardly a useful message for the rest of us presumed mere mortals. Centenarians and supercentenarians, specifically because they defy all odds and seem resistant to all kinds of lifespan-shortening diets and behaviors, are likely not the best role models of a healthy lifestyle.

Instead, perhaps we should look to specific cultures and large groups of people who are generally long-lived to understand what lifestyle factors might be important. Michel Poulain, Giovanni Pes, and others have identified "Blue Zones," areas of the world that are home to the longest-lived populations.[8]

These include Okinawa, Japan; Sardinia, Italy; Ikaria, Greece; the Nicoya Peninsula, Costa Rica; and Seventh-Day Adventists in Central California. Longevity in these Blue Zones has been attributed to healthy diets, relaxed lifestyles, good social habits, and positive attitudes. Much has been made of the Mediterranean diet, which is largely plant based and includes olive oil and red wine, while Okinawans' diet was low calorie. In addition to the Mediterranean diet, the longevity of Ikarians has been attributed to characteristics of their culture: a slower pace of life, low stress, lots of walking, and good social integration. Many of these differences are simple common sense, but are not necessarily the norm in many fast-paced modern cultures.

These Blue Zone observations led Dan Buettner, an endurance cyclist and National Geographic fellow, to suggest that one could live longer by making nine major shifts in behavior, which include increasing natural movement, like walking; not smoking; relieving stress; belonging to a community; having purpose to one's activities; shifting the diet toward fruits and vegetables; eating until one is only 80% full; and enjoying time with one's family ("family first") and one's friends ("Wine @Five").[9] An ongoing social experiment is taking place in Albert Lea, Minnesota, where the Blue Zone recommendations are being implemented on a citywide basis.* The project includes changes in traffic design and the addition of new pedestrian zones, bike paths, parks, and social spaces; new work-site and school programs; and changes to dietary offerings in restaurants and grocery stores. Since its inception, the city reports millions of dollars in savings in health-care costs, mostly due to decreased rates of smoking and weight-loss-related changes. Although life expectancy as a result of these changes is projected to be an added 2.9 years, even if no one lived longer, the community overall seems healthier and has increased its quality of life while reducing its medical costs. This last point, reduced medical costs, should be stressed, since such infrastructure changes on a citywide level should be considered a long-term investment rather than a frivolous expenditure.

Of course, centenarians are not limited to the Blue Zones; Japan, Italy, Portugal, and Spain all produce centenarians at a rate of 40–50 per 100,000 people. (Weirdly, though, an audit showed that as many as 200,000 Japanese thought to be over 100 are actually unaccounted for, meaning that these data may be skewed.) Nevertheless, living a full century is becoming less and less remarkable. Maybe lifestyle is not everything, a classic "nature vs. nurture"

* "Blue Zones Project," by Sharecare, accessed January 26, 2023, https://albertlea.bluezones project.com/.

dilemma. Perhaps some populations are long-lived—once infection and childhood disease are reduced—because they are *genetically* prone to long life, in which case their genomes might reveal the secret to long life. It has been estimated that somewhere between a quarter and a half of the variation in human longevity has a genetic basis.[10] In fact, the best predictor of whether you will live a long time is whether you have long-lived family members, suggesting a strong genetic component to longevity. Men with a centenarian sibling increase their odds of becoming centenarians about 17-fold—but of course you won't know whether that will benefit you until it's basically too late. One remarkable example of sibling longevity is the Kahn family. The four Kahn siblings—Irving (109), Helen "Happy" Reichert (109), Peter Keane (103), and Lee (101)—lived exceptionally long and were fairly healthy until the end, exemplifying the "compression of morbidity" that Fries and others idealize as the best-case scenario for long life. The genomes of such families may reveal important clues to longevity.

But how can we find those genetic clues? Centenarians' DNA has been collected and probed for sequences that they may share that differ from those of non-centenarians in genome-wide association studies, or GWAS. These studies rely on the collection of enough individuals to detect SNPs (single nucleotide polymorphisms, or changes of a single base in DNA) that occur in a higher frequency in the long-lived population than in a population with a normal lifespan. The GWAS approach has worked fairly well for some diseases that can be attributed to a small number of genes, but it is by no means straightforward.* GWAS are a bit like looking for a needle in a haystack—but if you have enough needles, you are more likely to find them, so increasing the frequency of the characteristic you are looking for helps a lot. For GWAS to work, there must be enough cases with and without the disease or phenotype one is looking for, so that there will be sufficient statistical power to decide that the DNA variant is actually associated with the disease, not just coincidentally. It can be a daunting task to collect enough individuals with the phenotype of interest, particularly those sharing the same DNA variant. Because there are no "control" cases of the same age in a centenarian study, often the offspring of centenarians are compared with appropriate age-matched controls.

---

* GWAS show correlations, not measurements of health or longevity. SNPs associated with diseases can give us only predictions, not diagnoses. Someone with a particular SNP may be more likely to live long (or develop a specific disease), but this is not a guarantee. This is different from diagnostics, which tell us that someone does or doesn't have a disease.

A second hurdle in the success of GWAS is the high cost of getting the relevant information. While SNP-chip DNA microarrays (arrays with probes representing the relevant single nucleotide polymorphisms to be probed, rather than using the whole genome) were the original GWAS tool, they were expensive, which limited the amount of data one could gather. Later, just using the small fraction (>5%) of the genome that gets translated into proteins, the "exome," was used, by capturing that portion of the genome and then just sequencing those regions—but, of course, that would miss any SNPs that are in noncoding, regulatory regions, which could be very important as well.[11] Whole-genome sequencing (WGS) was originally prohibitively expensive but has become used more frequently owing to the ever-dropping cost of sequencing technology, but SNP arrays are still used to find differences that affect longevity (or to determine if they even exist). Scientists have focused on centenarians with shared ancestry, such as Okinawans, Ashkenazi Jews, Germans, Danish twins, the highly related Icelandic population, and Finns.[12] Other studies have enrolled varied populations (e.g., the New England Centenarian Study,[13] Longevity Genes Project, Long Life Family Study, Religious Orders, Nurses Study, and Framingham Study) and looked for associations that might emerge even from more heterogeneous populations. (It is also worth asking what the relationship is between Blue Zone lifestyles and the genetics of the people living in the Blue Zones.) In any case, as more people live extremely long, and costs of sequencing genomes continue to drop, we will soon have much more information on the genetics underlying longevity.

So what do the genomes of the exceptionally long-lived reveal? First, let's be clear about what GWAS can and cannot tell us. The result of such a study is the identification of a consistent association between a phenotype (here, long life) with a particular DNA base in the genome (the SNP, or single nucleotide polymorphism). Consistency, or significance, is assessed by the p-value, which tells us how likely it is that this association is not real, but rather just a coincidence (so, very small p-values mean we are more confident that the association is real). We infer that a particular gene is implicated by its proximity to the identified SNP—it could be in the coding region of the gene, an intron between exons of the gene, or a regulatory region that controls something about the expression of the gene, and since the human genome is complex, it is also possible that a SNP is in a region of DNA that controls the expression of a gene that is quite far away (an "enhancer"), which sometimes complicates interpretation. We can have even more confidence that a gene is involved if multiple SNPs from several studies also identify that gene.

What GWAS cannot tell us is *how* that gene affects longevity—that is, what the underlying mechanism of that gene's effect on longevity might be. If the SNP (the difference in the DNA code at a particular site) causes something very obvious, like a mutation that shortens or "breaks" the protein, we can be fairly certain that that gene's function normally inhibits long lifespan. But it is rare to have such an obvious change, since that type of mutation is likely to also cause some deleterious—even fatal—effect early in life (remember the antagonistic pleiotropy theory from the last chapter), so what we are more likely to identify in longevity GWAS is something much subtler. Finally, DNA for the most part (unless it has mutated) is the "static" or stable component of the central dogma, unlike dynamically produced RNA and protein, so having a SNP that correlates with a disease doesn't tell us anything about *when* in one's life that might occur. Despite these caveats, GWAS can be a powerful method to start understanding the molecular basis of disease, and has begun to give us some clues to the genetic basis of longevity.

Early GWAS longevity studies indicated that variants of genes called *APOE* and *FOXO3A* are associated with long life, and these findings have been repeatedly confirmed. Later studies also pointed to the involvement of the CHRNA3/5 nicotinic acetylcholine receptor in longevity. These three genes are the most consistently identified candidate genes across all longevity GWAS studies.[14] More recently, a SNP whose gene function is unknown, *5q33.3*, has emerged consistently.[15] So what are these longevity-associated genes, and why might they be associated with lifespan? APOE might sound familiar to you because of its association with Alzheimer's disease. APOE is involved in regulation of lipids and levels of HDL, or high density lipoprotein; high levels of HDL (the "good cholesterol") are associated with better cardiovascular health and increased lifespan. APOE has several variants—ε2, ε3, and ε4. The most common variant is ε3, and the ε4 variant is most associated with increased levels of risk of developing Alzheimer's disease, while the ε2 variant has been associated with age-related macular degeneration and age-related hearing loss. However, despite its consistent association with differences in longevity in GWAS, exactly *how* APOE variants affect lifespan must still be unraveled.

APOE SNPs have been found to be in "linkage disequilibrium" with a gene called *TOMM40*—that is, the *APOE* and *TOMM40* SNPs are found together more often than you would expect by chance—and therefore some longevity association might actually operate through TOMM40, a mitochondrial protein. Because GWAS themselves do not reveal any information about the underlying molecular mechanisms, other types of experiments on these genes directly are necessary to understand how each variant might contribute to longevity.

Other genes may act in a similar manner to APOE, affecting cholesterol metabolism. Like APOE, the CETP cholesteryl ester transfer protein (in fact, the variant that the Kahn siblings had) is also associated with differences in longevity and may also affect HDL levels, in turn affecting cardiovascular disease risk and longevity.[16] *CETP* has been associated with reduced dementia and Alzheimer's disease risk, as well.[17] Lipoprotein(a) (LPA) is another GWAS-identified gene that may regulate aging and age-related disorders (diabetes, cardiovascular disease) through its role in cholesterol and lipid regulation. Whether, or how, these aging and Alzheimer's disease effects are related is not yet clear. The mysterious chromosome location 5q33.3 has also emerged in recent longevity GWAS and may have a role in blood pressure regulation and cardiovascular disease; further studies may clarify the gene function in question, though. (Lipid regulation is a common theme in aging research, as you will see; lipid dysregulation is associated with metabolic disease and obesity, while many of the conditions and mutants that regulate lifespan (see later chapters) in fact may do so through the regulation of lipids.) In any case, there is a strong connection between cholesterol metabolism and longevity, so understanding exactly how this works could have great benefits for human health.

The consistent identification of the *FOXO3A* gene in human longevity studies alongside *APOE* was a vindication of sorts, after years of model-system research identifying FOXO as a major regulator of lifespan. FOXO is a transcription factor—that is, a DNA-binding protein that directs the expression of mRNA of specific genes. Research first done in *C. elegans* on the FOXO homolog (that is, the worm version of the FOXO protein) called DAF-16, then in flies (dFOXO), then in mice showed that FOXO is required for insulin-signaling-mediated longevity in many species, and increasing the activity of FOXO homologs extends lifespan (more about this in chapter 6).[18] Variants of *FOXO3A* were found to be associated with human longevity in a study of long-lived men of Japanese descent, as well as a study of southern Italian centenarians, and other *FOXO3A* variants were associated with longevity in a mixed population of German centenarians.[19] Similarly, variants in the IGF1R (insulin-like growth factor 1 receptor) that regulates FOXOs in worms, flies, and mice are associated with human longevity,[20] paralleling the discovery—again, first in worms, then flies, then mice—that inhibition of the insulin/IGF-1 signaling pathway, which normally acts to block FOXO activity, extends lifespan. Specifically, IGF-1 receptor mutations are enriched in centenarians relative to controls in an Ashkenazi Jewish population. A later meta-analysis of Europeans also implicated circulating IGF-1 and IGF1BP-3 levels and FOXO3.[21] Together, these data suggest that the insulin/IGF-1 signaling pathway and its control of FOXO activity is likely responsible for at least some component of extreme human longevity. These small changes in DNA likely lead to small tweaks in the activity of the proteins they encode, but with significant enough effects that they can ultimately affect how long one lives.

The identification of the *CHRNA3/5* nicotinic acetylcholine receptor as a longevity candidate gene is particularly interesting because its link to aging might be in part through its regulation of a human behavior: smoking. Regression of parental lifespans in a large dataset (the UK Biobank) suggested that SNPs in the *CHRNA3/5* gene are associated with nicotine dependence, lung cancer, and chronic obstructive pulmonary disease—all diseases that are secondary to smoking—as well as schizophrenia,[22] a disease in which many patients self-medicate with nicotine. Does the *CHRNA3/5* nicotinic acetylcholine receptor alter lifespan through a variation in addiction to nicotine, which would in turn promote smoking? Future work may tease out these possible links between genetics, behavior, and longevity. The authors also incorporated lifestyle factors into their study, which revealed that an increase in body mass index (BMI) of one index unit reduces lifespan by 7 months, while each year of education adds 11 months to expected lifespan. These behavioral effects were thought to be mediated by diet and smoking, respectively, as BMI was associated with cardiovascular disease and obesity, and education was associated with a reduction in smoking-related diseases. The convergence of nature and nurture (environment) through genetic background and health behavior tendencies is important to understand, as genes do not operate in a vacuum—both our behavior and our genes are likely to have major effects on rates of aging and age-related diseases.

Inflammation, or the chronic activation of immune responses, is another consistent motif that has emerged from aging studies.[23] It was first noticed that there is a significant gene expression "signature" of inflammation in transcriptional studies of aging tissues, including the brain, and GWAS have further supported this finding, identifying inflammation gene SNPs that are associated with longevity differences. These genes include the major histocompatibility locus, *MHL*; the IL-6 inflammatory cytokine; and human leukocyte antigens, *HLA* (*HLA-DQA1/DRB1*);[24] which are all associated with aging and inflammatory disease. The study of "inflammaging" is a growing area of research partly owing to these findings. While regulated inflammation is an important part of the immune response, aberrant, unregulated inflammation is detrimental. In fact, part of the reason that senescent cells are so bad is not just that they die but that they secrete factors that exacerbate inflammation. Judith Campisi's lab has characterized this phenomenon, dubbed the senescence-associated secretory phenotype, SASP (which we will discuss in greater detail in chapter 13).[25]

Early GWAS were somewhat underpowered, simply because there were not enough centenarians to enroll in the work, but the results hinted at genes that might play some role in longevity. More recent work has utilized ever larger datasets. Although the rate of centenarians in China is not particularly high, the sheer size of the population ensures that a reasonable number of 100-year-old individuals can be found (although one could argue that then the signal-to-noise ratio should be lower, of course). A study of a Han Chinese population examined genomes from a whopping 2000 centenarians (!) and 2000 middle-aged controls,[26] allowing the identification of a larger list of candidate contributing genes with good certainty. The inflammatory gene *IL6* emerged from this work again, implicating inflammation's role in age-related decline, as did *APOE* and *5q33.3*, and eight SNPs overlapped with those previously identified in studies of US and EU populations. As in other studies, cholesterol regulation emerges as a theme: the authors zeroed in on the *APOE* ε4 allele/*TOMM40* gene, which plays a role in regulating HDL. A follow-up study took into account the psychological and cognitive characteristics of the large Han Chinese group, and found that "good psychological resilience and optimism are keys to the exceptional longevity enjoyed by centenarians"[27]— but you might be optimistic, too, if you'd managed to outlive everyone else!

Another approach to understanding exceptional longevity is to look at the most extreme cases—so back to our supercentenarians, people who live beyond the age of 105. James Clement, a lawyer by training who has become interested in understanding longevity, and George Church, the Broad Institute's sequencing maven, are taking a novel approach to cracking the longevity mystery: they have made the DNA sequences of supercentenarians available to any researchers who are interested, essentially crowdsourcing the problem.[28] Clement has taken it upon himself to identify and collect DNA from every supercentenarian he can, and Church's lab sequences the DNA and makes the data available to interested collaborators. While the "Betterhumans Supercentenarian Research Study" set is currently relatively small (about 60 supercentenarian DNA samples, plus an additional 19 samples that were previously collected and sequenced by Stuart Kim at Stanford),[29] the hope is that these extreme cases will highlight particularly important genes. This work is still ongoing, so it is too early to tell whether this approach will yield genes that previous centenarian studies have not; certainly the continuous addition of new samples to the set will make the conclusions stronger. One question we can ask, in addition to "Are there specific gene variants that are different in this population?" is whether these individuals with extremely long lifespans have lived so long simply because they don't have the common disease-causing gene alleles, so they are protected. My colleague Joshua Akey and co-workers examined the data from Clement's dataset to answer this question, and their preliminary analysis suggests that supercentenarians aren't simply less "burdened" with disease-causing genes; this means that they live long despite having all the same kinds of disease genes as the rest of us. This is an important hypothesis to test, because it makes it more likely that changes in single genes could help a person live long and avoid common illnesses, even if they are genetically predetermined to have those diseases. Another study of Italian semi- and supercentenarians came to the opposite conclusion, suggesting that DNA repair mechanisms must be more efficient, as they found lower somatic mutation patterns.[30]

Because GWAS of longevity are difficult, only a few have been carried out. Another approach to identify genes important in aging is to combine several GWAS into a meta-analysis—that is, an analysis of many studies together. By focusing on GWAS of *age-related diseases* rather than on the limited set of genome-wide studies based purely on longevity, Yousin Suh's group (then at Albert Einstein College of Medicine, now Columbia) was able to greatly expand the number of studies they could examine, with 410 GWAS and five age-related-disease categories.[31] They then used a very stringent significance cutoff ($p < 1 \times 10^{-5}$) for genes associated with the traits, to identify significantly associated SNPs with high confidence. They grouped the age-associated traits into cardiovascular disease, metabolic disease, cancer, neurodegenerative disease, and other age-related traits. All three of the genes that were represented in all categories were in the MHC locus (which is often overrepresented in GWAS). When this criterion was relaxed to include genes that were in at least three of the five age-related-disease categories, additional genes emerged, including *TOMM40/APOE*, inflammation genes, and the cell cycle/senescence regulator *p16INK4a*. Genetic pathway analysis of the genes identified nutrient-sensing signaling, lipoprotein metabolism, genome maintenance, proteostasis, and other canonical pathways of aging. Interestingly, while age-related diseases were most similar to metabolic diseases, suggesting that the two might share some mutual cause, the least overlap was between metabolic disease and neurodegenerative disease, yet both are associated with aging.

In another meta-analysis approach, Stuart Kim's group (Stanford) developed "informed GWAS" (iGWAS), which utilizes previous knowledge about age-related diseases combined with information from 14 meta-analyses of disease GWAS, along with a p-value weighting strategy to uncover new genes.[32] Basically, the idea is that SNPs that increase your age-related-disease risk would decrease your chances of becoming a centenarian. They found eight genes that associate with longevity. These age-related categories included not only the usual APOE/TOMM40 association with longevity (which is a kind of sanity check for longevity GWAS at this point), but also CDKN2B/ANRIL, an "antisense non-coding RNA in the INK4 locus" that may play a role in cellular senescence, the O blood group gene ABO, HLA loci (associated with immune function), and SH2B3/ATXN2, which is associated with neurological disease and has been shown to extend lifespan in *Drosophila*. The advantage of these types of meta-analyses is that they might reveal genes that do not meet the bar for significance in straight-up longevity studies, and they do not suffer from the small sample size that most centenarian studies do.

Despite all the progress identifying candidate genes associated with longevity, a monkey wrench has been recently thrown into the whole idea: a group from Calico, the Google-based longevity company, used data from Ancestry .com to explore genetic links to lifespan (note: not exceptional longevity) and uncovered none—but instead showed that an individual's lifespan is best correlated with the lifespan of their spouse, and even their in-laws.[33] The former result rules out genetic factors, and the latter result seems to rule out immediate environment as the most important factor in lifespan. This surprising result sounds odd until you think about the fact that people usually marry someone within their own "type" of social group—similar education level, religion, economic status, and lifestyle in general. Calico's results suggest that lifestyle and socioeconomic factors, after all, may outweigh most genetic links to lifespan, at least among people with normal rather than exceptional longevity.

Nevertheless, even within shared socioeconomic groups, there are people who do live exceptionally long. And there are people who live exceptionally long despite the odds not being ever in their favor. How can we understand what might be going on in their cells and tissues that allows them to live longer than their counterparts, even under the same living conditions? For that, we need to not just *identify* potential genes that could regulate lifespan but also *test* them—and to do this, we need experimental systems to help us untangle cause and effect, since we can't use humans for such studies. Specifically, we need to manipulate genes in short-lived, genetically tractable model systems, to figure out whether genes that are different are really causing changes in lifespan, or whether they just correlate with differences in lifespan. Maybe then we can finally unravel the secrets of supercentenarians.

4

Long-Lived Species and Longevity Mutants of Model Organisms

You have evolved from worm to man, but much within you is still worm.

—NIETZSCHE, *THUS SPAKE ZARATHUSTRA*

ONE OF THE BEAUTIFUL THINGS about evolution, from a scientist's perspective, is that many of the genes that we find affect longevity and aging in one animal are well conserved in others; as Jacques Monod observed, "What is true for *E. coli* is also true for the elephant."* This fact allows us to use fast-living animals to discover and carefully dissect new longevity mechanisms, and then later we can more quickly test our hypotheses in so-called "higher" (slower-lived, more complex, more expensive, and more ethically fraught) organisms, which are more difficult to use for unbiased, discovery-based science. In fact, aging research uniquely benefits from these short-lived model systems simply because of the time it takes to carry out longevity work in mice or primates, unlike fields such as developmental biology, where an embryo's formation might not take too long to study in mammals. Once we understand how these pathways work, it's easier to understand how they might function in our own cells.

---

* "Tout ce qui est vrai pour le Colibacille est vrai pour l'éléphant," quoted in Pallen, M., 2006, "*Escherichia coli*: From Genome Sequeces to Consequences (or 'Ceci n'est pas un éléphant . . .')," *Canadian Journal of Infectious Diseases and Medical Microbiology* 17:114–16, https://doi.org/10 .1155/2006/345319.

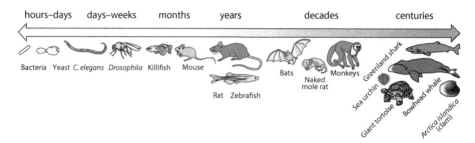

| hours–days | days–weeks | months | years | | decades | centuries |

FIGURE 2. Scales of lifespans range from hours to days for single-celled bacteria and yeast to the extremes of several centuries for giant tortoises, whales, sharks, and clams. While the super-long-lived animals are fascinating, it is difficult to experimentally test them; organisms at the faster end of the scale (from zebrafish downward) allow manipulation and rapid experimentation. All but bacteria (prokaryotic) are eukaryotic; those from *C. elegans* onward are multicellular.

How can we study genes that control rates of aging? One approach is to try to understand how extremely long-lived individuals, such as centenarians (chapter 3), or long-lived species, like the bowhead whale and naked mole rat, differ from their shorter-lived counterparts (figure 2).

While human genome-wide association studies (GWAS) and whole-genome sequencing have allowed us to identify genes that may be correlated with long human life, the inability to experimentally test these genes in humans and long-lived organisms—and the exceptionally long time it would take to carry out such an experiment if we could—suggests that identifying gene mutations that confer long life upon shorter-lived, well-studied model systems is a more powerful approach to study longevity. Groundbreaking longevity research done in yeast, *C. elegans*, *Drosophila*, and mice has been instrumental in making longevity studies feasible, ushering in our current era of aging research.

Comparative Biology: The "Longevity Quotient" and the Study of Long-Lived Animals to Understand Longevity

Humans are not the longest-lived creatures on earth, and it seems that we could learn important lessons from some of nature's Methuselahs. We can divide exceptionally long-lived organisms into two categories: those that are much longer lived than humans, and those that are long-lived *for their size*— that is, they have an extreme "longevity quotient," or LQ.[1] For the rest of this section we will focus on animals, but it should be noted that the longest-lived organisms on earth are in fact trees. This includes an 80,000-year-old, interconnected aspen colony, and several species of trees (mostly gymnosperms) that have been measured to be 4000–5000 years old: the Great Basin bristlecone pine, the Cypress of Abarkuh, and the Llangernyw Yew.

Many of the longest-lived animals on our planet are sea dwellers. Beside several species of whales, non-mammalian marine animals can be exceptionally long-lived; this ranges from the huge (think sharks) to relatively small (like clams and sea urchins). For example, the quahog clam *Arctica islandica* can survive for several centuries; in fact, an ocean quahog collected off the coast of Iceland in 2006 was named "Ming," for the dynasty during which it was originally thought to have been born.[2] Later carbon dating experiments suggested that Ming had been born in 1499, and lived a whopping 507 years before being frozen upon harvesting and then popped open to count its rings (oops). Of course, we don't know how long Ming might have lived if it had not suffered this unfortunate fate, given that its heart displayed no significant oxidative protein damage or other signs of senescence.

Vertebrate marine animals can also live extremely long: in 2016, the Greenland shark (*Somniosus microcephalus*) was found to live up to five centuries, the world's longest vertebrate lifespan.[3] Because proteins in the lens of the eye are deposited while the shark is still in its eggshell and are never replaced, these tissues can accurately reveal how long ago that eye was first made through radiocarbon dating (thanks to nuclear tests done in the 1950s and 1960s that provide radiocarbon bomb pulse markers). Analysis of the lens proteins of 28 female sharks' eyes allowed the researchers to estimate when each animal was born; their ages ranged from 272 to 512 years. This shark species is quite large and thus lacks predators, and lives in very cold water (37°–41°F), likely slowing its metabolism, two factors that may contribute to its extreme longevity.

But the cold, underwater life is not the only way to survive for centuries. Jonathan, a giant Seychelles land tortoise living on the island of Saint Helena (and commemorated on the island's five pence piece), hatched in 1832.* Jonathan was sent as a gift from the Seychelles to the government of Saint Helena when he was 50; he is currently the longest-living live terrestrial animal at 190 years. In January 2022 he surpassed the previous record,[4] which was held by Tu'l Malila, a tortoise who lived to the age of 189. Other tortoises are also

---

* Currently the world's oldest known land animal is Jonathan, an Aldabra giant tortoise who lives on the grounds of the governor's mansion in Saint Helena, an island off West Africa.

known to live long: Galapagos tortoises have been reported to have lifespans similar to Jonathan's (although I'm not sure my guide Mario on Santa Cruz island was accurate in claiming that Luis the Galapagos tortoise is 220 years old). How tortoises live so long is not understood, however—and it's unlikely (unwise and unethical) that we'll ever do experiments on these rare animals.

Long-lived mammals may hold some clues that are relevant for humans, but they simultaneously highlight the need for short-lived experimental models. For example, in 2015, the genome of the bowhead whale was sequenced.[5] Bowhead whales have been found to live over 200 years—and in at least one case certainly over 100 years, judging from a whaling harpoon injury—and are the longest-living mammals known. Multiple experimental and computational approaches were used to tease out candidate longevity genes; most powerfully, the authors compared the bowhead whale sequence to that of the minke, a related whale that lives only 50 years, as well as to those of the dolphin, cow, and other mammals. They looked for genes that were similar but had slight differences in their sequences, reasoning that these changes might account for the large difference in lifespan. Perhaps unsurprisingly, the genes that emerged from this work are mostly those we already knew: genes involved in cancer suppression, including FOXO3; DNA repair (ERCC3); and the signaling pathways that control growth, development, and longevity (TOR, FGF, and insulin/IGF-1; more to come about these genes in later chapters). Weirdly, CHRNA10, a nicotinic alpha 10 receptor, emerged from these comparisons, calling to mind the human CHRNA3/5 gene identified in GWAS data (maybe bowhead whales live so long because they don't smoke?). Despite the promise of identifying new genes from this long-lived mammal, few new longevity candidates emerged, although more work must be done to test this. The biggest problem with identifying "longevity" genes in long-lived organisms, just as with human GWAS results, is verification of candidate genes or SNPs, since one cannot do experiments in the whales; testing any new candidates would require using a short-lived model system or tissue culture. But the combination of candidate alleles with short-lived systems could be powerful, if harnessed correctly. The hope is that, eventually, new genes rather than those we have already identified in GWAS or model-system studies will eventually emerge from the sequences of long-lived animals.

Longevity increases roughly with size from species to species (think elephants and whales), but size dependence of longevity of some species is roughly inverted *within* species. For example, as I mentioned previously, small dogs live on average longer than large dogs; their lifespans are inversely correlated with their levels of the insulin-like growth factor IGF-1, which controls their size. This is somewhat like the long lifespan that is associated with human genetic variations that suppress growth. For example, Laron syndrome (LS) is a form of dwarfism that is also linked to a lack of IGF-1 signaling.[6] (You might have noticed that, again, the insulin growth factor pathway plays a large part in connecting growth to longevity.) People with LS have short stature but are at lower risk for cancer and diabetes, and possibly other age-related diseases. Patients with LS are the subject of endocrinology studies that aim to understand this interesting constellation of characteristics. LS was first identified among a cohort of patients in Israel, but LS patients can also be found around the world, including in several South American countries, such as in a remote area of Ecuador.[7] Interestingly, these Ecuadorian LS patients may have Sephardic Jewish ancestors who migrated from Spain during the post-Inquisition Spanish colonization of the Americas.

Perhaps the most interesting animals for longevity studies are those that live longer than one would expect them to, given their size—that is, they have an extreme "longevity quotient," or LQ.[8] An LQ of 1 means that the animal lives as long as you'd expect for an animal of its size, on average, while animals exceptionally long-lived for their size have LQs much greater than 1. For example, the little brown bat lives into its 30s, and Steve Austad has reported the longest bat lifespan yet, a 41-year-old male. Yet a bat's body size is comparable to that of a mouse, whose lifespan is on the order of two years. Thus, the brown bat's LQ is extremely high, almost 10. Marsupials, such as opossums, and shrews occupy the other end of the LQ range, living much shorter than one would expect for their size; it is thought that some species die quickly after breeding due to a hormonal surge, much like salmon.

The longevity of bats is still a bit of a conundrum, although new sequencing techniques are cracking the case. As I mentioned in the last chapter, while it would be easy to assume that this longevity is merely connected to metabolic shifts associated with hibernation, some nonhibernating bats also have long lifespans. Instead, at least part of their longevity may be due to their ability to escape from predators through flight, and benefits of hibernation might be due to this predator escape, as well. (Hibernation itself may have some benefit owing to its "overwintering" and association with slow life histories.) But how exactly this predation escape changed the animals *at the genetic and molecular level* to allow such extremes in longevity is not yet clear.[9] The telomeres (the ends of chromosomes, which shrink with every round of cell division) of bats appear to remain long with age,[10] but it's not clear that they have more

telomere-building (telomerase reverse transcriptase, or TERT) activity, relative to mice. Perhaps genome and SNP comparisons with mice, like the bowhead whale approach, will reveal other candidate genes that confer such striking longevity upon bats. Whole-genome analysis of the Brandt's bat revealed
unique sequence changes in insulin and growth factor pathway genes (GH
receptor and insulin-like growth factor 1 receptor),[11] suggesting that at least
part of this bat's high LQ is due to decreased IGF-1-like signaling, in addition
to low reproductive rates, hibernation, cave dwelling, and flight. Gene expression (mRNA) studies of specific tissues might also reveal differences that
could be important for bat longevity—and these candidate genes and SNPs
could be tested by putting the bat version of those genes into mice and seeing
how long the mice live.

If you have been following aging research for any length of time, by now
you have probably heard of naked mole rats, the ugly little rodents that are the
wonders of the longevity field. These rat-sized animals live into their 30s or
beyond, resulting in an LQ of at least 4. They are hairless, burrow dwelling,
and eusocial (that is, they have a queen and workers), and they can survive in
low-oxygen and toxic soil conditions—in a nutshell, they are weird, fascinating creatures, even without the fact that they can live more than three decades.
Moreover, unlike their mouse counterparts, they rarely develop cancer, and
their health, metabolism, and cellular physiology appear to change very little
with age—thus they experience the compression of morbidity that is the goal
of extended lifespan. How they achieve their absurdly long lifespan is the focus
of much of the work being done with naked mole rats. The animal's subterra-
nean lifestyle might protect it from predators and thus decrease its extrinsic
mortality, similar to the bat's ability to fly, allowing it to evolve a longer lifespan. It would also be less exposed to ultraviolet rays living underground, of
course. However, it also survives low oxygen and high carbon dioxide levels,
which must require other metabolic adjustments. Counter to the reactive oxygen species theory of aging, naked mole rats seem not to live long by increasing
their ability to deal with oxidative stress, as their proteins still show significant
levels of oxidative damage,[12] and antioxidant proteins are not increased.

Rochelle (Shelley) Buffenstein's lab has demonstrated that the naked mole
rat shows "negligible senescence"—that is, its mortality seems to not increase
with age, which has not been observed for any other mammal.[13] Nevertheless,
naked mole rats display many of the hallmarks of senescent cells,[14] including
the senescence-associated secretory phenotype (SASP), which can lead to
increased inflammation and other deleterious aging phenotypes, suggesting

that they do not avoid aging by eliminating SASP—although retaining hallmarks of senescent cells might explain their resistance to cancer over their long lives. However, they do have increased DNA repair, proteome maintenance, autophagy, and proteasome activity. This enhancement of stress resistance may be mediated in part by maintenance of RNA splicing,[15] which has been shown to change with age in other organisms (I'll explain more about RNA splicing with age in chapter 8). Comparative genomic analysis suggests that naked mole rats deal with aging by regulating the processes that repair DNA and proteins, insulin and TOR signaling, telomere maintenance, and tumor suppression,[16] as well as changes in mitochondrial uncoupling activity that might account for their differences in thermogenesis.[17]

A surprisingly simple and unexpected (at least to me) mechanism of low incidence of cancer in naked mole rats involves the building of long, high molecular weight (HMW) sugar chains called hyaluronic acid (HA), through the activity of the hyaluronic acid synthase gene, Has2,[18] which builds the HA chains. High levels of HA may allow the animals the highly elastic skin they need to squeeze through tight tunnels. Vera Gorbunova and Andrei Seluanov's lab showed that HMW-HA may also be responsible for the increased ability of cells to be resistant to malignant transformation. To test whether this is just a weird naked mole rat phenomenon, mice making the naked mole rat version of the Has2 gene, which synthesizes the HMW version of HA, should be tested for longevity and cancer prevention, assuming that it will also trigger the early contact inhibition phenotype that naked mole rats use to prevent cancer incidence. Because of its simplicity, expression of Has2 may be a promising avenue that could affect other organisms and systems.

Like naked mole rats, other eusocial animals (that is, species with castes of workers and queens but a single shared genome) have interesting longevity characteristics. Ant and bee species (e.g., *Apis mellifera*) have queens and social castes with different roles in the hive and correspondingly different lifespans. Eusocial insects present a particularly vexing example for those who cling to the reproduction/longevity trade-off idea, because the queens are both very long-lived and remarkably reproductive. They may achieve these differences through compartmentalization of tasks and regulation of the insulin signaling pathway (chapter 6) and lipid metabolism,[19] while some molecular mechanisms maintain gene expression that is specific for each social caste. (We will talk more about these "epigenetic" mechanisms that regulate aging in a later chapter.) Eusocial model systems may offer insights into aspects of longevity that humans could also use.

New Model Systems for Aging Research

While exciting, the pace of discoveries in many long-lived animals has lagged behind traditional model systems because the former systems still lack genetic tools that allow easy manipulations and gene studies—but that drawback is changing rapidly with the advent of genomic analyses, antisense and RNA interference techniques, and CRISPR gene-editing approaches. The ability to knock down, replace, and tag genes in these eusocial animals might help us understand not only how gene activity affects lifespan but also how social structures and cooperation might affect longevity of individuals and groups, which in turn should help us understand how the same factors affect human longevity.

Every year in my Aging course, I ask students to pick a new "model system" and write a brief essay on (1) what question in aging research this model might be ideal to help answer, and (2) what would have to be done to develop the system into a true model system for lab research. I've been pleasantly surprised at the range of animals the students have proposed, which have included salmon, Blanding's turtles, New Zealand white rabbits, gray mouse lemurs (a short-lived primate), bottlenose dolphins, marmosets (*Callithrix jacchus*), honeybees (*Apis mellifera*), tardigrades (*Ramazzottius varieornatus*), sea squirts (*Ciona intestinalis*), planarias, and brown marsupial mice (*Antechinus stuartii*). Each of these animals may be useful in addressing a particular aspect of longevity regulation, such as the true cause of semelparous death (salmon, marsupial mouse), close homology to humans for the study of cognitive decline (lemurs, marmosets), exceptional stress resistance (tardigrades), exceptional longevity (turtles), sleep (white rabbit), and regeneration (*Ciona*, planaria). The new genetic tools becoming available make it easier to imagine these will all become tractable models at some point soon, and I've been excited about the creativity the students exhibit when picking out these still-developing model systems for the study of aging.

How Invertebrate Model Systems Became the Kings (and Queens) of Aging Research

As I pointed out above, one shortcoming of human studies is that we could never carry out tests in humans of the genes identified in genome-wide association studies to *verify* their roles, and so we must infer their importance from other information. The same problem exists for the study of candidate longevity genes in other long-lived animals, and many of these interesting animals lack the genetic tools that would allow us to study the genes that have been uncovered—or live so long it would be hard to finish the experiment in a reasonable time. So how can we test the genes that we discover through human GWAS, or by comparative genome studies?

One idea is to artificially increase or decrease that gene's level in human cells in vitro (i.e., in cell culture) and then determine whether the gene in question affects how long the cells survive or divide. Cells can divide only a limited number of times before becoming sick and dying; this limit was first recognized by Leonard Hayflick at the Wistar Institute in Philadelphia in the early 1960s, and thus is now known as the Hayflick limit. The cells become unable to divide because they are undergoing cellular senescence, which has specific metabolic and cellular properties, including shortening of telomeres and the secretion of toxic metabolites that poison other cells (see chapter 10 for more information). In tissue culture, we can manipulate the genes in question using molecular approaches and then test the gene's effect on cellular senescence and other characteristics. Of course, this lets us know what is going on at the cellular level, but doesn't explain how things might work in a whole animal.

Another, much more laborious but potentially more informative way to test candidate GWAS genes is to mimic the candidate genetic change in mice (that is, tweak the mouse version of the human gene in question), and then study the mouse's lifespan and other health- and age-related phenotypes—but, of course, testing the effect of a gene on mouse lifespan takes several years, so it would be best to know a lot more about the gene before embarking on such a path. Both of these approaches will benefit from new CRISPR-based genome-editing methods that allow exact modifications to be made, rather than making inferences from extreme changes, such as gene deletions or overexpression. Another option is to simply blast ahead and look for drugs that affect the pathway identified by the genetics; in the case of APOE, CETP, and other genes that we know affect HDL/LDL levels, we may already know enough from other studies to take this approach, particularly since cardiovascular disease is a useful age-related end point that can be evaluated instead of lifespan.

Yet another way to study aging is to instead use simpler, genetically tractable animals to test these candidate genes and pathways. One of the first "model-system" longevity experiments was the surprising discovery in 1935 of caloric-restriction-induced longevity (chapter 7). Clive McCay was subjecting rats to different levels of diet to understand the possible effects of nutrient deprivation during wartime. It was expected that the rats in the study with the lowest level of diet would fare worse, so it was surprising to find that they in fact seemed healthier and more youthful. This observation opened up a new field of research into dietary restriction and its effects on longevity. Mice have become the mammalian model of choice for genetics (because they are smaller and easier to house, and thus more genetic tools have been developed for mice), although some nutrient studies are still done in rats.

The best genetic model systems are short-lived, so scientists can do many tests quickly, but still recapitulate aspects of aging that are shared with humans, allowing testing of lifespan effects of gene manipulation in old animals. For these types of questions, we turn to small invertebrate genetic model systems, the workhorses (so to speak) of the longevity field. Although worms and flies seem very different from us, in fact they share most of our genes. This means that we can study the processes we want to understand in humans in a fast-living organism, so that we can learn a great deal more about the gene and its interactions. These tiny animals have allowed insights into the cellular roles of aging genes that we might otherwise have had to guess about. And instead of just looking at how many times a cell can divide, we can examine how an animal behaves, and how long it is healthy, all within a few weeks.

It's hard to convey how important invertebrate model systems have been specifically for longevity research. Without these small, short-lived, genetically tractable models, it's very likely we would not know much more about aging now than we did in 1985. Being able to change the levels at which a gene product (mRNA and protein) is made, or visualize proteins in a live animal while watching it age (or not), has completely changed how we think about every process in aging. That's because we can finally really test what is going on, instead of simply theorizing or tracking changes with age, without knowing which changes are correlative and which are causative.

Groundbreaking research done in *C. elegans* and *Drosophila* has paved the way for our understanding of how longevity might be regulated in mammals. Often, a gene is identified through a genome-wide screen for effects on longevity or stress response in one of these model systems, then later confirmed in mouse studies. This is a logical order of events, because the small model systems are cheap to grow and have short lifespans, and the field has developed toolboxes of genetic tricks that allow the manipulation and study of candidate genes. In short, we can take a lot more chances and study many more genes in these cheaper, shorter-lived animals; we have to be a lot more selective about what we study when it comes to knocking down a gene in an animal that takes a few years to study and is expensive to house during those years. If a gene is identified through human GWAS or genomic analysis of a particularly long-lived animal, we can turn to one of the small, short-lived model systems to better understand how that candidate gene functions to affect longevity. In worms and flies, a gene of interest can be knocked out or down, or overexpressed, or fluorescently tagged and observed for its cellular and subcellular localization, which one could also do in tissue culture. But in a live animal, we can also see how the gene functions in specific tissues, and how it affects behavior and other whole-animal characteristics, like lifespan, healthspan, or behavior. We can also carry out genetic screens to identify genes that might interact with the original candidate gene. Thus, we can understand a great deal about a novel gene in just a few weeks, before we ever invest in the two- to four-year-lifespan experiment in mice. In addition to expanding what we know, it also is a nice approach to focus on genes that are worth studying further in mice.

Historically, the fruit fly *Drosophila melanogaster* has been the premier model system for genetic analyses, as it was first developed specifically for genetic studies by Thomas Hunt Morgan and colleagues over a century ago. Not surprisingly, some of the earliest genetic longevity experiments were done in flies. For example, in a long-running experiment started in 1981, Michael Rose selected the last progeny of the brood each generation; over the course of many generations, he found that the final generation had an extended lifespan compared with the original strain.[20] Surprisingly, these flies also lay more eggs all through life (but are reported to have other detrimental phenotypes). Because this experiment was done before the current era of whole-genome sequencing, the underlying genetic cause of the difference was not reported, and may be complicated by "gene × environment" effects—but it would be fascinating to know what genes have driven these differences that have resulted in a quadrupling of lifespan from the original batch of flies.

Dame Linda Partridge's work ushered in a new era of *Drosophila* studies of longevity, first studying the effects of mating on lifespan,[21] and then moving toward genetics and dietary effects on lifespan,[22] some of the important contributions of fly longevity work. Flies have a short lifespan compared with mice—only about 40–50 days—and a rich history of research on development, behavior, and neural activity that complements aging work in flies. Flies have been the stars of the genetics field for decades, and as such, many genetic tools have been developed to manipulate their genes, making this system extremely tractable. Despite their small size, they have real organs, such as a heart, gut, and a complex brain, so the findings in this organism often have very obvious correlations with human biology.

But my favorite longevity model system is the nematode *Caenorhabditis elegans*, for reasons that will become clear soon. Geneticist Sydney Brenner, after learning about work on free-living nematodes by Ellsworth Dougherty, Victor Nigon, Margaret Briggs, and others, very deliberately developed *C. elegans* as a model to study nervous-system development in the late 1960s, because of its balance of simplicity and complexity that seemed genetically tractable. This little worm led to the discovery and genetic analyses of many of the basic cellular processes that are cornerstones of molecular biology today. For their discoveries in *C. elegans* of the "genetic regulation of organ development and programmed cell death," Brenner, H. Robert Horvitz, and John E. Sulston were awarded the Nobel Prize in Physiology or Medicine in 2002, and the related nematode *C. brenneri* was later named in Brenner's honor.

Worms are simple to raise in the lab, as they eat bacteria, hundreds can live in a single petri dish, and they develop from an egg to an adult in only two days. (You can also freeze them, giving them a bit of an edge over flies in the experimental practicality department.) *C. elegans* adult hermaphrodites have only 959 cells, and those cells do not turn over or get replaced, so anything that lives long does so by keeping those exact same cells healthy longer, a concept that will become important as we go further. Those 959 cells make up just a few major tissues (neurons, intestine, hypodermis, and muscle), and the only cells that divide in the adult are the germline stem cells. The entire animal develops in a stereotypical fashion (that is, the same way every time), which revealed the genetics behind programmed cell death. Because it has a simple nervous system of only 302 neurons, the connections of those neurons were possible to map by electron microscopy, revealing the first complete animal brain "connectome."

But the true value of *C. elegans* in aging and longevity research is its very short lifespan. A worm's adult lifespan is only about three weeks long, which is slow enough that you can actually watch them age (and quantify it!), but fast enough to do many experiments quickly. By the 1980s, Michael Klass had identified a set of long-lived mutants that presaged all later genetic longevity studies,[23] as the mutants ended up falling into categories that have become the canonical longevity mutant pathways that we will learn much more about: insulin signaling ("dauer" formation mutants that couldn't properly regulate a larval arrest stage), caloric restriction ("*eat*") mutants that can't grind up their bacterial food, "clock" (*clk*) mutants (they just grew slowly), and even sensory loss mutants that extend lifespan (which we'll discuss in chapter 13). (Dauer is a long-lasting but reversible alternative larval state that the worms enter when their environment will not support further growth and reproduction, and Klass also studied this phase.) It's quite remarkable to look back at Klass's original paper and see the major longevity pathways appear in this simple genetic screen for long-lived animals—even before these genes could be easily sequenced and identified at the molecular level.

The next few years would change the longevity field, owing to the impact of *C. elegans* researchers. In 1988, Tom Johnson (University of Colorado) found a long-living mutant that he designated *age-1*,[24] which Gary Ruvkun's lab (Harvard) later cloned and found to encode a gene that acts downstream of the insulin/IGF-1 receptor, called PI3 kinase[25]—a gene that was known to be important in cancer. Before that gene was cloned, however, another longevity gene was identified that ended up cracking open the field. In the course of preparing to perform a screen to discover longevity mutants in 1993, Cynthia Kenyon, a professor at UCSF, found that a mutant called *daf-2* doubled lifespan, while its known downstream gene *daf-16* suppressed its lifespan extension.[26] ("Daf" means **da**uer **f**ormation defective; *daf-2* mutants at moderately high temperature go into dauer when they shouldn't, while *daf-16* mutants don't make dauers when they should.) When the Ruvkun lab cloned *daf-2* in 1997,[27] it turned out to encode the worm's only insulin/IGF-1-like receptor—and since we have all heard of insulin, this was exciting because one could then imagine that a similar change might also affect human lifespan. When *daf-16* was cloned,[28] it was revealed to encode the worm's only FOXO transcription factor, which you might remember later turned up in human longevity GWAS. The genes in the rest of the insulin signaling pathway, from the receptor through the regulatory kinases and phosphatases,[29] to the mediators of FOXO's translocation in and out of the nucleus,[30] were cloned one by one, and their roles in regulating lifespan in worms, then later other organisms, were established. These findings set in motion the current era of longevity research, as they proved that single-gene mutations could extend lifespan, a clear repudiation of Williams's notion that longevity extension could not be achieved through single changes. (I will explain the loophole in this "single gene" change soon.) Nevertheless, skepticism remained because people assumed that dauer formation—the reversible arrest in a pre-reproductive larval state—a worm-specific event, was required for long lifespan.[31] Later work showed that these events could be temporally separated;[32] as I mentioned earlier, *daf-2* knocked down in adulthood still increases lifespan, meaning the animals don't have to go through dauer or even decrease reproduction in order to live long. Echoing Klass's early work, Siegfried Hekimi identified mutants that mimicked caloric restriction (*eat* mutants) and slowed them down (*clk* and other mitochondrial mutations),[33] rounding out the set of longevity mutants that Klass's work had presaged.

Soon after the flurry of worm longevity papers, the insulin signaling pathway was also implicated in longevity regulation in *Drosophila*, when Linda Partridge's and Marc Tatar's labs found that the long-lived *chico* mutant encodes the IRS gene just downstream of the insulin/IGF-1 receptor,[34] and that the fly FOXO homolog acted downstream of insulin signaling to control longevity.[35] Since flies don't go into dauer, the idea started finally taking hold that we could learn something about human longevity from studies in invertebrate model systems—even worms. One by one, the genes that connected insulin/IGF-1 signaling with FOXO activity in both worms and flies were found to control longevity. Later, these genes were shown to have lifespan-regulating activity in mice as well, although it is clear that the story is more complicated in mammals.

When Craig Mello and Andrew Fire described RNA interference, or RNAi, in the 1990s (another Nobel Prize–winning discovery),[36] it made worm genetic work even easier—you could inject worms with RNA, soak them in it, or, best of all, just feed worms bacteria that expressed the double-stranded RNA of your gene of interest, and you would essentially end up with a mutant worm. The discovery of RNAi opened up the possibility of carrying out large-scale "forward" genetics—that is, going from a known gene knockdown to the discovery of a characteristic ("phenotype"). In 2002, Julie Ahringer's lab built a whole-genome RNAi library (one gene per bacterial stock that could be fed to the worms),[37] enabling worm researchers to carry out comprehensive genome screens for whatever phenotype they could measure, including longevity. Feeding bacterial RNAi, while it had some downsides, was so easy and fast (no cloning of mutants necessary) that the worm longevity field exploded at this point. Whole-genome RNAi screens quickly identified both new and known longevity genes, ushering in the study of mitochondrial genes, autophagy, heat shock proteins, protein translation factors, and more.[38] (I will explain how each of these processes affects longevity in more depth in coming chapters.) Combining mutants with RNAi suppressor screens enabled the description and ordering of new pathways. Engineering worms to allow RNA interference to function only in particular cell types, and the engineering of transgenic animals with expression of a gene in only one tissue, helped researchers determine exactly where genes were acting, unravelling tissue specificity and cell autonomy questions that are harder to address with traditional genetic techniques that affect the whole body. The typical discovery pathway soon became: (1) perform a genetic or RNAi screen in worms, describing the genetic mechanism in detail there; (2) verify in flies and uncover more organ specificity; and (3) verify in mice.* The development of automated lifespan machines has further increased the throughput and reproducibility of C. elegans lifespan analyses.[39]

In any case, the ability to rapidly screen in invertebrate model systems has greatly accelerated longevity-regulating-gene discovery. Current methods move fluidly between GWAS and gene network predictions to screens in C. elegans, back to examining gene expression levels in patients,[40] as we recently did for a study of Parkinson's disease. The ability to experiment in multiple systems increases our confidence that the genes and pathways we have identified will be relevant for humans.

––––––

I will focus mostly on worms and flies in this book because they have contributed a great deal to what we understand about how aging is regulated, particularly about genetic mechanisms. You may have noticed I missed the two extremes of lifespan here—extremely long-lived (primates) and extremely short-lived (single-cell) model systems—but both of those ends have been important in longevity research, as well.

At the very long end of lifespan studies are those done in primates. The obvious advantage to studying primates is that they are very closely related to humans; if something works to extend the lifespan of a rhesus monkey, it's likely to work for us, as well. The downside, however, is that they live a very long time (an average of 25 years), and thus longevity studies can take decades—far outlasting the time that a graduate student, postdoc, or even an untenured professor can spare. Even that would be fine, unless you realize that your initial experimental conditions might be slightly off—which has happened. Groups at both the National Institute on Aging and at the University of Wisconsin, Madison, set up long-running caloric restriction studies, and at the end (after 30+ years) they compared their results.[41] The NIA group did

---

* And, cynically: publish a paper on the mouse and then imply in university press releases that it was first found in mice . . .

not see a statistically significant effect on longevity, but the UW group did. Admirably, the two groups worked together to discover the source of the discrepancy. The UW group may have fed its control group a less healthy (higher calorie) diet than the NIA group did, causing the UW controls to live shorter than the NIA animals, leading to a bigger difference with their calorically restricted animals. Despite the study differences, the data that continue to be gathered from the monkey tissues will be valuable—and one could even argue that the overfeeding regimen is more like a typical Western diet, anyway.

Because the lifespan of a rhesus monkey is so long, it's worth asking whether other, shorter-lived nonhuman primates could be useful, including lemurs, marmosets, and other small nonhuman primates. The South American common marmoset (*Callithrix jacchus*) has an average lifespan of only 5–7 years,[42] develops much more quickly to reproductive adulthood (1.5 vs. 3–4 years) and has more offspring, is much smaller than rhesus monkeys (only 300–500 g body weight as opposed to 6–8 kg), and displays several humanlike age-related disorders. Its small size means it is easier to house, as well. Although it is not a commonly used model yet, these advantages, plus the possibility of carrying out genetics using new CRISPR approaches, suggest that marmosets and perhaps other small primates will offer promise for aging research.

At the other extreme are the very short-lived organisms, such as yeast and bacteria. You might be surprised that bacteria "age"—but, logically, as soon as you establish any type of cell asymmetry (that is, when cells don't divide exactly in half), you will have aging, one cell that is older than the one it spawned. An elegant tracking experiment showed that *Escherichia coli* (*E. coli*) bacteria with older poles (ends) tended to stop reproducing earlier than the bacteria with young poles, a form of replicative senescence. These results suggest that one daughter cell always has more damaged components (that is, protein and DNA) than the other, leading to asymmetry. Such asymmetry is even more readily evident in *Caulobacter crescentus*, which has an end that affixes itself to a rock in a flowing stream, while the other end buds off. Eventually, the affixed cell will stop dividing, while newer cells that bud off continue to reproduce.

But, of course, bacteria are less like us because they are single-celled prokaryotes, not eukaryotes. Eukaryotes like us have a true nucleus and a few other cell features that are shared, like organelles. (Eukaryotes likely arose from the engulfment of one prokaryote by another, giving rise to a symbiotic relationship between the two, and the evolution of our current-day mitochondria.) Yeast (*Saccharomyces cerevisiae*) is a very commonly used genetic system because it is a simple eukaryote, with fewer than 6000 genes and very tractable genetics, allowing the use of homologous recombination to create specific mutants, for example. Because it is a single-celled organism, there is of course a limit to the kinds of questions that can be asked, but yeast has given us great insights into many of the basic questions in cell biology, genetics, and metabolism.

The lifespan of yeast has been measured in two different ways: you can ask how long a population can survive, particularly after a period of starvation, which is known as *chronological* lifespan. A few genes, such as SCH9, a homolog of one of the kinase proteins in the insulin signaling pathway (Akt), were identified in a search for genes that extended the period that yeast could survive once in "stationary phase," the stage after exponential growth has ended. SCH9, in turn, regulates the superoxide dismutase, one of the proteins that fight against oxidative stress. Caloric restriction extends yeast chronological lifespan, just as it does for other animals. After the first genome-wide RNAi longevity screens were performed in C. elegans,[43] a genome-wide screen of the yeast deletion library (that is, a set of strains in which an individual gene is knocked out) identified factors that are required for chronological lifespan, as well as testing the effects of amino acid starvation, and inhibition of TOR (target of rapamycin).[44] TOR was found to be the main determinant of chronological lifespan, which also depends on genes like AMPK that function in energy-level sensing and act in dietary-restriction-induced longevity in other organisms, as well (see chapter 6 for more information on dietary restriction and longevity).

A second approach to studying lifespan in yeast is to measure how many times the cells can divide, which is called the *replicative* lifespan. This can be done either by tracking a single mother cell as she divides or by counting the "bud scars" that are left behind. In fact, budding of a daughter cell from a mother cell can lead to retention of damaged proteins in the old mother,[45] and sorting of pristine proteins into the new daughter cells—an excellent mechanism to guarantee that the best version of the organism will continue in perpetuity. The longevity effects of increasing the activity of a protein called Sir2 and its regulation of $NAD^+/NADH$ energy metabolism in longevity (more about this in later chapters) was discovered through yeast replicative lifespan studies. Yeast's claim to fame, at least in the popular press, was David Sinclair's discovery that a chemical known as resveratrol increased replicative lifespan. Resveratrol can be found in grape skin, and thus wine was celebrated as a longevity tonic (despite the fact that the amounts of wine that would be required to affect lifespan would be toxic many times over). The identification of the TOR nutrient signaling system in yeast and its roles in growth and longevity has become hugely important not just for lifespan, but for the understanding of how cells regulate growth. The fact that yeast cells are able to partition new proteins to their daughters while retaining old, damaged proteins in the mother suggests that a relatively sophisticated sorting system is at work to preserve the existence of a pristine copy, ensuring survival of the species for generations to come. This mechanism in yeast requires the activity of Sir2, although Hugo Aguilaniu's lab later showed that the analogous process of protein rejuvenation in *C. elegans* oocytes does not.

One downside to using yeast for longevity work is the fact that replicative lifespan experiments are quite laborious—much more work than moving worms to new plates every day, for example. While researchers have developed tricks to make it easier (like refrigerating the cells to slow things down overnight, and staining of bud scars in lieu of counting daughters), this problem may have slowed progress beyond the unbiased discovery of a handful of longevity genes, like Sir2 and other NAD^+/NADH pathway components, for many years. In fact, a full-genome yeast genetic screen for replicative longevity was not completed until 2015, more than a decade after such screens were performed in more "complicated" organisms such as *C. elegans*. This is the opposite of the general research trajectory in other fields of genetics and cell biology, where yeast research generally leads the way because of its ease of use. Yeast is great for most genetics and has been the real workhorse for genetics, genomics, and cell biology, and no other model systems would be successful without it—as a field, we always use yeast to work out new methods, and then we extend that work to harder systems, like worms.

Nevertheless, yeast, for all of its benefits, is actually quite a difficult model for lifespan studies. (I will lose friends, and probably future funding, for writing this.) I found this out the hard way. I had been working in worms for several years before David Botstein asked me to teach the Project Lab course at Princeton, and to use yeast for the course, since we had a lot of reagents for yeast experiments, and the other instructors were yeast experts. I decided that I should at least do something that I'm interested in, so I got some of the students to try to use yeast mutants for replicative lifespan studies, which I had never done before. My students and I quickly discovered that this assay, for lack of a better word, sucks. You had to pick the daughters away from the mothers every few hours, or you could delay it for a while by putting the plates into a refrigerator, greatly slowing the experiment—not a great option since we had only a few weeks to do everything. The fact that it was more difficult and slower to do a longevity assay in yeast—*the whole point of using yeast is that it's fast and easy!*—was demoralizing to the students, so we switched to an easier assay for their work, worms, and I felt grateful to be working with worms in my own lab (and we used worms for future classes).

Recently, however, high-throughput approaches (that is, rapid parallel testing), such as Dan Gottschling's "mother enrichment program," corresponding "daughter enrichment programs,"[46] and the development of microfluidic devices that allow automated tracking of dividing cells, have addressed the manual-labor problem of replicative lifespan, allowing labs to carry out discovery analyses and more precise experiments. The ability to isolate young and old yeast populations is particularly powerful for the study of cell biology and genomic and epigenomic studies. For example, Gottschling's lab found that as mother yeast cells grow older, their lysosomes function less well, leading to pH changes in the cell.[47] Lysosomes are required for efficient "cleanup" of proteins in the cell through a special type of protein recycling called "autophagy," so the failure of lysosomes leads to the inability of the cells to divide further.

The ability to sort young and old yeast cells is also enabling the exploration of epigenetic differences (chapter 15); for example, age-stratified yeast collection can be used to identify epigenetic marks that are associated with aging. Despite its small size and single-cellularity, yeast is a powerful model system to study aging, especially when linked to its strengths in genetics and cell biology.

————

Finally, a relatively new aging model, the African turquoise killifish (*Nothobranchius furzeri*), has blazed onto the aging research scene in the past few years. Killifish live in the transient pools that form in the rainy season in Zimbabwe and Mozambique; during the dry season, they survive in a diapause state, but can then become reproductive in just a few weeks. Because they must go through their entire life cycle within the rainy months, their lifespan is only 3–4 months. This animal is the best of many worlds for longevity research: it is a vertebrate, so it will share most of the processes that we care about in humans; it can be raised in the lab, much as zebrafish are housed; tools that have been developed to genetically manipulate zebrafish, including CRISPR, can be applied to killifish; and, finally, killifish are short-lived, so experiments can be done very quickly, or at least within a more reasonable time than even mouse experiments (2–4 years).

This fish was first brought to the forefront of longevity research by Anne Brunet's lab (Stanford), when postdoc Dario Valenzano, who had captured and studied killifish as a graduate student, worked on developing killifish tools for aging research.[48] The Brunet lab sequenced the genome, which allows the whole community to understand the organism better.[49] Brunet's team also carried out comparative aging studies on two strains that were captured from separate pools, with varying longevity due to their varying diapause durations in the dry season. Perhaps not surprisingly, in addition to a few other genes, they found that the insulin/IGF-1 signaling pathway contributes to the longevity of killifish. The fish also of course has a nervous system and can be used for neurodegeneration and behavioral studies, including the recent development of learning assays;[50] the development of CRISPR genome editing has made killifish genetics possible.[51] Because different killifish strains can be found in different regions of Zimbabwe and Mozambique,[52] the relationship between ecology and genetics of the strains and longevity can be studied, as well. And, excitingly, Valenzano's lab showed that transfer of gut microbiota from young to middle-aged fish can extend lifespan,[53] an intriguing finding that suggests that the microbiome can have large effects on longevity. While still in its infancy as a genetic model for longevity research, the numerous advantages of killifish will likely make it an ideal model for both bridging invertebrate and vertebrate findings, and for discovery research in its own right.

————

If you gain only one thing from this book, I hope it will be an understanding of the importance of diverse model systems and the utility of basic research in the study of biology. While a scientific discovery often makes the news only once a gene's function has been verified in mammals or humans, in truth, the work to get to that point usually started at least a decade before in some other, less-celebrated organism, like worms or flies or yeast or fish. No single model system is sufficient to address all aging questions, but several systems can complement one another. The work done in these systems is usually the reason that we know what we do about any longevity mechanisms, and in the coming chapters I will tell you about the experiments that have taken us this far.

5

What Is Aging
(and How Can We Measure It)?

BIOMARKERS OF AGING AND
"QUALITY OF LIFE" METRICS

Maybe it's true that life begins at fifty. But everything else starts to wear out, fall out, or spread out.

—PHYLLIS DILLER

AGING IS THE MAJOR RISK FACTOR for many chronic and neurological diseases, prolonging the low-quality stage of life. Therefore, it is not enough to know how to extend lifespan; we must also understand how to maintain *quality of life* as we age. Although we all see aging, developing standards to understand and quantitate components of the aging process can help us understand how aging proceeds. Measuring this "healthspan" will allow us to determine whether lifespan-extending measures also extend quality of life, or merely extend the low-quality fraction of life. Model systems can even be used to model human quality of life characteristics, including cognitive decline (loss of learning and memory with age), which may offer a powerful approach to address human aging issues.

Look around in a crowd sometime. We are good at guessing someone's age at just a glance; we see it in wrinkled faces and in graying hair, in how a person walks, in how agile a person is . . . and weekend warriors feel it themselves every Monday morning. Like Supreme Court Justice Potter Stewart's assessment of obscenity, we all know aging when we see it. If aging is so obvious to us, it seems we should also be able to measure and quantitate it—but how?

Several large observational longitudinal studies have been set up to track and measure aging, distinct from tracking disease, as has been done in the long-running Framingham Heart Study. These aging studies include several that are funded by the NIA. For example, the Baltimore Longitudinal Study of Aging (BLSA) was established by Nathan Shock, who led the National Institutes of Health's Gerontology Research Center and was one of the earliest to appreciate the value of studying aging as its own field. Shock, William Peter, and Arthur Norris initiated the BLSA in 1958 (women were included starting in 1978), with the aim of following healthy people over time; they aimed to "observe and document the physical, mental, and emotional effects of the aging process in healthy, active people."* The logic of the BSLA methodology is that following *individuals over time*, rather than simply comparing groups of people at different ages, helps control for environmental factors, since one's life experiences and environment can have large effects on one's health. Not only does the BLSA follow physical changes with age but it also tracks cognitive changes and the development of chronic disease, which are both major factors in aging.[1] Other changes, such as behavioral and environmental factors, are also noted.

Another longitudinal study, the Rush Religious Orders study, was started in 1993, with the goal of understanding changes in memory and movement with age.[2] By assessing the cognitive abilities and post-mortem brain tissue of more than a thousand nuns, priests, and brothers from all over the United States, the study will provide information that will contribute to the understanding of age-related cognitive decline. Other studies, such as the InCHIANTI study,[3] which focuses on elderly individuals in Tuscany, and long-running epidemiological studies of the health of registered nurses, have other emphases, but each longitudinal study is helpful in understanding the relationship between aging and a variety of diseases.

Other longitudinal studies focus on or enroll groups that might reveal more information specifically about aging. For example, the Long Life Family Study (LLFS) is a longitudinal study of about 5000 people from the United States and Denmark that follows people from families with long-lived members.[4] This study includes a wide range of ages—from 25 to 110 years—and both sexes, and has more aged individuals than most available studies; it focuses on

---

* "BLSA History," National Institute on Aging, accessed January 26, 2023, https://www.nia .nih.gov/research/labs/blsa/history.

"healthy agers" in order to find clues about what might help a person live long and avoid disease along the way. In these studies, the goal is to collect enough longitudinal data on long-lived and potentially long-lived individuals (because of their family histories) to learn general trends and to make correlations.[5] Importantly, in addition to behavioral analyses and health information, blood samples have been taken, allowing the identification of biomarkers of biological age.[6] Finally, large studies and databases including the UK BioBank collect genomic and medical information on a huge scale, but without any bias for longevity or other characteristics. By analyzing genetic and medical data on the subjects themselves, as well as knowing family history data (such as the age at death of parents), correlations can begin to be drawn. The more of this information we collect, such as the underlying genetic backgrounds and blood markers that accompany changes in cardiovascular health, cognitive performance, and disease, the more likely it is that we will be able to understand what aging really is, at the genetic and behavioral levels.

Many people have conveyed to me their concern that by extending longevity we will just stretch out the frail part of human life, which of course would be a disaster on every level. In fact, it's one of the first questions I'm asked when I give public lectures—if I forget to show movies of those lively, long-lived mutant worms at an age when all of their normal cohorts are dead, some in the audience assume we're talking about just adding on years of poor-quality life. It's worth asking, do centenarians live better longer, or just longer? To answer this important question, Ismail and colleagues (2016) examined the health of long-lived individuals, analyzing the frequency of various age-related diseases (cancer, cardiovascular disease, hypertension, osteoporosis, and stroke) in two longevity cohorts, the New England Centenarian Study and the Longevity Genes Project, compared to the normal-lived controls.[7] By plotting the duration of "disease free survival," they found that long-lived individuals experience a significant delay in the onset of age-related diseases. That is, they stretched out the disease-free part of their lives, they did not just live longer in an aged, debilitated state. This was an important finding; it argues strongly against the gloomy assumption that a long life is always a worse life. In fact, these long-lived individuals are experiencing the "compression of morbidity" that Jim Fries and others have argued is the optimal manner in which to extend lifespan. At least for centenarians or near-centenarians, extended lifespan does not come at the cost of poor health. Thus, an important point that has come out of studies of exceptionally long-lived humans seems evident, but is worth stating clearly: ***long-lived people live longer by staying healthier longer***. While this may seem obvious, it's perhaps worth reminding skeptics of this point, since there is a lot of hand-wringing about extending the frail part of life if we try to extend human lifespan. If you are an optimist about the value of longevity research, this might have been exactly what you had expected, but it's good to confirm instead of just assuming it's true. Now if we can just understand *how* centenarians' longevity is extended, then applying that to other humans should not only extend lifespan but also extend healthy, disease-free survival. That, in a nutshell, is the entire goal of the longevity and aging research field.

When we think of how our own relatives have aged, we may think of loss of cognitive ability, motility, and independence. (The term "frailty" is used clinically to describe this loss of robustness with age.) Balance, strength, and endurance are all lost with age, contributing to frailty. As I mentioned, one's gait changes with age,[8] which in itself may greatly affect quality of life, as motility is key to one's independence. Gait is greatly affected by aging because it incorporates both muscle and neuronal abilities; older people often walk not just more slowly but also with a wider stance to combat a loss of balance, which is why we can so easily recognize an "old person's walk." Gerontologists noticed that gait speed might be a good proxy metric for overall health. With this in mind, clinicians developed a simple series of tests, called the Short Physical Performance Battery, or SPPB.[9] This series of simple tests includes rising from a chair multiple times, standing with the feet in different positions to measure balance, and measuring the time it takes to walk a short distance (eight feet to four meters, depending on the study). Despite its simplicity, the SPPB test is useful in predicting how long a patient will remain out of a nursing home, and can predict functional decline.[10] The SPPB assessment of participants enrolled in the InCHIANTI study was strongly predictive of loss of mobility and disability over three years.[11] More recent work from the BLSA quantitated 3D parameters of gait, largely corroborating the previous observations of reduced gait speed and overall joint flexibility with age.

Another important question is what the systemic factors are that change in aging people. The most consistently age-associated changes could then be considered "biomarkers" of aging. Paola Sebastiani and colleagues examined 38 selected circulating factors in the blood of healthy participants in the LLFS to see if there were any good aging biomarkers.[12] They picked this set of factors because they might be associated with aging-related diseases, based on previous research and clinical practice. (It is important to note that this kind of study may miss anything *not* previously thought to be associated with aging, because the particular facts are chosen in advance based on common clinical measurements;

this is different from *unbiased genome-wide* or *proteome-wide* tests, for example, which can discover new biomarkers—but then, of course, whether a protein exists in the blood and can be measured is a different issue.) Most (34 of the 38) of the circulating factors they measured showed some age-dependent change, and a few showed very strong changes with age. For example, a protein called the "N-terminal pro-hormone of brain natriuretic peptide," which is correlated with congestive heart failure, had the strongest age correlation. Other potential age biomarkers included the circulating factors cystatin, DHEA (dehydroepian-drosterone), and sRAGE (soluble receptor for advanced glycation end product). In the future, these biomarkers might be useful in studying age-related diseases, but this study didn't make a clear case for a single biomarker being "the one" for measuring aging. A more unbiased approach might uncover better biomarkers, but would also require a specific study to do so, rather than utilizing easily ob-tained, universally measured clinical factors.

What about more obvious, visual changes with age? In addition to judging by gait, we are very good at guessing how old other people are—or how fast they might age—just by looking at their faces. As we all know (or have unfor-tunately experienced ourselves), faces change a great deal with age. Why are faces so depressingly good at revealing age? Facial aging is due to both extrinsic and intrinsic factors. Extrinsic factors include excessive sun exposure (UV damage) and smoking, which we all know has damaging effects on skin. Stress can also accelerate facial aging. You might remember the piercing green-eyed gaze of a 12-year-old Afghan refugee, Sharbat Gula, staring out from an iconic *National Geographic* cover in 1985. In 2002, the same photographer, Steve McCurry, managed to track her down again; she had lived much of her life in poverty and under stress.* Despite the fact that it was only 17 years later, she had aged considerably, looking much older than a woman of 30. Her life had been a hard one, and her face showed it.

We are still discovering intrinsic factors, but those might include basic metabolic functions. As we age, not only does our skin wrinkle but we also lose the layer of fat underneath it, which gets redistributed, leading to dark circles under the eyes and sagging of the cheeks. Together, these extrinsic and intrinsic factors cause wrinkling, dark spots (age pigment, aka lipofuscin and advanced glycation end products, or AGEs), and fat leaving where you want it (under your eyes) and going where you don't (to your jowls). If you

---

* Almost 40 years after leaving Afghanistan as a 6-year-old orphan, she was given a home in Kabul by the Afghan government, as reported in 2017.

are at least in your 30s, the chances are you'll do a fairly good job of guessing someone's age just by looking at their face. (Don't ask a kid, though—to them, we are all old.) As a result of these changes, just a glance can reveal a person's true age. We've had a lifetime of "training" in this area, inadvertently gathering information about how old people are and storing it away in our brains, a lifelong real (as opposed to artificial) intelligence project.

But can we predict how long someone will live from *how they look right now*? In a remarkable psychology study, participants were asked to look at a group of 100 men's photos from a 1923 University of Toronto yearbook, and to then guess how long the person lived.[13] (No women's or minorities' photos were included in the study because there weren't enough in the yearbook, a reflection of university society at the time—so this all may need to be repeated later!) They also scored the photos for a variety of other factors, such as perceived wealth, perceived health, attractiveness, symmetry, adiposity, maturity, and likability. (One of my favorite notes in the methods section of this paper is that they were careful to control for moustaches, glasses, and hairstyles—something my lab has never had to do with worms.) Remarkably, the viewers were able to predict the age of death with some accuracy: "The model revealed that perceived age of death was a significant predictor of actual age of death." What might surprise you is that "Health/Attractiveness and Power" was not the best predictor; instead, "Perceived Wealth" correlated better with perceived age of death, maybe an indictment of our medical system more than anything else. In the end, a combination of these qualities, which together indicated socioeconomic status, best predicted age of death. Adiposity didn't correlate well with age of death, despite the fact that other studies showed a correlation between adiposity and likelihood of dying from heart disease. However, the authors note that this study group didn't have high adiposity scores, meaning that being overweight may have predictive power for death when people are heavier than the individuals in this yearbook study were. Another factor, "affect," which basically measured smiling, didn't correlate well—but the authors note that in this group, "the large majority of faces did not look very happy"! (And smiling has weird effects—it makes young people look older, and old people look slightly younger.)[14] In any case, the study may have uncovered what we already suspected, that wealth is correlated with longevity, likely because of access to good health care and lower environmental stress. Nevertheless, it would be interesting to see how one's longevity might be predicted even as early as in one's college years.

What if we could use these changes and other information our faces convey to predict *how fast we are aging*, basically as a biomarker of health? Jing-Dong

Jackie Han's group has used 3D imaging of faces to develop a "facial phenome"—that is, a set of facial phenotypes, such as distances between nose and mouth, corner eye slopes, asymmetry of the two eye slopes, lip thicknesses, and so on, to describe the face and how it changes with age.[15] Han's group reconstructed 10-year average face profiles from over 300 faces of Chinese origin. Using 17 landmarks, they aligned the faces and quantified the facial features. These phenotypes were then compared across ages, and clustered with blood biomarkers from the same individuals. Surprisingly, Han's group found that facial features have more correlations with chronological age than blood biomarkers do!

What is even more remarkable is that they then used machine learning to *predict* aging. Using the characteristics that are most informative, they were able to predict the age of an unknown 3D face with fairly good accuracy (correlation 0.85, with an average 6-year deviation). They then grouped the faces into fast agers, slow agers, and well-predicted samples (and it should be noted that there are more of the first two groups in the older samples), and examined the blood biomarker data. Indeed, the aging-associated blood health markers, such as albumin, LDL-C, and total cholesterol, agreed with the fast- and slow-aging facial metrics. In fact, the predicted ages correlated better with the blood biomarkers than with the chronological ages. And at least for Chinese faces, facial features are more general than blood biomarkers, which are to some degree sex-specific. This is an exciting area of work, because in principle one could do a 3D scan of the face and assess age-related health instantly, which could subsequently be used to quickly monitor the efficacy of aging interventions more rapidly than simply waiting for changes in lifespan.

Ideally, we would use all of this biomarker information—gait, blood factors, and facial features—to study rates of aging in people. We could use these human biomarkers to study the efficacy of treatments, such as diet or exercise, as aging interventions. But to carry out genetic experiments, for example, to test whether a gene actually affects the rate of aging, or to carry out chemical screening for possible drugs, we cannot use humans, for both practical and ethical reasons. So how can we apply what we know about human biomarkers to model systems, like the genetic model systems of worms and flies?

One approach is to study each animal for its own characteristics of aging, or aging biomarkers. In the early 2000s, soon after the discovery of several different longevity pathways, there was great interest in characterizing *C. elegans* aging phenotypes. In 2002, the labs of both Monica Driscoll (Rutgers) and Cynthia Kenyon (UCSF) used microscopy to describe characteristics of *C. elegans* aging, including morphological features.[16] Delia Garigan in the

Kenyon lab used fluorescence and light microscopy to describe such changes with age as tissue breakdown (e.g., germline), accumulation of bacteria in the gut and pharynx of the worm, and oil (fat) droplet accumulation. Just as in human skin, worms experience the accumulation of lipofuscin (age pigment) in old animals. Using these factors, Garigan and Kenyon scored individual worms for their quality, then compared the rates of aging between normal (wild-type) and long- and short-lived mutants. This nonparametric analysis suggested that long-lived daf-2 animals, even when 20 days old, look more like young (2- or 5-day old) wild-type worms, and that the short-lived worms looked like aged wild-type worms. Similarly, Laura Herndon and Monica Driscoll used locomotory behavior and fluorescence and electron microscopy with David Hall (Einstein College of Medicine) to examine the aging of C. elegans tissues. Their beautiful electron microscopy and fluorescence images showed that worms exhibit sarcopenia, the breakdown of muscles, in much the same way that humans do, and that the skin of worms also wrinkles with age. (Surprisingly, the integrity of the nervous system, at least touch and motor neurons, appeared to be intact, but later work showed that neurons do physically degrade with age.)[17] Herndon also used a qualitative scoring metric of behavior, and showed that individual worms age at different rates, but that a long-lived mutant slows some aspects of aging (sarcopenia) but not others (lipofuscin accumulation).

One aspect emphasized in the Herndon study was the stochasticity of aging—that is, even though the worms studied were isogenic (of one genotype), they exhibited individual variation in how quickly they aged. While this concept is obvious from any examination of those blocky, steplike drops in Kaplan Meier survival curves (otherwise, the worms would all drop dead on the same day), these authors were the first to detail the progression of individual animals from healthy to decrepit. Later work has quantified this in ever greater detail and on larger scales, enabling the identification of factors, such as sustained large size, that are associated with long life.[18] Spectrofluorimetry of lipofuscin—the same stuff that you see in "age spots" on your hands—and AGE accumulation revealed that not only did longer-lived mutants slow this lipofuscin accumulation but also reflected the previously described motility behavior classification.[19] This suggests that age pigments are a good biomarker of biological age. Perhaps this is true of humans as well, since we see the same problems on our sun-damaged skin.

This idea of stochasticity in aging is accompanied by the concept that isogenic populations can be divided into slow and fast agers, and if one could identify a biomarker that distinguishes them at an early age, one could effectively predict the future lifespan of the individual. Such a biomarker would be of great interest to all of us, since the equivalent type of biomarker in humans could help let us know how long we might live; while that information might be depressing, it might also tell us whether we need to use an aging intervention, of which several are being developed (see chapter 17). Tom Johnson's lab (University of Colorado) tested this concept in worms by examining the levels of a fluorescently tagged small heat shock protein (*hsp-16.2::gfp*) induced by a brief heat treatment on their first day of adulthood.[20] The group found that the worms that turned on expression of the heat shock protein went on to survive the longest in a thermotolerance assay and to live the longest. Of course, this analysis is a tiny bit circular, since the damage induced by heat would have been best managed by worms that can turn on heat shock chaperones the best, but the concept still holds, that the variability *already present* in an isogenic population early in adulthood can predict the eventual longevity of the individual. That is because the levels of *hsp-16.2* induction reflect that individual's ability to withstand stresses, likely due to the underlying differences in the levels or activation of stress response factors in different individuals. Thus, *hsp-16.2* induction, like age pigment levels, can act as a biomarker of aging that may help predict future longevity.

You might be asking yourself: "we know a lot about how people die, and these studies show morphologically how worms age, but how do worms actually *die*?" We used to joke about carrying out worm autopsies, but David Gems (University College London) is now basically a worm coroner. His lab has carried out several studies of worms at the moment of death to understand *exactly* how worms die, and identified two types of death, characterized by pathology specific to the grinder (pharynx) of the worm.[21] In one set of worms, the pharynx/anterior intestine was swollen owing to bacterial infection,* much as Garigan and colleagues had described.[22] These deaths appear to happen in mid-lifespan, and might correlate with pharyngeal pumping rates, as the slow-pumping *eat* mutants seem to have lower rates of this pharyngeal swelling, and killing bacteria first decreases this kind of death. Even more tantalizingly, Gems's lab also identified a fluorescent burst in the worm intestine, a "blue wave of death." The blue color turned out, surprisingly, to be

---

* Closer inspection by electron microscopy suggests that this is in fact a swollen intestinal cell that encircles the base of the pharynx, rather than the pharynx itself, and in fact, this anterior intestinal phenotype may exist only in the specific, unusual genetic background that Gems's lab used.

not lipofuscin but rather derivatives of tryptophan, anthranilic acid glucosyl esters, which are released in the intestine in response to a calcium signal that triggers necrotic cell death through a change in pH originating in lysosomes in the most anterior intestinal cells. This is basically like the worm having a heart attack or an aneurysm (or an internal burst of fireworks) from which it cannot recover. Another time-of-death phenotype was described a few years earlier by Nick Stroustrup, Walter Fontana, and colleagues (2013), who used a flatbed scanner to carry out high-throughput lifespan measurements; they observed that the worms shrank slowly for a day or so before death, and then rapidly expanded immediately after death.[23] This reminds me of our own work on mating-induced shrinking and the cuticle's resistance to osmotic shock— so maybe the worm loses the ability to regulate osmolarity upon death.[24]

The Herndon/Driscoll and Garigan/Kenyon studies did an excellent job describing morphological changes with age, and Gems's lab and Stroustrup and colleagues described the moment of death.[25] Additionally, these studies suggested that long-lived mutants slow aging and are healthier, but these were all mutants in the insulin signaling pathway, including mutants of the DAF-2 insulin/IGF-1 receptor and the AGE-1 PI3 kinase that acts just downstream of DAF-2; age-1 mutants are also long-lived and healthy. Other mutants that extend lifespan through completely independent mechanisms had also been discovered and characterized, including dietary restriction mutants (eat-2) and animals with reduced mitochondrial function (clk-1, atp-3, cco-1). All of these worms are long-lived, so it would be easy to assume that they might all do everything better with age. However, those of us who had worked with both daf-2 and mitochondrial mutants already knew that this was not the case; in general, the mitochondrial mutants are small, sickly, and don't reproduce well, while insulin signaling (daf-2 and age-1) mutants seem like superheroes, looking young and battling all stresses. But this difference was not usually discussed in the papers that revealed the long lifespans of mitochondrial mutants, perhaps leading many readers to believe that they would all be healthier, and equally so.

To determine whether different longevity mutants extend all their behaviors, in 2004 Kerry Kornfeld's lab (Washington University) carried out a systematic study of a set of longevity mutants—daf-2, daf-16 (the short-lived mutant that "undoes" daf-2's beneficial effects), eat-2, and clk-1—for a set of specific phenotypes beyond lifespan, including pharyngeal pumping, motility, and reproduction.[26] By examining how the wild-type and mutant worms performed each different behavior, these researchers showed that some longevity mutants extended all behaviors, and some affected only a few. They used motor activity to characterize different stages of worm lifespan, classifying from I to IV, and after normalizing to total lifespan, determined what fraction of life each longevity mutant remained in each stage. This analysis suggested that the dietary restriction worms (*eat-2*) had the longest *proportional* reproductive period (stage I) relative to its total lifespan, and that the reduced insulin signaling mutants had the longest proportional stage IV. Conversely, *daf-16* mutants, which are short-lived, look "healthier" by this assessment. This was somewhat surprising, as *daf-16* worms look much worse and die early, but because they die early, the *proportion* of healthy lifespan looks better. That is, if you are a *daf-16* mutant, you won't live long, but at least the end won't be protracted. Heidi Tissenbaum's lab (University of Massachusetts) carried out a similar set of assays a decade later and came to a similar conclusion, that not all longevity mutants are created equal.[27] This conclusion seemed obvious to those of us who had worked with and quantified the behavior of these mutants, but bore repeating, as it had somehow become expected that all long-lived animals would be healthier.

Tissenbaum's study put a slightly different spin on things, making the point that long life is not always better life—that is, lifespan had been uncoupled from healthspan. Although this point had been made at times previously, particularly when others had looked at characteristics of aging rather than simply lifespan, Tissenbaum made a point that others in the field had not pressed, which is that some long-lived mutants are just not healthy—which is the concern of many lay audiences when we talk about our longevity studies. For example, my lab found that worms can learn and remember associative cues (like Pavlov's dogs salivating for steak) and they lose this ability with age.[28] While *daf-2* mutants can do this for a long time and have great memory ability, *eat-2* mutants take longer to learn and have poor short-term memory, but can eventually learn and remember longer *with age*, while mitochondrial mutants are bad at learning and memory in general—so it's more complicated than just saying that if an animal lives long, its memory will be good. We also found that some long-lived animals have long reproductive spans, while others do not (chapter 8), and vice versa.[29] Therefore, the idea that not all longevity mutants extend quality of life had been shown previously, but because the idea that if an animal lives longer it *must* be healthier was pervasive in the longevity field, it was important to challenge this notion.

However, things went a little too far; in the lay press, soon the summary of Tissenbaum's findings morphed into the idea that *daf-2* mutants are actually *unhealthy*. This struck most C. *elegans* researchers as quite bizarre; *daf-2* worms are robust and healthy for a very long time, even when wild-type worms have already started to die. Beyond longevity, *daf-2* mutants survive longer than normal worms under a wide range of adverse conditions: high heat, oxidative stress, hypoxia, heavy metal exposure, infection—basically, you name it, *daf-2* worms do it better, except for their slightly reduced and delayed reproduction (more on this later, in chapter 8). So it was strange that the take-home message being peddled was that *daf-2* worms are *less* healthy. The key to this argument is that *daf-2* worms have such a long lifespan, so *proportionally* they might only do as well as wild-type. The message that *daf-2* worms might actually be unhealthy seemed to be met with glee by some mammalian researchers. This may have been because of the relatively recent arrival of "healthspan" on the mammalian scene in the mid-2010s (even though worm researchers, such as Tom Johnson and Michael Klass, had made this point and used the term "healthspan" as far back as the 1980s), or maybe because they were sick of *daf-2* getting all the attention. To check these results, Rose DiLoreto in my lab reanalyzed all of the Bansal/Tissenbaum 2014 and Huang/Kornfeld 2004 study data and found that *daf-2* mutants are proportionally just as good or even better than wild-type worms.[30] DiLoreto's analysis did support both the Huang 2004 and Bansal 2014 conclusions that other mutants, such as mitochondrial mutants, are less healthy overall, and so the larger point of Tissenbaum's work, that healthspan and lifespan can be uncoupled, still stands.*

But why should we care about worm-specific traits? I would argue that whenever we can, we should model worm healthspan assays on *human* aging characteristics instead. It's surprisingly easy to judge how old someone is just by how they walk, because walking incorporates information about muscle function (speed) and balance. You'll remember that gerontologists developed the SPPB (measuring gait speed, standing strength, and balance) as a quick way to determine a patient's health and predicted likelihood of returning to a nursing facility. With the SPPB in mind, Hong-Gil Nam (Daegu Gyeongbuk Institute of Science and Technology, South Korea) started from a more human perspective: Could we model the kind of physical decline that humans experience in the worm? While worms can't do repeated chair stands or balance tests, they *can* move. Nam's lab filmed worms crawling on a plate with no food for 30 seconds, then analyzed the movies not only for average speed but also for the

---

* The damage had been done, though: the idea that *daf-2* mutants are unhealthy still persists, is often mistakenly repeated in talks and in study section, and has cost several researchers, including new professors, their grants.

maximum velocity within that time frame. It's worth noting that this analysis differs a bit from all of the other studies I've told you about so far. First, these worms were moved off of their bacterial (food) plates and filmed while on agar plates without food—this is an important difference that I will explain in a moment. Secondly, most studies report either an average speed or, in the case of the Bansal 2014 study, the length of the track left by the worm in the bacteria; Nam's study was the first to track both average and maximum velocity. Thirdly, after recording them, they returned the worms to their food plates every day for the duration of their lives; this allowed them to report the motility characteristics of each individual for its entire lifespan. Finally, they tracked almost 200 individual worms every day, so they could actually report the behavior of an individual worm across its entire lifespan, a truly arduous study.[31]

Nam's analysis revealed several interesting aspects. First, *maximum* velocity changes with age more dramatically than does average speed. This makes sense—I can't sprint as fast as I could in high school, but I can still jog at about the same rate, and I suspect that's true for many people, so maximum speed is more informative about my age. Secondly, when they split the middle-aged (day 9 of adulthood) worms into two groups by their maximum velocity, the high-maximum-velocity worms ended up living about 30% longer than did the low-maximum-velocity worms. Thus, maximum velocity is not just a bio-marker of longevity but, in fact, also a great *predictor* of longevity. Nam's results also showed that other measurements, such as pharyngeal pumping or average speed, perform far worse or have little predictive power, which is important to note—*not all aging characteristics are equally informative or predictive*. Maximum velocity correlated well with mitochondrial integrity, which may tell us one important requirement for moving quickly. Remarkably, Nam's lab used maximum velocity in mid-life to predict the eventual lifespan of an individual. This would be like me using the 100-meter-sprint times of 50-year-olds to predict how long each person will live. It might work! A study of 1562 active career firefighters aged 21–66 in Indiana came to a similar conclusion, with an inverse correlation of push-up capacity and risk of cardiovascular disease in the following decade.[32] The authors acknowledge that this study group may be somewhat skewed, and the push-up test might not be generalizable, but you get the gist. I still like the idea of sprints, myself.

In addition to wild-type worms, Nam's group carried out this analysis for the long-lived *daf-2* (insulin signaling) and short-lived *daf-16;daf-2* mutants. They found that the maximum velocity largely mirrored the lifespan of the mutants, with each mutant's maximum velocity being proportional to its lifespan. By this metric, *daf-2*'s healthspan is much longer than wild type's health-span, and is proportionally even slightly better than wild type's when normalized to the total lifespan. So, we wondered, what could be different between these metrics and those so famously reported by the Tissenbaum study?

This is where the fact that the Nam lab tracked the worms' velocity on food-less plates became so important. A worm's main job in life is to make more baby worms; to do so, it must find food so that it can take in enough resources to reproduce. When you put a worm onto a bacterial plate, it will crawl around for a while but then slow down and stop when it realizes it is on food, and start eating. *daf-2* worms seem to realize this much more quickly (I mentioned they are smarter, right?), so on bacteria they stop almost immediately and begin to eat, and so their tracks are shorter—leading to the idea that they aren't as active. But on "no food" plates, as Nam's lab performed, they know to look for food, and keep doing so. We hypothesized that *daf-2* worms might sense food better because they make more of a food-sensing receptor called ODR-10, and that if they had less of that receptor, they might act more like wild-type worms and keep looking for food. (We guessed this because it had been previously shown by Douglas Portman's lab [University of Rochester] that males have less of this ODR-10 receptor, leading them to explore more, increasing their likelihood of finding a mate—"choosing sex over food," as the headlines said. Boosting the level of ODR-10 in males caused them to make the wrong choice—decreasing foraging—and thus decreasing their mating success.)[33] This food receptor, ODR-10, turned out to be a good guess as the reason for the difference in veloc-ity on plates. We found that *daf-2* worms have about 20-fold higher levels of this receptor, explaining why they might sense food more rapidly and then slow down immediately when placed onto bacteria, as Tissenbaum's lab reported. We then tested this idea by asking what happens if they can't sense their food anymore, but they can still move as fast as they want to. To do this, we fed the *daf-2* worms RNAi bacteria that express double-stranded RNA of the *odr-10* gene, to knock it down. When we knocked down *odr-10*, the *daf-2* worms kept crawling on the bacteria for at least 10 minutes (which was when we finally stopped filming)—perhaps because they didn't "know" they were already on food. But they moved fast and continuously, showing that they are perfectly *able* to move. So by measuring the maximum velocity on "no food" plates, Nam's lab had measured the worms' true *ability* to move, rather than their *desire* to search for food—and that difference alone could account for much of the mistaken conclusion that *daf-2* worms are somehow low performing. This would be like looking at pictures of Usain Bolt lounging by the track, and coming to the conclusion that he is a slow guy. Needless to say, being able to distinguish average from maximum velocity is incredibly informative.

In fact, the field already had important information about how well *daf-2* mutants perform compared to wild-type worms, particularly in neuronal functions. The worm's neuromuscular junction is preserved longer in *daf-2* worms,[34] allowing them the ability to move longer, and at least part of this is mediated by the kinesin motor protein UNC-104. Kinesins carry neurotransmitter vesicles in the neurons. *unc-104*/kinesin is expressed at higher levels in young and *daf-2* worms.[35] *daf-2* mutation also prevents the development of the morphological defects in the neurons—like blebs, waviness, and unregulated outgrowths—that arise in wild-type worms with age.[36] We found that *daf-2* worms have better memory with age,[37] and maintain their ability to regenerate axons with age,[38] unlike wild-type worms. All of these characteristics suggest not only that *daf-2* mutants are able to live longer but that many of the functions that we care about most in humans—our abilities to think, to remember, and to move—are maintained by *daf-2* worms better with age.[39]

———

But, of course, healthspan analyses are not limited to people or worms. Flies are kept in vials, and their ability to walk upward is an obvious sign of health;[40] anecdotally, Linda Partridge's lab reported that calorically restricted flies not only lived longer but better maintained their motility with age.[41] David Walker's group developed an assay to measure the integrity of the intestinal barrier using an ingestible blue dye that spreads if the intestinal barrier is compromised (the "Smurf" assay).[42] With this tool they showed that other markers of health, such as changes in metabolism and markers of inflammation, including antimicrobial peptides, correlate with this visible assay and can be used to predict lifespan. Because flies have real organs, including brains and hearts, their functions can be assessed as well; for example, Rolf Bodmer's group studies fly cardiac function and its changes with age.[43] Molecular marks, such as glycation damage (AGEs) have been tracked in both worms and flies, and correlate with aging and poor health outcome, but can be delayed by dietary restriction.[44]

Josh Shaevitz's lab (Princeton) developed an unbiased method to observe fly behavior and changes with age in flies. His group videoed both male and female flies in a chamber and used machine vision to summarize the activity as a set of postures.[45] By clustering these postures, they were able to capture the flies' various behaviors, such as walking, grooming, and so on, to create a "behavioral map." By repeating this procedure as the flies grew older, they could assess changes in behavior through changes in the behavioral map. As expected, more-sedentary behaviors increased with age, but surprisingly this was after a burst of activity in mid-life. Shaevitz's group also used this approach to study models of neurodegeneration, including Parkinson's and Alzheimer's disease. Strikingly, at younger ages, these models best matched with aged wild-type flies, suggesting that neurodegeneration not just causes pathology but accelerates many aspects of normal aging. Whether these changes in behavior can be matched with specific molecular markers is not yet known, but this unbiased approach to visualize and measure aging is a powerful tool, and could be used to assess efficacy of treatments. A similar metric in humans might also be useful for prediction of neurodegenerative disorders.

Of course, much of what we extrapolate from invertebrates like worms and flies to humans is through the lens of testing in small mammals, particularly mice. Healthspan is quite visible in mice, and a wide variety of metrics are available to assess health, from behavior to tissue function to blood biomarkers. The mouse counterpart to the Baltimore Longitudinal Study of Aging in humans, the Study of Longitudinal Aging in Mice (SLAM) was established very recently (2021).[46] SLAM will assess both sexes of C57BL/6J (a commonly used mouse strain) and another strain of outbred mice for various phenotypes with age, and will collect tissue samples. These kinds of measurements in both mice and nonhuman primates offer obvious parallels with human healthspan and can more easily be extrapolated—but of course take much longer to do than studies in invertebrates. Developing work on healthspan metrics in the turquoise killifish may help bridge this temporal gap.

Finding biomarkers of aging and healthspan is a major focus of ongoing research not only in humans but also in model systems. In chapter 7, I will describe cellular biomarkers, nucleolar size in particular,[47] that correlate well with lifespan. Nucleolar size, which is related to protein translation, scales inversely with lifespan; dietary restriction and exercise are correlated with shrinking nucleoli from worms to flies to mice to humans, so this is one marker that seems universal. Cellular and molecular biomarkers will help us more rapidly test therapeutic interventions in both model systems and humans and will perhaps one day function as diagnostic tools. Developing new assays in model organisms that best model human degenerative characteristics will help us better understand how to maintain the best quality of life, even as we age.

6

Insulin Signaling, FOXO Targets, and the Regulation of Longevity and Reproduction

There is only one difference between a long life and a good dinner: that, in the dinner, the sweets come last.

—ROBERT LOUIS STEVENSON, *WILL O' THE MILL*

EARLIER, I MENTIONED my entrée into the field of aging: I had heard Cynthia Kenyon explain that single-gene mutations in the insulin/IGF-1 signaling receptor could increase the lifespan and youthfulness of *C. elegans*. After I heard her talk, I knew what I wanted to do—find out how those mutant worms were so healthy—and I had an idea of how to do it. But before I tell you about that, let's learn a bit more about how these worms shed light on aging and, in particular, how the insulin signaling pathway regulates longevity.

C. elegans has been a fruitful subject for the study of aging since its establishment as a model system in the early 1970s. While it was originally used to study development and neurobiology, *C. elegans*'s short lifespan—about two weeks—was first described in the mid-1970s. Within a few years, the field established some important aging-relevant facts: that nutrient conditions can affect worms' lifespan, that arrest in diapause ("dauer") extends lifespan, and that parental age can affect progeny characteristics.

In 1983, Michael Klass published the identification of several long-lived *C. elegans* mutants.[1] This work was remarkably prescient for its discovery of what we now recognize as the insulin signaling pathway (see below), caloric restriction (chapter 7), and chemosensory mutants (see chapter 13). None of the genes were cloned at the time (gene identification was much harder in the pre-genome era, especially for lifespan genes), so their identity and functions were unknown, but the principle of the existence of longevity mutants was firmly established. Tom Johnson picked up where Klass left off, first describing the long lifespan of the *age-1* mutant in 1988, with follow-up, single-author *Science* papers (rarely if ever seen these days!) in 1990 on this mutant's slowing of the rate of aging.[2] *age-1* mutants were reported to have low fertility, which was thought to confirm the idea of a "trade-off," that long life came at the cost of fertility.

Around the same time, Cynthia Kenyon, a young UCSF professor who had already established herself with her work on developmental gene regulation, had become interested in aging, and she set out to do a screen for longevity mutants. Reasoning that she would need a way to distinguish long-lived mutant parents from their (contaminating) offspring, she used a genetic trick: by carefully choosing a "background" (starting) strain that would prevent progeny from ever developing into adults, she could track the age of the parents to reveal any long-lived mutants. This was a key idea that would be hard to replicate in another animal, because worms have a superpower: the ability to push the pause button when times are tough. That is, when there is not enough food to go around, they feel too crowded, or the temperature gets too high, *C. elegans* forms an alternative larval state known as "dauer" (German for "lasting"). This spore-like stage allows a pre-reproductive animal to pause development until times are better, then later resume normal development when it detects better conditions (more food, lower temperatures). This helps the worms survive poor conditions in the wild.

Kenyon realized that using a temperature-sensitive mutant that got stuck in dauer even when it shouldn't make that decision—that is, a mutant that would go into dauer even when food is around ("dauer constitutive")—would be useful to visually distinguish aging parents from their offspring. Raising these dauer formation (*daf*) mutants at a normal temperature should allow parents to develop as usual, but when she shifted the plates to a higher temperature, their offspring would arrest in this dormant state as small larvae, allowing the screener to easily identify any long-lived parents, which would be obvious just from their size. But before she even carried out the screen, she discovered that her background strain, *daf-2* (which you have already read a bit about), was already incredibly long-lived—*daf-2* worms lived twice as long as their wild-type counterparts! On top of that, because the genetics of dauer formation had been well studied, "dauer defective" (Daf-d) mutants were also known to act downstream of ("epistatically to") *daf-2*—that is, loss of a second Daf-d gene would essentially "turn off" this dauer-forming process. Kenyon found that another gene, the Daf-d mutant *daf-16*, was required not only for *daf-2*'s dauer formation but also for its long life. Moreover, *daf-2* mutants were fairly fertile, turning the accepted notions of the field on its head. (This *Nature* paper also had an interesting author list: because Kenyon could not convince lab members other than rotation students to work on aging, she did much of the work herself and was the first author, an unusual position for a principal investigator in biology.) Kenyon's work established not only that *daf-2* worms are long-lived but also the epistasis of a longevity pathway from *daf-2* to *daf-16*.

At first the two longevity genes, *daf-2* and *age-1*, were thought to act in different pathways because of their differences in fertility and because the latter had been reported to not form dauers. But later it was found that the *age-1* mutant also contained a nearby sperm-formation mutation that was likely responsible for its low fertility. Kenyon's lab soon showed that the two genes acted in the same longevity pathway (that is, putting both the mutated genes into one worm—the *daf-2;age-1* double mutant—didn't cause them to live longer than the single mutants). Meanwhile, Gary Ruvkun's lab, along with the labs of Don Riddle and Jim Thomas, was investigating dauer gene epistasis, and found that *daf-2* and *daf-23*, another dauer-formation gene, act at a similar point in the dauer decision pathway.[3] (*age-1* had been reported to extend lifespan but not form dauers, suggesting that it acted in a different pathway, but Ruvkun's lab found that *age-1* was already known by a different name, *daf-23*; it turned out that *age-1 would* form dauers if put at high enough temperatures.)

At this point, cloning of the genes so that the pathway could be better understood was critical—and probably necessary for you readers to not give up on reading the boring worm nomenclature that was established by the *C. elegans* pioneers in response to flippant *Drosophila* naming—sorry about that, really not my fault. In 1997, Ruvkun's lab solved the *daf-2* mystery: the gene encoded an insulin-like receptor.[4] To me, this result fundamentally changed how we think of long-lived worms—their long lifespan was not just a weird worm phenomenon, but might have real implications for humans. This was, of course, exciting not just because the field could finally start to place the genes into a cellular context (and we could stop saying "dauer"), but also because of the obvious parallels with mammalian physiology. Once you hear "insulin/IGF-1 receptor" it's hard not to think that these findings might have some impact on people. This is not a trivial point; the fact that the genes all participated in dauer formation was convenient for the genetics, and interesting to worm folks, but made it difficult to convince non-worm researchers that this longevity pathway was not specific to worms. Identifying the worms' true genetic basis and the conservation with mammalian signaling pathways was a real breakthrough for the field. Up until this point, it would have been easy to assume that all of the longevity effects observed in worms were due to their ability to form dauers, a stage that many animals do not have. (I should probably mention here that we now know that hibernating bears and squirrels share some of the important genes regulated by the insulin pathway.) The existence of a nutrient-sensing pathway using insulin signaling that determined lifespan was harder to ignore. Because of the obvious parallels with mammalian physiology, the possible impact of longevity gene identification in *C. elegans* took on more importance.

The Ruvkun lab ran practically a cottage industry in the late 1990s of cloning the dauer/lifespan genes, including the insulin/IGF-1 receptor (*daf-2*), PI3 kinase (*age-1/daf-23*), insulin/IGF-1 receptor substrate (*irs-1*), PTEN phosphatase (*daf-18*), Akt kinase (*akt-1*), and PDK kinases (*pdk-1, -2*).[5] The Ruvkun and Kenyon labs raced to clone *daf-16*,[6] which turned out to encode a forkhead family O (FOXO) transcription factor that was already known in the mammalian literature, further increasing the relevance of this pathway to humans[7]—and, of course, you have read about the later GWAS findings about FOXO and centenarians. (PTEN was already known to play a critical role in cancer, as well.) The fact that DAF-16 is a transcription factor—that is, regulates the expression of many downstream genes—is critical to our story. Ruvkun later also identified almost 40 insulin-like peptide genes through a clever bioinformatics similarity search, adding an unexpected layer of complexity to an otherwise well-studied pathway.[8] Other components were added to the pathway,[9] but the flurry of gene characterization in the late 1990s unraveled many of the mysteries of the signaling pathway that controls longevity regulation.

Ironically, the double duty of the insulin signaling genes in dauer formation, which for technical reasons greatly accelerated their cloning and identification, may have inadvertently slowed down the acceptance of *C. elegans* as a general model for aging. On occasion, I still hear non-worm researchers opine that *daf-2* worms are not relevant for the understanding of lifespan in higher organisms because *daf-2* also regulates dauer formation. Many people missed the fact that in Kenyon's 1993 paper, daf-2 worms lived longer even when the temperature was shifted *after* the stage when worms would make the choice to go into dauer, or even after the larval stage (L3) altogether. This means that the worms did not need to go through a dauer state to live longer. Still, I guess one could have argued that the worms had a mutation in their background, so were still kind of different. It was harder to make this argument after Andrew Dillin, then a postdoc in Kenyon's lab, put wild-type (normal) worms onto bacteria producing double-stranded RNA of daf-2 *only* in adults—thus knocking down daf-2's gene expression via the RNA interference pathway well after development was done—and the animals *still* lived exceptionally long.[10] By putting the worms onto daf-2 RNAi at later and later stages of life, Dillin and Kenyon were able to firmly uncouple lifespan regulation from both dauer formation and fertility (which was affected by RNAi application before the third larval stage, L3), killing two annoying birds with one stone. Degradation of the DAF-2 protein very late in life, well past reproduction, still extends lifespan, in fact.[11] Still, the myth lives on—but at least *you* will know better!

Interestingly, because of the way that the original lifespans were done, another potential lifespan/dauer pathway was checked but missed. These days, when we carry out lifespan experiments in worms, we "censor" (delete from counting, on that day) worms that disappear if they die from unnatural causes. Sometimes the worms crawl up on the side of the plate, get stuck, and dry out; other times they wander off and you just can't find them! The worst is when they can't lay their eggs properly, so the mothers die of "matricide": they "bag," meaning that the eggs hatch and crawl around inside the mothers (super gross) or "explode" (yes, gruesome, I know). In the early lifespan experiments, some worms who probably actually died of matricide owing to defects in their ability to lay eggs might have been counted as natural deaths, so the mutants appeared to have normal or short lifespans, and were pretty much ignored after this point. The TGF-beta "dauer" pathway mutants were a key example of this: they were checked for longevity by Kenyon and others, and entered the worm canon as "*dauer forming but not affecting lifespan.*" My lab later discovered that TGF-beta dauer mutants in fact live long if this tendency to explode during reproduction is avoided. We found this in two ways: first, the whole-genome expression profiles of these TGF-beta mutants as adults looked too much like those of daf-2 to be unrelated; and secondly, when we were doing controls for another project on reproductive aging (see chapter 11), we found that when we blocked their reproduction and thus prevented matricide, all of the mutants in the TGF-beta dauer pathway were long-lived. In fact, they achieve this by regulating two insulins, *ins-7* and *ins-18*, which then act on the *daf-2/daf-16* pathway. So in fact both the insulin signaling and TGF-beta dauer pathways regulate dauer formation in early larval stages, and longevity later in life, and ultimately act through the insulin pathway.

The fact that longevity regulation by insulin signaling was not some weird worm thing became even clearer when the labs of Marc Tatar and Linda Partridge each reported the same longevity phenomenon in *Drosophila* mutants in 2001.[12] In fact, the fly field had been studying aging much longer, since it had been established as a model genetic system almost a century before Sydney Brenner started working with *C. elegans,* so fly researchers had a bit of a head start. The effect of temperature (1968) and mating status (1984) on fly lifespan had been noted—and Linda Partridge and Mariana Wolfner first studied the reduction in female lifespan caused by sperm and seminal fluid (a theme that we will revisit in chapter 12), and Marc Tatar had studied heat stress as a hormetic mechanism to extend lifespan. In the late 1980s, Michael Rose famously carried out a multigenerational selection for late reproduction that resulted in extended lifespan, but the exact genetic changes that allowed such differences were unknown and at the time were expected to be multigenic.[13] By contrast, Partridge found that mutants with a single-gene mutation, *chico* mutants, are small owing to a defect in the insulin/IGF-1 receptor substrate (IRS) just downstream of the insulin/IGF-1 receptor, and they are long-lived.[14] Similarly, Tatar observed that fly insulin/IGF-1 receptor (InR) mutants are long-lived.[15] These findings suggested that worms and flies shared this aspect of longevity regulation, establishing insulin signaling as an evolutionarily conserved mechanism to control lifespan. Tatar's lab then showed in 2004 that dFOXO, the fly homolog of DAF-16 and FOXO, regulates longevity through its activity in the fly's brain and its fat-storing tissue (fat body),[16] paralleling the Kenyon lab's discovery in 2003 that DAF-16 regulates lifespan through its activity in the intestine (which regulates fat) and neurons.[17] The description of *Drosophila* insulin-like peptides and other parts of the insulin signaling pathway further cemented the evolutionary parallels between worm and fly longevity regulation.[18]

But do these findings extend beyond invertebrates to mammals? The role of growth hormones on mouse lifespan had been investigated for several decades, as "dwarf" mutants that resulted from hypopituitary function—that is, lacking growth factors, including insulin-like growth factor (IGF-1)—were long-lived.[19] For a long time it was assumed that the longevity of these "dwarf" mutants was a result of their small size, which is not crazy, since many small animals, like dogs, are longer-lived than their large counterparts. But this size-longevity connection was uncoupled in mouse studies, as the influence of single-gene mutations in the insulin signaling pathway on lifespan was next found to be conserved through mammals. In 2003, Holzenberger and colleagues observed that mice heterozygotic for the IGF-1 receptor (that is, they have one wild-type and one mutant copy of the IGF-1R) are long-lived and, like flies, do not require small size for their longevity.[20] Reducing insulin signaling in the mouse adipose tissue (*fat insulin/IGF-1 receptor knock out*) also increased lifespan, just like dFOXO in the fat body of flies and DAF-16 in the worm's intestine, further strengthening the parallels with invertebrates.[21] Reducing IRS2 (the insulin/IGF-1 receptor substrate) in mice decreased food intake and reduced reproduction, coupling energy status to fertility.[22] Together, these results suggested that the same pathways were at work in worms up through mammals, making it harder to believe this was just an invertebrate phenomenon. Even more incredibly, the genes that Yousin Suh later identified in GWAS studies of centenarians (see chapter 3) were also in this pathway, specifically the insulin/IGF-1 receptor (InR), IGF-1, and FOXO3a.[23] The possibility of using this pathway as a therapeutic target has only recently been shown: mid-life treatment with an inhibitor of phosphoinositide 3-kinase (that's what *age-1* encodes) increased mouse lifespan 10% and improved some healthspan metrics, as well.[24]

So how *do daf-2* worms live so long? This brings us back to my own story. By 1999, when I heard Cynthia speak, it was clear that insulin signaling was doing something through the activity of DAF-16 to cause the worms to live long. Since DAF-16 had recently been revealed to encode a FOXO transcription factor, this genetic result implied that the major way that *daf-2* animals live long was through some sort of transcriptional change directed by DAF-16. When I heard this problem, I knew exactly how I wanted to solve it.

That's because I was in the right place at the right time. Stanford Biochemistry was one of the birthplaces of the growing field of genomics, with Joe DeRisi in Pat Brown's lab developing spotted microarrays,* Michael Eisen in David Botstein's lab developing new analysis and visualization methods, and statisticians like Rob Tibshirani developing new methods to analyze the data. Notably, this group of researchers accelerated the adoption of array technology by model-organism researchers because they openly shared methods, data, and

---

* Pat Brown later left Stanford to found Impossible Foods, the maker of Impossible Burgers, to help decrease global warming.

Why Isn't Decreased *daf-2* Activity Deleterious, like Diabetes?

The fact that *loss* of insulin/IGF-1 receptor activity extends lifespan and healthspan can sound counterintuitive, since the reduction of response to insulin results in diabetes. In fact, people with Laron syndrome (see chapter 2) have reduced IGF-1 signaling (hence the dwarfism phenotype) but are more resistant to diabetes, as are centenarians with mutations in this pathway. Perhaps the best way to think of it is that *daf-2* mutants are a bit like intermittently fasted animals. In worms, intermittent fasting (IF) activates DAF-16,[a] just as low insulin signaling does, and increases the stress response, making the animals healthier. *daf-2* animals seem to interpret the low activity of the DAF-2 receptor as a sign of a slight nutrient stress, and respond accordingly, slowing aging to allow the animals to reproduce longer. Excess glucose kills worms faster,[b] which is more like a human diabetic situation.

You might also wonder why *daf-2* worms don't take over the population, if they are so great and live so long. Wild type is wild type for a reason: *daf-2* is "stuck" in this IF-like state and as a result grows more slowly than wild type, gets to reproductive maturity later, spreads its reproduction out over a longer period than wild type, and has slightly fewer progeny in every generation than wild type does. All of this adds up to "*loser*" in evolutionary terms—within a few generations, wild-type worms will outcompete *daf-2* worms, since they have more progeny faster, over and over. Sure, if it gets hot or there's no food, *daf-2* will survive better, but in non-stressful conditions, wild type wins out.

tools, in direct contrast to commercial microarray companies, which primarily catered to mammalian researchers and medical consumers. This openness allowed researchers working on smaller organisms to build their own arrays. My classmate Joe DeRisi was a student in Pat Brown's lab at the time. Joe and Pat popularized the DIY approach starting in the late 1990s by teaching a course at Cold Spring Harbor: in only two weeks, the students built a microarrayer machine from parts, copied and amplified all 6000 yeast genes in 96-well plates, printed those pieces of DNA onto glass slides, then performed their own yeast microarray experiments. Joe was a real evangelist, convincing everyone within a 100-foot radius of his bench to use microarrays in their own work, including me.* Even though my graduate studies were in no way related to expression analysis, I had already learned a lot about them just by working in the same room as Joe.

By then, Joe was heading to UCSF for a prestigious fellow's position, and I decided that I would try to work in Cynthia's lab at UCSF and get Joe's advice on how to build my own *C. elegans* microarrays so that I could figure out how *daf-2* worms live so long.† That is because all of *daf-2*'s superpowers are due to the activity of DAF-16/FOXO, the transcription factor that "turns on" gene expression in response to *daf-2* signaling. I knew that if I could build a microarray that contained all of the worm's genes, then I could compare the long-lived *daf-2* insulin signaling mutants to short-lived animals, and we would have a readout of what exactly was going on inside the healthy animals. It was a great way to answer a big question, and the perfect time to do it.

I should note that Cynthia was generous to let me start this project. It required a lot of money and some belief that I would actually get it done in a reasonable amount of time, which was not at all guaranteed. The money was to buy a set of ~20,000 pairs of primers (short stretches of DNA).‡ Each pair of primers could be used to PCR (polymerase chain reaction) amplify one of the nearly 20,000 genes in *C. elegans*, a project that was made possible by the sequencing of the *C. elegans* genome. Those pieces of DNA would then be "arrayed" on glass slides in an ordered fashion, one spot per gene. Once the slides were made, they could be used for hybridization, the binding of fluorescently labeled experimental DNA samples to the spots, and then scanned for the fluorescence intensity at each spot, to give a readout of the relative levels of gene expression from each sample.

Building the arrays took a couple of months, with plenty of advice from Joe. The only opportunity I had to use any of my biochemistry PhD training was

* Mixing of labs was an oddity of Stanford Biochem at the time that I think could really benefit research but has been hard to implement in other places.

† Stuart Kim and John Kim (Stanford) had already built an array for worms, but I wanted to use a ton of arrays for my experiment (60–80), and at the time, there was a limit (understandably) on the number of arrays that could be used for collaborations, something like 4. I didn't realize until later that Stuart was interested in aging (see chapter on supercentenarians), as well.

‡ Julie Ahringer's lab used the same set of primers to build the RNAi library, one bacterial colony per gene, that the whole worm field later used for whole-genome screening for lots of phenotypes, including longevity.

in the first few weeks of my postdoc, when I purified tons (that is, 11 mg) of Taq, the polymerase used for PCR reactions that are necessary to make copies of the pieces of DNA. I did PCRs around the clock for a few weeks until all of the 20,000 genes had been DNA amplified.* Once I used the arrayer to print the purified PCR products onto polylysine-treated slides, I had my first microarrays, and I could start the experiments. I used *daf-2*, *daf-16*, and control RNAi clones to treat the worms, and I collected samples at different time points through mid-adulthood so that I could see what kind of transcriptional changes happen with age and which ones are really due to the insulin signaling pathway.† I also collected some samples of different *daf-2* mutant alleles, *age-1* mutants, and wild-type worms, so I could do a more standard analysis. I massively overestimated how many worms I'd need, filling up an entire incubator with 10-inch-diameter plates, 15 plates for each time point, with 10 time points for each time course for each strain. (This is why I sound like an annoying old lady recounting walking uphill both ways in the snow when my lab people call an experiment with a couple of small plates "big.") By labeling the mRNA from *daf-2* and *daf-16;daf-2* animals with two different fluorescent dyes on the same microarray, I could find the genes that are differentially expressed—that is, see genes that were up- and downregulated—and thus distinguish aging from youthful worms.

In any case, it was worth it. By collecting so many time points and mutants and hybridizing 80 two-color arrays in total, I could be very sure that when a gene was up in a *daf-2* mutant or RNAi, and down when *daf-16* was knocked out, that it was likely to be a target of the pathway, a gene that both *daf-2* and *daf-16* regulated. Using software that Mike Eisen had written, it was easy to see right away that a fairly large group of genes were blazing bright in the long-lived worms, and they clustered together by their pattern of gene expression. The genes I was looking for—that is, "good" genes that might help *daf-2*

---

* One unanticipated advantage to the project was that because Cynthia's lab didn't do this kind of work, I had to go around and borrow equipment from Joe DeRisi's lab, Peter Walter's lab, and others. I got to know most of the other people at UCSF who were also building their own microarrays, which was a great opportunity that most postdocs don't get.

† More precisely, I used Andy Dillin's RNAi clones to feed a sterile strain (*fem-1;fer-15*) vector control, *daf-2*, or *daf-2* + *daf-16* RNAi bacteria. I collected both a short time course (48 hours) and a long time course (day 1 to day 8 of adulthood), each of 10 time points, plus a few alleles of *daf-2* and *age-1* mutants, for a total of 80 arrays. I chose day 8 because that is before the worms start to die, since I wanted to focus on genes that *predict* lifespan, not genes that change due to death.

worms live long—were bright red (high) in *daf-2* worms, but green (down) in *daf-16* worms.

Right away, I could see genes I expected to find, like *sod-2*, a superoxide dismutase that was already known to be expressed at high levels in *daf-2*, and *ctl-2*, the catalase that works with *sod-2*. These two genes were some of the few that had been previously proposed to work downstream of *daf-2* and *daf-16*, so it was a relief to see these positive controls doing what I expected.

But in fact there were dozens of other genes that were expressed as highly as *sod-3* and *ctl-2* were, suggesting that those two genes were only part of the answer. Gratifyingly, most of these *daf-2*-upregulated genes made a lot of sense.[25] This set included heat shock proteins, which "chaperone" proteins while they fold, particularly under stressful conditions; "proteostasis" (that is, protein homeostasis) and proteasome components, which mark and degrade damaged proteins, so that proteins remain unblemished and functional; a metallothionein that acts in metal detoxification, and *daf-2* worms are champions at resisting metal toxicity; antimicrobial peptides, which you might expect in a mutant that fights infection well; fat metabolism genes, which fits with the fact that *daf-2* worms store more lipids; TCA cycle (citric acid or Krebs cycle), trehalose metabolism, and glyoxylate cycle genes, suggesting a shift in carbohydrate metabolism to increased fat storage; cytochrome P450s, which can act in detoxification but also are a component in the building of steroid ligands for nuclear hormone receptors; and autophagy ("self-eating") components, which help the cell to recycle its proteins. Since this experiment, the *daf-2*-upregulated gene list has become a veritable *Who's Who* of aging mechanisms (and we will discuss many of them in the upcoming chapters).

Conversely, there were genes that were down in *daf-2* mutants and up in *daf-16* animals; these genes are primarily involved in development and metabolism, with some functioning in the regulation of fat metabolism shifts that may benefit developing oocytes, which helps explain why *daf-2* worms make slightly fewer progeny.*[26] Many of the genes had only numerical codes as names, and we provisionally dubbed them "*dod*" genes (**do**wnstream of **D**AF-16) until their functions were better described, and many were renamed later.

I had figured out how to do the array work and find the genes, but I had no good way of testing the genes for their contribution to *daf-2* longevity, which

---

* Later we discovered the "anti-DAF-16," a factor called PQM-1, that turns on these developmental genes only when DAF-16 is not in the nucleus.

is what I really wanted to know—a list might tell you something about the underlying biology, but testing the genes would determine which of them are actually important. Testing genes can also distinguish between different hypotheses; one model was that just one single gene was the key regulator of lifespan downstream of DAF-16, while the opposite was also possible. Traditional genetics would be very slow, since that would entail crossing a mutant of each gene with *daf-2* and then measuring the lifespan of the double mutant. (While certainly possible, I am not a patient individual, and postdoc positions don't last forever.) Moreover, that would work only for the genes for which there were already mutants, but many of the genes were only predicted from their sequences in the genome, so no mutants yet existed (and this was long before CRISPR was discovered). We had already had good results in the lab with using RNAi, so it seemed clear that this was the way to go. But building RNAi constructs for the long list of genes in front of me posed a daunting task, even if I prioritized and pared it down to the top 50 or so.

At this point, I got really lucky: Andy Fraser and others in Julie Ahringer's lab (University of Cambridge) had used that same set of primers to build RNAi constructs for all of the genes on chromosome I, then later the rest of the genome, and Ahringer kindly sent this RNAi "library" to Cynthia. This allowed me to go very quickly from just having a list of genes to testing the functions of the genes, simply by putting the *daf-2* worms onto RNAi bacteria for each candidate gene and measuring its lifespan. Soon it became clear that many of the genes were required for the full longevity of the *daf-2* mutants, but *no single gene on its own had a large effect.* This result directly countered the prevailing hypothesis in the field, that combatting reactive oxygen species is the most important mechanism of slowing aging. In fact, loss of the superoxide dismutase *sod-3* and its partner catalase *ctl-2* had only a 5%–10% effect on *daf-2*'s lifespan, as did many of the other genes.

These results had several big implications for the regulation of longevity by insulin signaling. First, it eliminated the model that DAF-16 only regulated the expression of two or three genes, which is why a suppressor screen for DAF-16 never would have worked. Perhaps more importantly, it also killed the idea that *daf-2* required only elimination of oxidative stress to live long, the prevailing hypothesis (the ROS theory) in the aging field. Instead, the fact that *many different kinds of genes* are upregulated in a coordinated fashion, and each one makes a small contribution to longevity, explains the logic of using a transcription factor as the target of the nutrient signaling pathway: upon a change in insulin signaling that might accompany a stressful situation, DAF-16 can be rapidly deployed to turn on hundreds of genes. In turn, each of those genes contributes to the protection of the cell in a different way: carbohydrate and fat metabolism is shifted to help the animal survive long-term starvation, heat shock proteins help stabilize the proteins that are subjected to heat-induced unfolding, autophagy proteins start to help the cell utilize its proteins for energy and recycling, antimicrobial peptides protect the gut from invading pathogens, and still other proteins help the animal withstand a whole host of stresses (heat, metals, hypoxia, xenobiotics, etc.). Each one of these proteins and enzymes carries out an important job, and each is necessary for worms to be able to survive longer. Excitingly, most of these components are also expressed in humans, so there is hope that we can understand how to turn them on and make our cells live longer, too.

This type of coordinated response and additive effect on lifespan would have been difficult or impossible to find without the use of a high-throughput genomics approach such as expression microarrays, and high-throughput functional genomics using RNAi was critical for testing their contributions to longevity. Without these two tools, each *dod* gene might have been identified, tested, and a paper published on it wrongly claiming it was the most important gene downstream of DAF-16. (Of course, this still happens, even with genes that cause a 5%–10% effect.)

Another, perhaps more esoteric point that has been largely overlooked but is no less important was our use of the expression data to distinguish *aging-dependent* from *insulin-signaling-dependent* gene expression changes. I was interested in this question because for most of the history of aging research, investigators had used two opposing arguments to explain their data: that either a gene changed with age *as a result of* aging, or a gene changed with age *in response to* aging, a compensatory response. Before the discovery of longevity mutants, it was hard to distinguish cause from effect, and authors seemed to use whichever logic they felt made sense at the time, based entirely on their knowledge of the name of the gene in question. For example, a gene that is expressed more highly in aged animals might be neutral, or might cause a deleterious aging effect. But when you find out that the name of the gene is "heat shock protein" then the explanation magically becomes "this is a compensatory response to fight against aging." Maybe, maybe not—and this is the problem with using only changes *with age* to understand aging. Longevity mutants help us unravel and distinguish these confusing models.

I wanted to make sure we could tell the difference between aging and longevity regulation. Steven McCarroll, a graduate student in Cori Bargmann's
lab at the time, got interested in the project and helped me with another type
of analysis.* By ordering the expression data as a time course, we could ask
whether *daf-2* worms at the *whole-transcriptome* level looked overall younger
than the control worms or *daf-16* worms. We did this by treating the entire
transcriptome at each time point as a single datapoint (vector), then comparing each time point to itself or to all the other time points (aka a Pearson correlation). A perfect match, such as a comparison of a sample to itself, gets the
maximum score of 1, and is colored white; as the transcriptomes become more
different, their correlation score goes down (grayer to black if completely opposite, or −1). We ordered these sample points by time and calculated their
correlations to all of the other time points. Just by looking at where the perfect
matches were, we could infer how quickly the samples were changing from
one another. Remarkably, even though we could see that this was true in the
short time course, in the long time course, all three worms (control, *daf-2*
RNAi, and *daf-2+daf-16* RNAi) had similar *overall* transcriptomes. That means
that even though a large number of genes (about 500) are regulated by *daf-2*
and *daf-16*, on the *whole-genome* scale, it's a relatively small set; *daf-2*-regulated
genes are only about 2.5% of the total worm genome (about 20,000 genes).
The total aging transcriptome is largely unaffected by insulin signaling, even
though the *daf-2* worms slow down their rate of aging twofold. Another way to
think of it is that *daf-2* worms, by affecting a small *but very important* set of
genes, can essentially override the global aging gene expression changes that
they are experiencing. I found this remarkable—lots of gene expression was
still changing with age, but *daf-2* worms seem immune to those effects for
quite a while. That means if we found the right *daf-2*-like gene to tweak in us,
we might do the same—and centenarians might have done this already. With
these data from long-lived mutants, we could distinguish causal from non-
causal genes, and finally solve the cause/effect confusion problem that plagues
studies that look only at changes with age.

Another question we could answer only by having transcriptional data was
"What piece of DNA does DAF-16 actually bind to?" Like all transcription

---

* Steve McCarroll is now a professor of genetics at Harvard, and one of the leaders in the
single-cell sequencing field, having developed new experimental and computational methods
to tackle this big challenge.

factors, DAF-16 does its job by recognizing and binding to a particular sequence of DNA in the promoter region of a gene and attracting RNA polymerase II, the enzyme that does the actual job of making the messenger RNA of that gene. In vitro approaches suggested that DAF-16 homolog FOXO transcription factors bind to the DNA sequence TTGTTTAC, and it was dubbed the "DAF-16 binding element," or DBE.[27] Using this information, Sylvia Lee and Gary Ruvkun searched the C. elegans genome for genes that had this exact sequence in their promoters, and found that at least some (17) of these candidates have a role in lifespan and metabolism.[28]

I used the inverse reasoning: if all of the dod gene promoters are examined, are there sequences that appear more frequently than they should randomly? This unbiased approach, called enrichment analysis, can tell us whether there is a particular DNA sequence that appears statistically more often than one would expect, without us having to tell the algorithm what sequence to look for. Again, I benefitted from the development of new genomic analysis tools that used different algorithms (hidden Markov models [HMM] and oligo-analysis) to build up consensus sequences from large sets of data. I found a few interesting things. First, there was some variation in the DBE motif; while the TTT was at the core of the sequence (as for all FOXO transcription factors), DAF-16 bound to sequences that were similar but not identical to the DBE, which explained why using the strict TTGTTTAC sequence wouldn't have found all of the true DAF-16 targets.*

Surprisingly, a second motif popped up with this analysis, with strong representation in the downregulated targets. This "non-canonical" DAF-16 binding motif looks basically like another motif known as a "GATA motif," CTTATCA, and was later named the "DAF-16 associated element," or DAE.[29] It seemed clear that there was no way that DAF-16 itself was binding to this

---

* I didn't "search for" this motif in the genes' promoters at the time, since the overrepresentation analysis revealed them. A later paper asserted that only 11% of our genes had the DBE, because those authors assumed that the list of 58 genes that I tested were the only ones with the DBE motif—which is wrong; I just happened to test the top 58. In fact, 98% of our dod genes have the DBE in their promoters. In sum, any genomics approach needs to account for the rate of hits you would get at background—the rate you'd get by randomly picking genes in the genome. The fact is, the DBE is pretty ubiquitous, so cannot tell the whole story about what gets turned on and when. But understanding the concept of enrichment above background is pretty key here, and I'd love it if more biologists got it.

motif, but I couldn't understand the logic of the motif either. It didn't appear with any sort of regular distance from the DBE, and primarily (but not exclusively) was present in the promoters of DAF-16-downregulated genes. This was another clue that it was unlikely to be bound by DAF-16 itself, since these genes are turned on in mutants lacking DAF-16. All in all, the appearance of the DAE in the promoters of *dod* genes strongly suggested to me that there was a new, mysterious player in the insulin signaling pathway. The identity of this factor really bothered me, as it meant that we were overlooking an important part of the pathway.

A year later I was looking for a faculty job, and after giving a talk in the Biology Department at Columbia University, two men approached me. I immediately made the incorrect assumption that the older one was the professor—but in fact the younger one was physicist-turned-bioinformaticist Harmen Bussemaker, and the older one was his graduate student Ron Tepper. Ron was a fascinating character; he had gotten interested in bioinformatics from taking a few classes after selling his company and retiring, after his kids had graduated from college. Even though he could have spent all of his time on his boat in the Bahamas (which he occasionally did), he embarked on a new life as a graduate student, and then as a postdoc, in Harmen's lab. He was probably the only true "gentleman scientist" I will ever meet. (I usually forgot about Ron's lucrative past because we always just talked about science, but it would become clear later when the rest of us would stay in the UCLA dorms for the biannual International *C. elegans* Meeting, and he would stay at the Four Seasons.)

Ron had become interested in applying a novel genomic analysis technique to aging studies, so he and Harmen developed a new way to classify all the DAF-16-dependent genes from different studies, mine plus a few other *daf-2* expression studies, into one set of results—a tough job because different types of data from different microarray platforms can be difficult to integrate. Ron's thesis project was to develop a ranking algorithm to order all of the genes from highest in *daf-2* to lowest. His results firmed up most of my previous findings, including the presence of the DAE. Most importantly, when the modENCODE consortium published its ChIP-seq (direct binding of transcription factors to DNA) analyses of 22 *C. elegans* transcription factors, we hit the jackpot: it was clear that a completely unstudied zinc-finger transcription factor, PQM-1, bound to the mysterious DAE.[30]

My lab then tested PQM-1's role in *daf-2* signaling. We could see from lifespan analyses that it was important, but the real "Aha!" moment was when

Jasmine Ashraf, a research scientist in my lab, e-mailed to tell me the results of an experiment where she used RNAi to knock down the insulin pathway. Amazingly, she found that PQM-1 is regulated in the *opposite* way to DAF-16—that is, instead of going into the nucleus when *daf-2* is knocked down, it traffics out of the nucleus, and when *daf-18* (the PTEN phosphatase that normally prevents DAF-16 from going into the nucleus) is knocked down, PQM-1 goes back into the nucleus. In fact, knocking down *daf-16* drove PQM-1 more into the nucleus, and *pqm-1* knockdown drove DAF-16 towards the nucleus. I still remember seeing the results; my family was driving from Denver to Aspen, and as we approached the Continental Divide, I read her e-mail; seeing the photos she sent, I could finally understand both how PQM-1 works and why we couldn't see it all those years—it was basically DAF-16's shadow, the "anti-FOXO." Whenever DAF-16 is in the nucleus, doing its stress resistance job, PQM-1 is out of the way, sequestered in the nucleus, promoting the expression of developmental genes (figure 3).* In sum, this means that insulin signaling can direct worms to either grow and develop, or to resist stress—but not both at the same time.

In the years since these initial findings, the whole field has better fleshed out both the processes going on downstream of DAF-16—particularly those of proteostasis and autophagy—and we have discovered new cofactors of DAF-16 that regulate its activity or cooperate with DAF-16. These include HSF-1, which activates expression of heat shock proteins and chaperones; SKN-1/Nrf2, which acts particularly in response to toxic stresses; MXL/Mondo, which acts in concert with DAF-16 and HLH-30/TFEB to initiate autophagic responses; and SWI/SNF, which was identified in a DAF-16 co-immunoprecipitation approach.[31] We also have a better understanding of the complexities of DAF-16 regulation, which involves not only the Akt and PDK-1 and -2 kinases mentioned earlier, but also SGK-1 (which we will see again in the TOR section) and JNK-1, which seems to act in opposition to the other kinases in its regulation of DAF-16.[32]

---

* This antagonism between PQM-1 and DAF-16 also explained why the DAF-16 overexpression strain Malene Hansen (a fellow postdoc at the time, now at the Buck Institute) and I had made while in Cynthia's lab arrested in development: having too much DAF-16 in the nucleus might "push out" PQM-1, preventing developmental genes from being expressed. At the time, we overcame the problem by putting the worms onto *daf-16* RNAi during development; now we could understand why that worked.

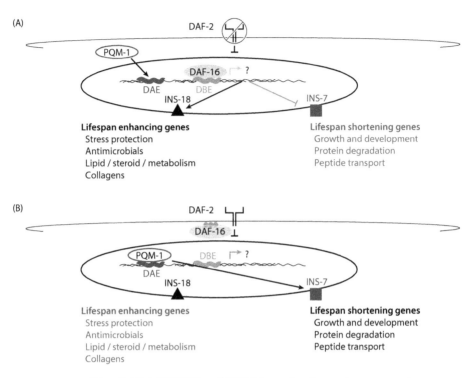

FIGURE 3. How DAF-16/FOXO and PQM-1 regulate longevity, stress resistance, and growth and development. (A) When the DAF-2 insulin receptor is mutated or not functional, the signaling pathway that normally keeps DAF-16 out of the nucleus is off, and DAF-16 enters the nucleus, where its transcription factor activity results in the transcription of genes that help keep the cell functioning (lifespan enhancing genes). (B) When DAF-2 insulin signaling is "on" DAF-16 is phosphorylated and excluded from the nucleus, and PQM-1 enters the nucleus to promote expression of growth and development genes (lifespan shortening genes). The mutual exclusivity of DAF-16 and PQM-1 allows cells to choose between stress resistance and growth/development (Tepper et al. 2013).

How Does This Insulin Signaling Pathway Actually Regulate Longevity?

Imagine a happily fed animal, say a worm on a rotting apple in an orchard. This worm has plenty of food, thanks to the ready supply of bacteria that flourishes as the apple decays. With plenty of nutrients at its disposal, the right "decision" for the worm is to develop as quickly as possible and to have as many children as it can while times are good. Under these circumstances, in response to the food cues that the worm senses, an insulin that binds to and activates the insulin/IGF-1 receptor is released from neurons and circulates, turning all of the worm's DAF-2 receptors to the "on" position. DAF-2 activates the AGE-1 PI3 kinase, which in turn activates the Akt and PDK kinases to phosphorylate DAF-16/FOXO. In this phosphorylated state, two "14-3-3" proteins, PAR-5 and FTT-2, bind to DAF-16 and prevent it from entering the nucleus when it is not needed.[33] This might allow PQM-1 to stay in the nucleus, activating a developmental program that lets the young worm develop into adulthood.[34]

Now imagine this same worm, or more likely one of its unlucky progeny, that is still around when the apple has been devoured and the bacterial supply is starting to run out. If the worm manages to avoid starving in the first larval stage (L1), it won't go into dauer, but it may get pretty hungry while it's growing. There's still hope: maybe the worm can crawl to a new apple and find another source of bacteria. If this is likely, then it's a better strategy for the worm to buy itself some time before reproducing. When the worm's neurons sense that there is a low amount of food but it is not yet starving, the insulin signaling pathway essentially turns "off," causing DAF-16 to enter the nucleus (and, concomitantly, PQM-1 to leave) and to start turning on all kinds of protective genes that help the worm's body survive longer while it looks for food. Simultaneously, the worm's germline gets the message to slow down reproduction until times are better (see chapter 8), protecting its oocytes until there is enough food to support progeny. If the worm gets to a new source of food, then its neurons will release an insulin that will communicate to the rest of the animal that it's time to turn off its stress response and resume reproduction. If there's no food, that hungry L1 will develop into a dauer animal, which is the state that most worms are in when they're recovered from the wild.

Of course, worms are not the only animals to experience times of hunger, and the insulin signaling and FOXO components I've told you about seem to act in the same way in flies and in mammalian cells. In all of these systems, a low nutrient state results in the slowing of aging and better stress responses. While the aging and longevity field likes to think of this as a way to "slow aging," in fact that idea makes no sense evolutionarily. The only reason to have any mechanism of responding to nutrient status is to help the animal "win" the evolutionary race—that is, to be sure that more of its own genome gets propagated through time. That means that it doesn't really matter how long that individual lives but, rather, how well its progeny survive and go on to make more progeny. And for that to happen, the individual must make the "right" choice about reproduction. When times are good, it makes sense to have progeny early and quickly. (One interesting recent finding is that early progeny are most likely to be phenotypically diverse, which may help in surviving all kinds of conditions.)[35] But if those progeny have no food to eat, then it makes more sense to slow down reproduction, look for more food, and start things up again later. In order for the animal to be *able* to reproduce, the mother's body must still be in good enough shape to do so without dying in childbirth. So all of this business about slowing down aging isn't really to help the hungry individual live to the ripe old age of three to four weeks, but rather to make sure that the next generation has the best chance of survival.

It's somewhat humbling to have this viewpoint—our own survival until old age may not be that important, and in fact our longevity may instead simply be a side effect of the need to be in good enough shape to have kids in in midlife. Everything after that may just be gravy. Finding the genes that *daf-2* and *daf-16* regulate and understanding how this signaling is conserved in higher organisms has helped us better understand how we might harness some of the most powerful "anti-aging" cellular processes—and I'm happy I had the chance to peer into the worms' transcriptome to learn more about longevity regulation.

7

Dietary Restriction

NUTRIENT AND GENETIC REGULATION OF LONGEVITY AND REPRODUCTION

You could have done anything. But you're just like your monkeys, locked up and underfed. The point is that you have to live a lesser life in order to live a longer one.

—LUCY TO VARYA IN *THE IMMORTALISTS*, BY CHLOE BENJAMIN

I SPENT MOST OF MY childhood as a serious gymnast. I wasn't that talented, but I was strong and driven, which, together with a good work ethic, almost hid my deficiencies, and I got pretty far before a broken ankle eventually forced me to quit. My parents weren't totally on the gymnastics bandwagon and always viewed the sport dubiously, for good reason: in addition to the constant injuries, anorexia nervosa was a growing problem among gymnasts in the 1980s. Extreme diets were encouraged by coaches because it delayed puberty and kept the girls light and small, all the better for flipping. Many top-level gymnasts suffered from some sort of eating disorder. (Sadly, one of the gymnasts on my regional circuit, Christy Henrich, eventually died of anorexia.)* In response to these problems, my mom drilled it into my head that I should *always, always eat.*

---

* Joan Ryan's 1995 book *Little Girls in Pretty Boxes* (New York: Warner Books) details the deleterious effects of extreme training conditions of high-level gymnasts and ice skaters, including Henrich's death (she was coached by Al Fong in neighboring Missouri) and conditions in Béla Károlyi's gym in Texas, where some girls from my team in Kansas transferred. Eating disorders were common in gymnasts, at least while I was competing.

I mention this to explain my personal aversion to strict diets in general, and the notion of caloric restriction in order to live longer in particular. Eating only 60%–70% of one's normal intake, which is the level required to induce a 10%–30% increase in longevity, strikes me as a horrible life choice. I'm also spoiled scientifically, having come from the *C. elegans* insulin signaling field, where we typically see lifespan increases of 200%, making an increase of only 10%–30% seem meager. Nevertheless, *even I* will admit that the data on the influence of caloric restriction on longevity and many aspects of healthy aging are compelling. There is an ever-growing body of work suggesting that dietary restriction might be an effective way to increase lifespan, reduce disease, and improve healthspan. Moreover, although it is challenging, it's an approach that almost anyone *could* adopt (but we are going to dig into this point as well), unlike strenuous exercise regimens or expensive drug treatments, so knowing whether or not dietary restriction could help humans live long and avoid age-related disease is an important question. Understanding the molecular pathways downstream of caloric restriction that trigger these beneficial effects could help us find ways to turn on those effects without the difficulty and deleterious effects of caloric restriction itself.

A Brief History of Caloric Restriction

At various times throughout history, people have been forced to undergo some form of dietary restriction, usually due to wartime rationing.[1] Later study of these forced but "natural experiments" suggested the possible advantages of dietary restriction, as a two-year reduction in calories in Denmark (1917) and four years of 20% caloric restriction in Oslo (1941–45)—both carefully managed to maintain sufficient supplies of nutritious food—correlated with greater than 30% reductions in mortality.[2] Okinawans in the last century had a 17% lower caloric intake than Japanese and 40% lower (and a less meat-based diet) than average Americans, and Okinawa produced centenarians at the highest rate in the world: 50 per 100,000, or almost five times the rate anywhere else.[3] However, the import of a Western diet in recent decades has now reversed the high life expectancy that Okinawans once enjoyed. In any case, low nutrient levels have been observed to correlate with improvements in several mortality factors.

The history of the experimental study of caloric restriction (CR; also known as dietary restriction, or DR) has been described extensively elsewhere, so I'll be brief. As far as I know, the first mention of linking limited nutrients, slowed growth (aka "stunting"), and extended lifespan was in a 1917 *Science* paper studying a small set of nutrient-deprived rats; Osborne and colleagues used reproduction as a proxy for lifespan, reasoning that it would be more practical as a measurement. In fact, reproduction in some of these "stunted" rats (25–31 months) lasted well beyond the average control reproductive span (15–18 months) and even beyond the average rat lifespan (about two years).[4] The stunted rats went on to die of "lung disease" by 32 months.

The father of the CR field, though, is generally considered to be Clive McCay (whom we'll hear more about in chapter 10, on parabiosis), a biochemist and professor of animal husbandry at Cornell University who made a series of nutrition breakthroughs in his career.[5] His work uncovered the deleterious effects of phosphoric acid in cola, and decreased calcium absorption into older women's bones. He also developed irradiation-based food preservation, and, with his wife Jeanette, a high-nutrient bread known as Cornell Formula Bread. In the early 1930s, while investigating the effects of nutrition levels that might be experienced during food shortages, McCay made the surprising discovery that food-deprived rats not only were able to survive but in fact both lived longer and suffered less from late-life disease. This was the first real inkling that restricting calories could be beneficial to health beyond its slowing of reproduction.

Although we generally regard McCay as the first CR researcher, Scott Pletcher, who studies dietary restriction in flies at the University of Michigan, recently pointed out to me that in addition to the Osborne study in rats (1917), the Polish researcher Stefan Kopec published his finding that intermittent starvation prolonged lifespan in *Drosophila* in 1928,[6] preceding McCay's much better known and oft-cited finding on caloric restriction in rats. Since these early studies, the longevity effect of caloric restriction has been shown in mice, worms, flies, monkeys, and even single-celled yeast, making it the first and perhaps only "longevity treatment" that applies to every organism tested. Later work established that caloric restriction can confer other health benefits on animals, including lower rates of tumor formation.

What Exactly Is Dietary Restriction?

One problem that plagued the DR/longevity field was the lack of agreement about what DR really *is*, exactly, even though the "restriction of calories to 60%–70% of a normal diet" is the definition that is commonly used. That sounds great, until you try to figure out what that means for a worm or a fly or a monkey in the lab, given that we don't really know what or how much most animals eat in the wild. In fact, our normal lab conditions are often the equivalent of eating McDonald's every day, a never-ending buffet of lab chow (or, in the case of worms, bacteria). *How do you properly restrict a diet?* became the burning question in the field in the early 2000s. As a postdoc, I watched at Cold Spring Harbor Laboratory Molecular Genetics of Aging meetings as first the yeast field (*what % glucose is real CR?*) and then later fly, worm, and mouse researchers argued vigorously about the proper way to carry out DR experiments, and what chemicals might or might not induce the effect. (To be honest, this is another reason why I stayed away from DR research in my own lab—*daf-2* genetics seemed much cleaner and less debatable.) It seemed clear that simply removing food for a time extended lifespan in all organisms, but the exact timing, amount to restrict, dietary components, and the molecular pathways that interpreted the DR signal and turned it into longer life needed to be discovered. In fact, the genetic and regulatory mechanisms that control the longevity response to diet were not well studied for a long time, since it was assumed that everything must be controlled through the insulin pathway and that *daf-2* was just a caloric restriction model—which was at least in part incorrect. In 2009, Eric Greer and Anne Brunet nicely showed that at least five different types of DR could be induced in worms and act through at least seven different molecular pathways,[7] illustrating how complicated DR can be (and how everyone arguing passionately about DR was probably at least partially correct but seeing only a fraction of the picture.)

Mice are used extensively in the study of DR, and because they are mammals some of these findings might be more applicable to humans than what we find in flies or worms, but mouse studies are of course much slower than similar studies in invertebrates. Combinations of genetics with DR conditions (e.g., testing to see whether long-lived growth hormone receptor mutants live even longer when dietarily restricted) to figure out which genes are necessary for the effects have been compelling. Mice lacking growth hormone receptor (GHR) do not experience further lifespan extension from DR, suggesting GHR and DR may act in the same pathway.[8] Similar experiments combining genetic mutants and DR conditions have helped tease out the molecular pathways that are necessary to sense decreased nutrients and convey this information across the whole system (as I'll describe below). Identifying these components is critical in unraveling the puzzle of dietary input and the physiological output.

One of the best presentations on dietary restriction that I've ever seen was a tag-team talk from Rozalyn Anderson (University of Wisconsin, Madison) and Julie Mattison (National Institute on Aging) to explain their seemingly contradictory results on caloric restriction in primates from a pair of long-running studies done in rhesus macaques.[9] These experiments are difficult to carry out, as you might imagine, simply because of the length of the study: in order to find out whether caloric restriction affects the longevity of monkeys, it takes around four *decades*; the studies were originally started in the late 1980s by Donald Ingram (NIA) and Richard Weindruch (UW). The UW study had reported a significant increase in lifespan due to dietary restriction, but the NIA study failed to find such a strong effect. To their credit, rather than fighting over which result was "right," the two groups carried out a separate, detailed analysis of the design of the two studies to discover why the results might have differed.[10] Through discussion and analysis of the differences in their results and methodology, the two labs were able to identify the source of the discrepancies. This analysis revealed that there were differences in the sources of the monkeys, the timing of the onset of the DR conditions, the nutrient composition of the diets, the feeding timing and practices, and the body weights of the animals. Although there were many differences, the one that likely had the biggest impact was the slightly different control diet for the monkeys. The Wisconsin study's control diet was essentially more food than they would have eaten in the wild, so the matched DR-treated animals did much better by comparison. That is, if the control diet of one study is not the same as that of the other, then it will affect the interpretation of the entire experiment. While we can criticize this choice of control conditions, it is hard not to draw a parallel with our own calorie-laden Western diet and conclude that perhaps we, too, would benefit from some DR, just as the monkeys whose control food was on the more generous side did. In addition to generating valuable information for the DR field, the authors presented an excellent model for open communication in science: rather than arguing about who was "right," the two groups analyzed all of the data to zero in on the truth.

Flip It and Reverse It: The Fleeting Benefits of Dietary Restriction

How quickly does DR have an effect? Dame Linda Partridge's lab (University College London) reported that DR dynamically affects *Drosophila* mortality—that is, the likelihood of a fly dying at any point changed immediately in response to DR. Adult flies fed a normal diet and then switched to DR immediately slowed their mortality rates; in fact, the effect seemed almost instantaneous.[11] This is very different from just stretching out how long an animal lives, having a cumulative effect over time, or taking a long time to get started. (Unfortunately, the paper does not show any other healthspan data, like motility, which would have been interesting to see.) There is a flip side, though: disturbingly, mortality rates *increased* again almost immediately upon DR cessation, suggesting that the effects of DR might be short-lasting. The starting and stopping of DR effects might reflect the kinds of flexibility that the animal has evolved to rapidly respond to fluctuations in the availability of nutrients in the wild.

The idea that DR has fairly rapid effects—but which can be undone quickly, as well—is important, although more data on healthspan would be necessary to fully understand whether overall DR is beneficial even if not practiced continuously. The fleeting nature of the effect of DR is a bit concerning—in Partridge's *Drosophila* study, the benefit didn't seem to last or even to slow aging once DR was stopped. Translated into human terms, this would be like practicing DR for 30 years and then stopping one day, and then having a heart attack the next week and dying—terrifying. Clinical studies would be difficult, but necessary, to determine whether there are benefits or dangers to halting DR in humans.

The Cost of Food Deprivation: Slowed Brain Function

The hard-core DR fans out there won't like this, but anyone who's had brain fog from missing a few meals will not be surprised to hear that the mammalian brain, which uses about 20% of the body's calories (despite the fact that it makes up only 2% of its mass), works less well under reduced nutrient conditions. When mice were fed 15% less than their ad libitum counterparts, resulting in a 15% reduction in body mass, conductance of the AMPA receptor in the visual cortex decreased, even though all the mice ate normally before testing, suggesting it was not a short-term effect of hunger.[12] Decreasing AMPA receptor conductance saved 29% of the animals' ATP (adenosine triphosphate, a molecule that is a source and store of energy at the cellular level), so there is logic to the system. Compensatory increases in input resistance and depolarization maintained the neurons' excitability, but response variability was increased and coding precision was decreased. That is, the mice had impaired visual discrimination as a result of their long-term decrease in energy availability. Restoring leptin levels restored their visual function, suggesting that the decreased nutrient levels were a signal to the neurons, rather than acting through a direct metabolic regulation of the brain's AMPA receptors. These results in the mouse visual system echo our findings in worms, where we saw that a model of DR (*eat-2* mutants) performed less well in memory assays;[13] nutrient restriction also deleteriously affected long-term memory in *Drosophila* and blowflies.[14] The reduction of metabolically costly functions to save brain power when nutrients are restricted long-term seems to be an evolutionarily conserved mechanism to promote survival under extreme conditions. DR, as attractive as it might be, does not come without a cost—particularly to neuronal function, apparently.

A Calorie Is Not Just a Calorie

The complex mixtures in most animals' diets make it difficult to immediately know the proper way to restrict their diets. Which component is the most important part to cut down—is total calories the important factor? Or is it fat, or sugar (carbohydrates), or total protein, or particular amino acids? The roles that these specific components play in lifespan and aging rates are being tested in every organism. Of course, practitioners of keto and paleo diets have their preferences, and diet trends seem to fluctuate. Because some studies suggest that particular components, such as sugars or protein, rather than total calories, are most likely to control lifespan, we refer to the general treatment now as "dietary restriction."

To study dietary requirements in *Drosophila*, Scott Pletcher's lab (University of Michigan) carried out a careful analysis of fly lifespan with respect to the relative contributions of the three major components of the fly diet (protein, sugar, and fat), creating a sort of "phase diagram" of dietary components to find the optimal ratio to increase lifespan.[15] The optimal diet for fecundity (that is, reproduction) was high protein and low sugar, while lifespan was best with a more balanced diet; and it may be unsurprising that the flies fed high sugar were measurably more obese.

While there has been a focus on the obvious culprits—fats and sugar—there have also been studies of specific amino acid restrictions in several different animal models, revealing that restriction of proteins in general, and specific amino acids in particular, might extend lifespan and have positive effects on other health metrics, such as motor and cognitive function.[16] Methionine is

Don't We Already Know What's Bad for Us? Why It's Important to Know When Researchers Have Conflicts of Interest

If it seems like nutrition advice goes in cycles, it's not your imagination. Every few years the news trumpets a new viewpoint about what we should eat (and, more importantly, not eat). New diets become popular but then fade—low fat, high protein, low sugar, "detox cleanses," Atkins, paleo, keto, South Beach, . . . on and on, but in actuality some of these are repackaged old diets. Americans are also bigger than we used to be: since the 1950s obesity rates have more than tripled,[a] and waists have grown more than *six inches*—all despite heightened awareness of diet and exercise. Of course, some of this is no surprise: portion sizes in restaurants have increased, daily caloric intake has increased by almost 400 calories, and standard Western diets contain more meat and cheese but less milk and fewer eggs than 50 years ago, and 40% more sugar than in the 1950s, while our behaviors include more sedentary time (e.g., screen time and car travel) and less manual labor.

How did we get here? A damning report in 2016 revealed that at least some of these deleterious shifts in our diet are due to the sugar industry's purposeful suppression of results suggesting that sugar is responsible for cardiovascular disease.[b] Stanley Glantz is a UCSF statistician who was also involved in exposing the tobacco industry's cover-up of the known risks of smoking (think "The Insider") and the Koch brothers' tobacco industry financing of the Tea Party;[c] Glantz and colleagues found that the sugar industry had systematically covered up data showing that sugar is unhealthy, and had influenced the nutrition field since the 1960s. Studies in the 1950s of dietary factors correlating with high rates of cardiovascular and coronary heart disease (CVD and CHD) mortality led to a link with added dietary sugars, and a separate link with dietary fats (total fat, saturated fat, and dietary cholesterol). But in the 1960s the Sugar Research Foundation used its influence (money) in the nutrition research field to blunt any criticism of high sugar diets—in fact, the sugar industry realized this was an opportunity to gain *more* of the market by shifting the blame to fats, promoting a low-fat—but high-sugar—diet as "healthy." To do so, researchers were paid to bury evidence linking CVD to heart disease, and to write review articles criticizing studies that linked sucrose to CVD.[b] This campaign was so successful that in the following decades the thought that sugars were to blame for CHD had all but disappeared, with fat bearing the brunt of the blame for CVD risk. As a result, people were advised to stop eating eggs and high-fat foods, and "low-fat" became the diet buzzword and remained so for decades. Meanwhile, high-fructose corn syrup (HFCS), a cheap sugar abundant in the United States, has been added to almost every American food product to increase sweetness and alter other characteristics—in fact, "low fat" diet foods have lots of HFCS, making it even more difficult to avoid a high-sugar diet. Our consumption of HFCS has increased from 37 grams (1977) to over 55 grams (2008) per day. No wonder people are confused; if nutrition researchers have conflicts of interest because they've been paid by that industry to influence the litera-ture, influencing the contents of food products, then consumers lose out, even when they are trying their best to be healthy.

interesting as it is the first amino acid in most proteins because of the start codon (that is, the RNA code that signals the "start" of a protein is also the signal to attach a methionine and start a peptide chain), so it is a relatively good measure of the number of mRNAs that are translated into protein; restriction of methionine has been reported to increase mouse lifespan and resistance to oxidative stress.[17] Furthermore, methionine alone was sufficient to rescue the fecundity of both flies and worms subjected to DR without decreasing lifespan, suggesting that the amino acid signal, rather than nutrients themselves, con-trolled the physiological response.[18] Restriction of the branched-chain amino acids (BCAAs) leucine, valine, and isoleucine has also been found to extend male mouse lifespan even when applied in mid-life,[19] largely acting through the TOR pathway, which senses amino acids and other nutrient components. High-protein, low-carbohydrate diets have been associated with increased car-diovascular mortality[20]—kind of surprising given the current popular notion that high-protein, low-carb diets are good for health and weight loss. In the end, switches from animal- to plant-based diets might be beneficial not because of the quality or nature of the specific amino acids in each but rather the decrease in total protein typically ingested in plant-based diets.[21]

Some specifics of DR might be "private"[22]—that is, specific to a particular organism—because the natural diet of each animal is different, but the sensors that detect DR and signal to the organism and the ultimate results of DR might be "public" (well conserved across many organisms). Asking these questions in a variety of animals will help us understand the whole molecular process of DR regulation of health and longevity. This is particularly important to clarify as people strive to eat better to become healthy, lose weight, and live longer— it sounds great, but the devil is in the details if you are trying to choose between low fat, low sugar, low carb/keto (high protein and fat), or other popular fads. Inconsistent public messages plague the human nutrition world, or at least the fraction that makes it to the public consciousness. One point that is often missed is that the goals of specific diets are often in conflict with one another. Building muscle is obviously not going to require the same diet as losing weight, or becoming a good long-distance runner, or living long. What was eaten in "paleo" times might be optimal for short periods, not necessarily for longevity or long-term health, and the young men in my Molecular Genetics of Aging class who lift weights and take BCAA supplements on a regular basis were not happy to realize that their diets might shorten their lifespans, if the current research holds. It would not be surprising to find that diets that offer short-term benefits may be deleterious in the long term, but we need more research to test these theories.

Timing Is Everything: Intermittent Fasting and Time-Restricted Feeding

We've talked about the benefits of limiting what you eat, but what about *when* you eat? Because dietary restriction is psychologically challenging, more re- cently the effects of intermittent fasting (IF) and time-restricted feeding (TRF) have been explored.[23] These approaches often do not involve ingesting fewer calories but rather restricting eating to short intervals, or limiting fasting to a few days of the month. This may be much more achievable than severe dietary restriction. Perhaps not surprisingly, various forms of IF have become popular, ranging from whole days of "off vs. on" eating per week to restricting the hours that one eats each day. For example, the "5:2 diet" involves eating only 500– 600 calories on two non-sequential days of the week, and has been very popu- lar in the United Kingdom. Other similar IF diets are more extreme versions, including alternating days of fasting. Another route is to provide the nutrients that make up a DR diet: Valter Longo (University of Southern California) has developed a broth (ProLon) that mimics fasting. (There is enough popular literature on all of these diets, including books by their inventors, that more explanation is not necessary here.)

Rather than extreme restriction for full days, restricting the hours that one eats, aka time-restricted feeding, may be much easier to do. Research on circadian rhythms and "chronobiology" suggests that there are optimal times to be awake, to take medicines, and to eat. Tying caloric intake to the circadian rhythm may lead to optimal health;[24] eating most of our food during the day rather than at night might be optimal, as glucose tolerance appears to be lower at night.[25] There are plenty of data showing that the opposite—ignoring our circadian rhythms—is deleterious, as evidenced by the poor health, higher cancer rates, and shorter lifespans that night-shift workers commonly experience. The flip side, that restricting meals to a mid-day window of 8 to 10 hours, is based on the idea that the benefits of fasting could be induced by matching the time of eating to optimal hours of metabolic expenditure.

What exactly are the effects of IF on animals? Valter Longo's group studied the effects of a periodic diet that mimics fasting (periodic fasting-mimicking diet, or FMD) on many different organs of mice and found that this diet was pretty miraculous—it made the animals lean, slowed cancer development, and improved tissue regeneration, organ function, and cognitive function.[26] While it did not increase maximum lifespan, the periodic FMD did increase median lifespan—so if you were a mouse on this periodic FMD, you might not live longer, but more of your mice buddies will stick around the same length of time. Sounds pretty awesome.

Time-restricted eating (TRE) seems like it would be the easiest and most practical approach to any of these methods to induce DR benefits in humans, and therefore might have the greatest impact. But there is evidence that the timing of the TRF has to be just right: at least in flies, only nighttime fasting induced autophagy (cellular recycling)—and without nighttime autophagy, there was no benefit to TRF.[27] It would really stink if you went to all the effort to starve yourself and didn't get anything out of it. In fact, TRE in humans was dealt a blow in late 2020. Ethan Weiss (UCSF), whose group studies metabolism, was interested in testing TRE in a well-controlled study, after becoming interested in the results that Satchin Panda (Salk Institute) had reported in animals. (Weiss reports that, at the time, he himself was such a believer that he was practicing 16:8 TRE—that is, eating only between noon and 8 p.m. each day.) Weiss's group carried out a decently sized randomized clinical trial of 16:8 TRE (116 participants, 60.3% men, aged 46.5 years on average, overweight with BMI from

27 to 43 kg/m^2).[28] Importantly, their comparison wasn't a simple "before and after" treatment; instead, they used a control group ("continuous or consistent meal timing") who also ate three meals a day plus snacks, but in no specific time window and without a fasting period. There was also no particular diet or exercise plan that either of the groups had to follow; they just had to report their weight changes, and a subset of the group were measured in person for various metabolic parameters. After 12 weeks, the shocking result was that *TRE did not lead to any statistically significant differences from the controls!* How can that be? As the authors point out, having a clinical control arm to the study was key: if there had been no control arm, the study would have concluded that TRE causes weight loss, but since the control arm *also* showed an average weight loss (and in both groups some people gained weight!), there was no statistically significant difference between the 16:8 group and the control group. This was also true for most other parameters they measured (insulin, glucose, lipids, sleep, activity, energy expenditure, or fat mass). More disturbingly, the TRE group lost more lean mass—the kind of mass you don't want to lose.

The importance of this study cannot be overstated: because Weiss's group included a proper control—which is hard to do in human nutrition studies—they were able to distinguish the effects of TRE from those of just being enrolled in a clinical trial in general. Perhaps the only thing we can take away from this is that having to report your weight each day to someone else might make one more mindful of one's diet, regardless of when one eats. Not exactly a gold star for TRE in the absence of caloric reduction, but that's how science sometimes goes.

Caloric Restriction in Humans— Who Chooses to Restrict Their Diet?

All the genetic and molecular information we currently have about the mechanisms of dietary restriction is the result of studies done in model organisms, but some clinical trials of DR-like regimens are underway. Obviously, lifespan is difficult to test, but health can be measured periodically to assess any improvements. One study is the Comprehensive Assessment of Long-Term Effects of Reducing Calorie Intake, aka CALERIE, which is taking place at several different institutions.[29] These studies are in fact all somewhat different from one another, using different levels and duration of calorie restriction, and enrolling people of different ages and weights. Notably, several of the studies are aimed at overweight individuals, rather than DR of people at normal weights, so whether DR is actually being studied is debatable. Instead, we are more likely to better understand how to reduce obesity from these studies.

However, there are some true believers who are voluntarily subjecting themselves to CR, members of the Caloric Restriction Society. Relative to healthy age-matched controls, a 2007 study of CRS members (average age of 50, practicing CR for about six years) showed that they have lower BMI (19.6 vs. 25.9), better "good" cholesterol, better fasting plasma insulin and glucose levels, lower blood pressure, and lower chronic inflammation markers.[30] These are all very positive health metrics, and the data to this point suggest that their incidence of type 2 diabetes, atherosclerosis, and other age-related diseases is likely decreased, as well.

While these health improvements are all great, it is notable that the ratio of men to women in the CRS is 29:4, suggesting that this extreme diet is not equally attractive to men and women. When I travel around to give seminars, I've asked DR researchers I meet whether men or women are more likely to enroll in studies of one of these extreme diets and stick with it—that is, is the drop-out rate from these DR studies sex biased? More than once, I have been met with blank looks, suggesting to me that not enough people are paying attention to this question—but it's important, as the effects of DR have been shown to be quite different in male and female mice, and there are sex-specific effects of DR mimetics on lifespan as well. If there is no benefit to be had in one sex, then the field has an obligation to make that clear to people who are interested in subjecting themselves to some dietary regimen. And as I mentioned before, we should also pay some attention to the risk of eating disorders when discussing the benefits of CR, rather than blithely suggesting it as a cure-all, particularly if one sex is more susceptible.

I also think we need to know how DR affects moods. While it's harder to measure in flies, worms, and yeast, the authors report anecdotally that dietarily restricted primates are, understandably, quite cranky. To be fair, there are reports on this point historically: 36 (already lean) World War II conscientious objectors became subjects in a 1950 study called the "Minnesota Starvation Experiment" and had abnormal psychological behaviors after six weeks of an extreme CR diet.[31] Interestingly, the members of this group ended up living eight years longer than the 1920 life expectancy, which, given what we know about the reversibility of DR, is unlikely to be due to this restriction, but instead due to other factors in these men's lives, those same factors that led them to be conscientious objectors to war in the first place. Another study using a less extreme diet reported no adverse effects on mood or cognition and even some improvements,[32] although both studies reported negative effects on libido (note again that the subjects were all men). We don't know if there are certain types of people who are more likely to voluntarily adopt such extreme diets, but I think that answer in itself would be interesting to find out. Anecdotally, I know far more men who subject themselves to some sort of DR or IF—and brag about it accordingly—but the best way to assess whether there is a gender bias in choosing to carry out DR will come from analyses of drop-out rates from studies that start out with equal numbers of men and women. My hope is that eventually we will know not only the benefits of DR but also whether these choices and effects are sex-specific, and the psychological challenges to the maintenance of DR long-term. Anyone who has ever dieted can probably tell you how hard it is to maintain *any* diet, much less one that reduces your calories by 30% or is extreme in any way; long-term weight maintenance is extremely challenging and the lost weight is usually regained within a few years,[33] so it would not be surprising if most DR regimens are too difficult to maintain.

One well-known CR researcher and advocate, Dr. Roy Walford, studied the effects of DR on aging in mice and practiced CR himself, even going so far as to suggest that his fellow occupants of Biosphere 2 also practice CR in response to an alarmingly low crop yield.[34] While he had spent most of his career studying and promoting DR, sadly Walford did not live long enough to reap the potential longevity effects: he died from complications of ALS (amyotrophic lateral sclerosis) at the age of 79. While Walford surmised that the air conditions within Biosphere 2 and possible resulting hypoxia might have contributed to his development of ALS,[35] at least one study of DR treatment of a mouse model of ALS suggested that CR might transiently improve motor performance but ultimately hasten onset of disease[36]—if true, Walford's decades of DR might have had a deleterious influence on his development of ALS. On the other hand, it's impossible to know this without the proper genetic analyses, and many cases of ALS are spontaneous. In fact, one could just as easily argue that Walford might have *staved off* ALS for decades because he calorically restricted—it is impossible to know now. In any case, the transgenic ALS model studies do warrant caution in applying DR to every possible disease; the possibility that DR might be deleterious should also be considered. And our own findings on worm models of Parkinson's implicate low BCAA signaling in Parkinson's-like symptoms, while our work on worm learning and memory suggests that DR models take longer to learn and have worse memory—but they lose it less quickly when they get old, suggesting that maybe waiting until the last possible minute to practice DR might be a wise strategy. Knowing exactly what benefits and costs DR presents in many different aging and disease contexts will be increasingly important as more people try to apply different versions of it to their own lives.

The Genetics of Dietary Restriction— Finding the Regulators

How do cells get the message that they are being calorically restricted so that they do the "right" thing? In the past few years, the identification of the molecular pathways that mediate these responses to DR has accelerated. Identifying and characterizing these regulatory components has helped us understand how low nutrient levels are sensed, and how this information is ultimately turned into a longevity "decision" that is then executed at the biochemical level in cells.

The First Dietary Restriction Genes

Of course, I'm going to start here with worms. In chapter 6 I mentioned the work Michael Klass did back in the 1980s to identify long-lived mutant worms.[37] At that time, the exact genes that caused the long-lived phenotypes were not identified, but you might remember that a few of the mutants essentially couldn't "chew" their food—that is, their pharyngeal muscles could not be activated to grind up their bacterial food—so it was clear that they had limited dietary intake. Thus, decreasing food intake extended lifespan of wild-type worms, just as it had in other organisms. In 1996, Siegfried Hekimi's lab published that "*eat*" mutants, which have defective pharyngeal pumping and therefore cannot take in bacteria effectively, are long-lived, just as Klass had found.[38] *eat-2* turned out to encode the nicotinic acetylcholine receptor, which is required for the pharynx to respond to acetylcholine to regulate its pumping rate. Indeed, if worms are fed smaller bacteria that are easier for the worm to chew, caloric restriction is "undone" and the worms no longer live longer.[39] Thus, *eat-2* is a genetic mutant that is effectively calorically restricted because it can't eat well—like if I had my jaws wired shut and had to eat through a straw—but that doesn't tell us how DR works at the molecular level, it just provides worm researchers a nice genetic tool to cause dietary restriction in worms.

Other than the identification of *eat* mutants, and Hekimi's identification of "clock" genes (some of which turned out to regulate mitochondrial function, as we will discuss later),[40] the genetic dissection of dietary restriction in *C. elegans* lagged behind its counterpart, insulin/IGF signaling (IIS), where the components of the pathway were being identified one by one. As I mentioned earlier, the cloning of IIS mutants was greatly aided by the additional role of IIS in dauer formation, which simplified the isolation of the mutations for cloning. The cloning of lifespan mutants, at least prior to whole-genome-sequencing technology, required the researcher to carry out many crosses and months-long lifespan experiments, and then they had to go back to the families of the animals for additional experiments to narrow down the genetic region— so lifespan extension is not an easy phenotype to track through traditional genetic approaches.* By contrast, dauer formation is quick and is an obvious and easily followed phenotype that is visible in single worms (as opposed to population phenotypes like lifespan), and because dauer is a pre-reproduction stage, the mutant animals can go on to have progeny, making the genetics easy to follow.

Despite the fact that IIS and DR were assumed to be the same pathway in mammalian studies, it's important to note I'm treating these as two different mechanisms because of genetic results in *C. elegans* demonstrating their independence: the reduction of *daf-2* extends *eat-2*'s lifespan,[41] so they appear to be additive⁺—that is, they do not act in the same pathway. More critically, loss of *daf-16*, which you'll recall is absolutely required for *daf-2*'s extended longevity, did not alter *eat-2*'s lifespan. Said differently, *daf-16* mutants can still have their lifespan extended when crossed with *eat-2* mutants or when dietarily restricted. These results suggest that the DR pathway and insulin signaling are genetically separate mechanisms, at least at the regulatory level; the downstream pathways have some overlapping cellular outputs, however.

---

* This problem was solved when the RNAi library came out. It still wasn't easy, but researchers could know within a few weeks whether each of the 20,000 genes affected lifespan.

+ Sort of—hypomorphs are mutants that lose some but not all function (unlike null alleles). So actually you can add two hypomorphs to get a greater effect. In fact, RNAi knockdown of *daf-2* in a *daf-2* hypomorph mutant can lead to a longer lifespan, illustrating this point. *daf-16* requirement is a much better test.

Sir2, NAD⁺, and Resveratrol

Some of you longevity aficionados might be wondering why I haven't yet mentioned the yeast gene *Sir2*. For several years, the Sir2 story made the news almost constantly. In fact, Sir2 and the "red wine drug" resveratrol,[42] which was thought to activate Sir2, may have been the only thing you've heard about as far as molecular genetics of aging goes.*

A few years after *daf-2* was found to control worm lifespan, Sir2 was dubbed a "lifespan gene," after it was discovered that Sir2 was required for yeast to have a long lifespan under CR conditions (that is, in a mutant that blocked the ability of yeast to use glucose).[43] Then, a few years later, the small molecule resveratrol was reported to extend lifespan by activating Sir2.[44] The idea of having found an "answer" to aging, and the idea that one compound might fix it all, was very appealing, almost intoxicating. Different underlying mechanisms were attributed to Sir2's pro-longevity function: regulation of repetitive ribosomal DNA (rDNA) copy levels, then metabolism (NAD^+ levels), transcription of stress genes, telomere function—and with almost maddening frequency, every new paper declaring each new component "the master regulator" of aging. More recent studies of Sir2 suggest that its effect on longevity is due to its role in NAD^+/NADH metabolism, its influence on FOXO activity, and its activation of the mitochondrial unfolded protein response (UPR), which we will address soon. In the early days of Sir2, there was a lot of excitement and an almost equal amount of arguing about how it might control lifespan, what the correct dietary restriction conditions are, whether replicative or chronological lifespans were "real lifespans," and whether resveratrol does anything to extend lifespan or not.

Let's first review how one measures lifespan in yeast. There are two main approaches, replicative lifespan (RLS) and chronological lifespan (CLS). In RLS, the number of divisions (daughters) that a mother yeast cell can produce before it wears out (senesces) is counted, while CLS measures how long a cell can survive post-mitotically (after division). Basically, RLS measures how many times cells can divide, and CLS measures how long nondividing cells can survive. Some of the genes that regulate RLS also affect CLS, but many are separate. There is some debate about which of these best model aging in higher organisms, but CR extends both of types of lifespan. CLS is

---

* One of the goals of this book is to alert readers to the fact that there is a whole world of aging research that has nothing to do with Sir2.

easier to carry out, since one just needs to take a culture that is sitting with restricted conditions and see if it can be revived, whereas RLS requires micromanipulation to move each daughter away from the mother after each division, and is very challenging. However, RLS has more evolutionarily conserved regulators, so is generally considered the better analog to lifespan in higher organisms.

OK, so what is Sir2, exactly? Named as a "**s**ilent **i**nformation **r**egulator" for its first discovered role, that of transcriptional silencing of "cryptic mating loci" in yeast,[45] Sir2 was eventually found to be an NAD^+-dependent protein deacetylase. This means that it has an enzymatic activity, removing acetyl groups from proteins, that requires nicotinamide adenine dinucleotide (NAD^+), which is a coenzyme and metabolic product of the NAD biosynthetic pathway. Sir2 is a member of a larger group of enzymes that in mammals are known as sirtuins, which are all NAD^+-dependent protein deacetylases that are responsible for regulating a huge range of important cellular processes.[46] NAD^+ is so important to human health that its deficit can cause a disease known as pellagra, and prevention of pellagra is the reason that foods are often fortified with niacin (aka vitamin B3), a precursor to NAD^+. NAD^+ is critical for the function of several different enzymes, including the sirtuins, which utilize NAD^+ in their enzymatic removal of acetyl and acyl groups, and poly(ADP-ribose) polymerases, which among other things are important for DNA repair and ribosomal RNA biogenesis, two important processes in aging and longevity. Because NAD^+ is used at such high rates in the cell, it is both synthesized (starting from tryptophan) and salvaged from nicotinamide through different enzymatic pathways. Levels of NAD^+ in the cell decline with age, and since it is important for many cellular functions, a sensor for NAD^+ levels would naturally be a good candidate for an aging regulator. While Sir2's original substrate (target) was thought to be histones, which made sense in the context of chromosomal regulation, later it was found to also deacetylate other proteins, including DAF-16. "14-3-3" proteins bind to DAF-16 depending on its acetylation status and regulate nuclear entry, so SIR-2 essentially acts in the insulin signaling pathway in worms.[47]

Sir2 also plays a role in repressing extrachromosomal rDNA circles,[48] which helps then extend lifespan; this might be through Sir2's shift to the nucleolus, where it participates in rDNA silencing. Aguilaniu and Nyström showed that Sir2 is also critical for the fascinating process of asymmetric distribution of protein between mother and daughter cells[49]—a selective process that keeps daughter cells as pristine as possible by sorting damaged and new proteins into the mother and daughter, respectively. This is truly an amazing system for keeping each daughter cell brand-spanking new—not inheriting old, damaged protein is a great way to keep new cells functioning, even as the mothers continue to age. (Interestingly, Aguilaniu's lab later found that *sir-2* is *not* required for an analogous protein rejuvenation process in *C. elegans* oocytes,[50] suggesting that *sir-2* is not the root of all youthful processes, and is not even conserved in this selective sorting mechanism.) Sir2 was found to be necessary for RLS extension by caloric restriction, while Sch9, the S6 kinase that regulates TOR function, is important for CLS extension.[51] CLS was used later to study the role of the TOR pathway in DR-induced longevity.[52]

Probably due to the difficulty of manually performing yeast RLS assays, the discovery of new genetic factors of longevity in yeast outside of the SIR-2 pathway for many years lagged in comparison to the booming genetic longevity discoveries happening in worms and flies, which benefitted both from traditional genetic approaches and particularly whole-genome RNAi screens. A comprehensive analysis of reproductive lifespan in 2015 revealed more genetic players in yeast longevity regulation,[53] more than a decade after whole-genome lifespan studies had been carried out in *C. elegans*.[54] Recently the problem of RLS slow throughput may have been solved through clever genetics that mark either the daughter or the mother, such as the "mother enrichment program" and microfluidic approaches that physically separate the mothers and daughters.[55] These new techniques have solved the throughput bottleneck in replicative aging studies in yeast, so there is likely to be a resurgence in yeast longevity work that is not solely focused on DR and Sir2.

In worms, *sir-2* was studied first in a large chromosomal duplication that included other genes that might have in fact caused some of the lifespan increase, but Lenny Guarente's lab (MIT) later made a low-copy SIR-2 overexpression strain that also increased lifespan. Using that old duplication strain, David Gems published a headline-grabbing title in *Nature* stating that there is "no evidence for lifespan extension," while his and other studies using the low-copy SIR-2 overexpression strain consistently found that overexpression of *sir-2* in worms causes about a 15% increase in lifespan. Let's be clear: a 15% increase in lifespan is not *nothing*; you'd be happy with that in your own life, and much of the lifespan field—particularly the DR field—*does* focus on lifespan extensions of this magnitude. Of course, in comparison to the effect sizes caused by mutants like *daf-2* or by loss of the germline, which can double lifespan, this might seem small, but even DR usually maximally causes only a 30% increase in lifespan. So a 15% extension caused by overexpression of

C. elegans SIR-2 is simultaneously statistically significant, real, and yet not huge compared to other interventions or mutations, so its total dismissal because of a study done in a suboptimal duplication strain seems a bit disingenuous.

Because Sir2 activity was necessary for CR-induced longevity, and its over-expression extended lifespan, it seemed logical that an activator of sirtuins might be an effective longevity drug. To that end, David Sinclair's lab looked for such compounds, and resveratrol, a compound present in the skin of red grapes and berries, seemed to have this activity. His group found that resveratrol and other related compounds could extend the lifespan of yeast.[56] Soon after the yeast longevity study, resveratrol was shown to increase the lifespan of flies, worms, and later mice, honeybees, silkworms, and two killifish species (*Nothobranchius furzeri* and *Nothobranchius guentheri*).[57]

Resveratrol was found to activate SIR-2 through the use of a fluorescence-based peptide assay, but the results were later challenged as an artifact ("resveratrol has no detectable effect on Sir2 activity in vivo").[58] Resveratrol was known to have antioxidant properties before it was identified as a sirtuin activator and may have non-sirtuin targets, so much of the study of resveratrol has focused on whether it truly acts through Sir2. In each case, controversy and conflicting findings ensued.[59] In mice and primates, it seems that resveratrol might be most effective in improving the health of animals fed a high-fat diet (like our Western diet), and may act not only through SIRT1, the mammalian homolog of Sir2 (a "sirtuin") but via other regulatory molecules, as well. Whether resveratrol and its related compounds act directly and only through sirtuins is still debated, and whether it benefits non-obese animals is unclear, making it difficult to move forward into the development of a human pharmaceutical.[60]

Sir2's star has faded a bit in the longevity field, as drugs meant to activate it, such as resveratrol, were not well supported in different labs, and other mechanisms were shown to play a greater role in regulating longevity. Sir2 is not the only sirtuin in the cell; since the discovery of Sir2 in yeast, seven mammalian sirtuins that function in different subcellular organelles and compartments (nuclear, cytoplasmic, and mitochondrial) have been identified.[61] These SIRT mammalian homologs do have important roles that I do not want to diminish, and the study of these sirtuins is a hot area of ongoing research. Sirtuins are critical regulators of cellular functions that were not even imagined when Sir2 was originally discovered, ranging from DNA repair to metabolic regulation, mitochondrial regulation, transcription, fat differentiation, insulin sensitivity, lipid regulation, and host-virus interactions. While it may not be

My own skepticism about resveratrol, after reading Sinclair's first yeast paper on it where he briefly mentions flies and worms at the very end, led me to test it myself; I dissolved resveratrol in ethanol and spread it on bacterial plates, and compared lifespans of these worms to ethanol-only control treated bacteria—and found that resveratrol caused a remarkable 45% increase in lifespan. Interestingly, I got these results only if the bacteria were old. The resveratrol effect shrank when I used fresh bacteria, suggesting that the bacteria were metabolizing the drug—but the original results seemed convincing.

the *only* and most important regulator of lifespan, Sir2 overexpression in yeast, worms, and flies does extend lifespan, and it influences mammalian stress responses, so it does play at least some role in regulating longevity. And resveratrol and Sir2 may play a very important role in reducing the deleterious effects of overfeeding—but it seems that Sir2 cannot be the end-all of the DR story, despite early enthusiasm about its potential importance.

Finding the Regulators: TOR and the Explosion of DR Regulators

In addition to Sir2, there are several different pathways that animals use to extend lifespan in the face of nutrient restriction.[62] Much of the field is focused on figuring out what those conditions are, identifying the downstream signaling pathways that convey these restricted conditions to the animal, and then, ultimately, the cell biological events that result in extension of lifespan.

The mid-2000s witnessed a slew of discoveries about the molecular basis of DR-mediated longevity. Several of these findings centered around the nutrient sensor and growth regulator TOR (target of rapamycin) in one way or another.[63] Rapamycin, or sirolimus, is a small molecule originally isolated by Suren Sehgal in 1972 from a soil bacterium found on Easter Island, also known as Rapa Nui (hence *rapamycin*). mTOR is the "mammalian" or "mechanistic" TOR and is a serine/threonine protein kinase (i.e., a signaling enzyme that phosphorylates—adds a phosphate group—to serine or threonine amino acids in a protein) that has been extensively studied for its antifungal, immunosuppressive, and anticancer functions.[64] The two complexes that TOR forms are mTOR complex 1 (mTORC1) and mTORC2—also known as Raptor and Rictor, respectively—and have different input sensitivities and cellular signaling outputs.

The main thing to know is that TOR had already been implicated in nutrient sensing and growth regulation through its effects on translation, so its connection to longevity and metabolism is logical. mTOR may be best described as a nutrient sensor, and it integrates information from multiple signaling pathways, including insulin and growth factor pathways. As usual, worms were the first animal in which the connection with longevity was shown, as reduction of TOR (known by the catchy name *let-363* in worms, as its complete loss is lethal) extended lifespan.[65] Soon after, Pankaj Kapahi in Seymour Benzer's lab (Caltech) found that overexpression of the TOR inhibitors TSC1 and TSC2, and inhibition of TOR and S6 kinase, extend fly lifespan in a nutrient-dependent manner.[66] These results in worms and flies showed that the TOR pathway was involved in longevity regulation, in addition to its previously known roles in growth and immunity. TOR reduction by loss or by rapamycin treatment was also shown to extend both RLS and CLS in yeast, while Malene Hansen in the Kenyon lab showed that reduction of translation mediates DR-controlled longevity in worms.[67] AMPK is a protein kinase that senses the levels of ATP and AMP, the high- and low-energy forms of the nucleotide, and through its subsequent phosphorylation activity, signals to other proteins to indicate these levels. Most importantly, AMPK activates TOR when there is insufficient ATP, through sensing of AMP-to-ATP ratios— that is, if energy levels are low, TOR turns on. The field was able to start connecting the dots from nutrient levels, to ATP levels and ratios, to regulators of transcription involved in longevity regulation.

Suspecting that a transcription factor similar to DAF-16/FOXO might be key to CR-mediated longevity, Andy Dillin's lab used RNAi to knock down *C. elegans*'s 15 forkhead proteins in *eat-2* mutants; they found that loss of the FoxA homolog PHA-4 completely blocked the long lifespan of both the DR-mimicking *eat-2* mutant and direct restriction of bacterial food. Susan Mango's lab later showed that TOR/*let-363* in worms and Raptor (*daf-15*) acts to block PHA-4 activity through regulation of *rsks-1*/S6 kinase, which leads to lifespan extension.[68] This put the pathway together, from nutrient activation of TOR and S6 kinase and AMPK signaling to PHA-4/FoxA activity;[69] lack of nutrients results in activation of PHA-4, which, like DAF-16, then increases lifespan. Later, Alex Soukas and Gary Ruvkun found that the Rictor/TORC2 complex also regulates it all—fat, growth, and lifespan.[70]

Anne Brunet's *C. elegans* findings at first seemed controversial because she showed that the energy sensor AMPK plays a role in DR-mediated longevity and requires DAF-16/FOXO to do so,[71] distinguishing it from the previously studied *eat-2* longevity pathway, which Andrew Dillin's lab had found required the FoxA transcription factor PHA-4. Eric Greer, a graduate student in the Brunet lab, went on to investigate the genetics of *C. elegans's* DR in greater detail; echoing some of the complexities of the DR diet, Greer and Brunet determined that there were several distinct dietary restriction pathways, depending on the source of the food, the genetic mutation (*eat-2*), dilution of bacteria in liquid or on plates, or timing of the restriction.[72] This careful analysis explained why different research groups were coming to seemingly different conclusions: each different dietary restriction regimen utilizes different molecular sensors and signaling molecules and different downstream transcriptional regulators (PHA-4, HSF-1, SKN-1, DAF-16) to execute DR-induced longevity.

How does intermittent fasting (IF) fit into the picture? The Nishida lab (Kyoto University) used an alternating feeding/fasting regimen to induce longevity in worms, and found that IF signals through TOR and a GTPase called RHEB-1, which in turn regulates TOR and then DAF-16.[73] IF also causes downregulation of *ins-7* (an insulin-like peptide that I had first identified to have a role in longevity regulation in my postdoctoral work),[74] which in turn signals through DAF-2 and DAF-16. This IF regimen resembled at least one of the DR regimens Greer and Brunet had studied, linking the insulin and DR pathways and blurring the lines between these mechanisms, which were previously thought to be genetically distinct. Thus, several different ways to induce longevity through nutrient restriction were mapped out through genetic experiments in *C. elegans*.

The Cell Biology of Dietary Restriction

Genome-wide Analyses of DR

But what does DR actually *do* in the cell, once these regulators have signaled that the organism is experiencing DR, to help the animal live longer? For that answer, we have to peer into the cell itself. Unlike the insulin/FOXO pathway, where the kinase cascade of insulin/IGF-1 receptor to PI3 kinase to phosphorylation regulation of DAF-16 was found and cloned before the transcriptional targets were determined, many of the downstream effects of DR were found before the regulators and signaling molecules were discovered. That is because various groups worked on identifying the downstream transcriptional changes induced by caloric restriction in flies, mice, and primates using newly developed microarray analysis.

In 1999, Tomas Prolla's group (University of Wisconsin, Madison) performed one of the very earliest microarray experiments in the aging field to examine the gene expression changes in aging skeletal muscle caused by caloric restriction.[75] They reasoned that muscle displays obvious age-related changes (sarcopenia, or muscle wasting) and because muscle cells are all postmitotic (meaning that they can't just renew themselves by replacement when they get old), transcriptional changes might be informative. Although the arrays they used were not full-genome—microarray technology was just getting rolling and thus many of the arrays at the time were incomplete—the kinds of genes that they found to be changed with age and prevented by CR were largely consistent with later findings. These included genes that affect energy metabolism, stress response, proteostasis (protein homeostasis), and damage repair—in general, the *categories* of genes that induce longevity in CR-treated animals are similar to those that the insulin pathway controls through FOXO-regulated transcription, even though the exact genes differ. Later CR transcriptional studies of other tissues, such as adipose and brain, emphasized the same themes, with some tissue-specific genes.

In flies, Pletcher and Partridge found similar sets of genes,[76] plus others involved in innate immunity and reproduction, to be regulated in response to caloric restriction. Importantly, Pletcher's analysis took advantage of the whole-genome nature of expression data to ask whether there was any evidence for specific parts of the genome to become defective, but the data suggested instead that there are specific changes that arise with age. (Steve McCarroll and I came to the same conclusion in our analysis of worms with age and in insulin mutants.)[77] Consistent with the results from direct DR expression experiments, Pandit and colleagues (2014) found that the gene targets of PHA-4 in the long-lived *eat-2* mutant include regulators of autophagy, proteostasis (ubiquitin-mediated protein degradation, protein folding), and metabolism, as well as microRNAs (miRNA) that form feed-forward signaling loops to propagate the signal.[78] Again, while the specific genes regulated by PHA-4 differ from those that are targeted by DAF-16, the gene categories are similar to those found in insulin signaling mutants.

Later, Jing-Dong Jackie Han's lab used a systems biology approach to identify three main transcriptional modules temporally regulated by DR and IF in *C. elegans*: one that involved *rheb-1* and TOR that responded early to DR; another using DAF-16 that regulated a transcriptional response; and a third involving AMP kinase (*aak-2*) and calcineurin (*tax-6*) that was associated with a "starvation" module.[79] By genetically manipulating these modules in an additive manner they were able to recapitulate the longevity response. This work reinforces the concept that Anne Brunet's lab had shown earlier through genetic analysis, that DR has multiple input and output pathways that are only partially overlapping, and act through regulation of both translation and transcription.

Cell-Nonautonomous Signaling of DR— How Systems Are Coordinated

How does an animal's cell know that it's being calorically restricted? While some effects are a direct result of metabolic changes, others are conveyed to the whole animal through inter-tissue signaling. In C. elegans, neurons that sense low nutrients may signal to the rest of the system to coordinate responses, particularly to large, metabolic tissues. For example, the SKN-1 transcription factor, which is required for development and oxidative stress response, was also found to be required for longevity induction by dietary restriction.[80] Remarkably, expression of SKN-1 in just two sensory neurons in the worm's head (the ASI neurons) is sufficient to rescue the DR lifespan response, suggesting that there is a systemic, cell-nonautonomous response coordinated by neuronal SKN-1 activity (although the downstream signals remain unknown). Similarly, AMPK regulation of a cofactor called CRTC-1 in C. elegans neurons signals to regulate metabolism in the rest of the animal through octopamine signaling to non-neuronal cells, resulting in healthier mitochondria in other cells and longer lifespan.[81]

In fasted mammals, the fibroblast growth factor 21 (FGF21) hormone (aka "the starvation hormone") is secreted by the liver and coordinates several responses across the whole system. FGF21 does so by signaling to several tissues, mediated by its uptake through the β-Klotho single-pass transmembrane receptor. (Klotho was one of the original "longevity" factors, named for one of the three Fates in Greek mythology, spinning the thread of human life.) FGF21/β-Klotho signaling controls insulin signaling to adipose tissue, where it in turn regulates metabolism via AMPK and TOR, increasing glucose uptake and insulin signaling through PGC-1α signaling. FGF21 also signals to the brain; FGF21 can cross the blood-brain barrier and is taken up into specific regions where β-Klotho is expressed, such as the superchiasmatic nucleus.[82]

Remarkably, in addition to roles in metabolism and circadian rhythm, FGF21 can affect preference for sweet tastes and alcohol, perhaps setting a feed-forward regulation of behavior in motion.[83] In this way, in addition to coordinating the response to starvation across tissues and inducing metabolic effects that are necessary to deal with fasting (such as stimulating glucose uptake and blocking nutrient-expensive growth functions), a rise in circulating FGF21 levels could cause the animal to seek out the nutrients it is lacking—namely sugar—to quickly resume normal conditions. Excitingly, transgenic overexpression of FGF21 significantly extended lifespan in both male and female mice without decreasing their food intake.[84] At least part of this lifespan extension seems to be due to its inhibition of insulin-like growth factor 1 (IGF-1) signaling in the liver—again linking a DR model with the insulin signaling pathway.

The Cell Biology of Dietary Restriction: Translation and the Nucleolus

What about *non-transcriptional* effects of DR? Some of the biggest clues came from the original RNAi screens for longevity, in which several different components of the DR pathway, as well as components necessary for protein synthesis, emerged (e.g., TOR/*let-363* and *rsks-1*/S6 kinase); these included ribosomal protein genes and translation machinery. In yeast, flies, and worms, knockdown of ribosomal proteins and translation initiation factors was found to extend lifespan.[85] These effects are independent of DAF-16 and are not further extended by DR or by TOR knockdown, suggesting that TOR and S6 kinase signaling ultimately impinges on protein translation. (In fact, an undergraduate in my lab, Daniel Liu, once collected all the longevity mutant gene expression data we had into one giant heat map; while *daf-2/daf-16*–regulated genes were off the chart, the one consistent factor across all long-lived C. *elegans* mutants, no matter what the genetic origin, was the reduction of ribosomal protein gene expression. This result suggests that all longevity pathways—even *daf-2* mutants—involve at least some protein translation inhibition.) Therefore, regardless of transcriptional outputs of the DR pathway, signaling that ends up reducing protein translation is likely to account for much of DR's effects.

Changes in translation rates are reflected in cell biological changes, as well. The nucleolus—a weird little organelle that is not as famous as the nucleus or mitochondria but is key to the regulation of longevity—is the site of the large and small ribosomal subunits that are responsible for translation of mRNA into proteins. Perhaps not surprisingly, inhibition of translation causes the nucleolus to shrink. Wu and colleagues (2018) connected the dots between translation, nucleoli, ribosomes, TOR, lipid metabolism, PHA-4, and starvation survival in C. *elegans*: by screening for mutants that have excess fat, they found a ribosomal processing protein called RRP-8 that localizes to the nucleolus.[86] RRP-8 is required for ribosomal RNA (rRNA) processing, which is perturbed by nucleolar stress; nucleolar stress activates PHA-4 transcription via TOR signaling, and PHA-4 subsequently induces expression of lipid biosynthesis genes. Thus, PHA-4 appears to be a "nucleolar stress sensor" that promotes survival under starvation conditions through lipid metabolism. PHA-4 has additional transcriptional targets that affect other stress-resisting cellular functions.[87] In fact, the correlation of small nucleolar size—that is, a reduced function of the ribosomal proteins in the nucleolus—with long lifespan fits with the idea that reducing the levels of ribosomal proteins decreases protein translation. Adam Antebi's lab showed that the (inverse) size of the nucleolus may be a good indicator of healthspan and lifespan, from worms to flies to mice and even human cells—that is, if the nucleolus is small, the animal is likely to be healthy and long-lived, but if it's large, it's a sign of poor outcome.*[88] So the size of the nucleolus, an oft-ignored organelle, might actually be the most informative indicator of health, owing to its critical role in the regulation of translation and sensing of nutrient stress.

Putting It All Together:
How Dietary Restriction Slows Aging

When an animal lacks nutrients, the logical thing to do is to (1) immediately stop carrying out energetically expensive cellular functions, like growth and reproduction, and eventually memory; (2) save nutrients for the most important survival functions, like food sensing; and (3) figure out how to get nutrients again by changing behavior. To do these things, it is critical to stop making new proteins (by blocking translation and downregulating ribosomal components), maintain the ones you have (through proteostasis and other damage repair), get energy from wherever you can (like recycling materials through autophagy), coordinate the responses at the cellular level (via TOR) and across the whole system (through FGF21 signaling), and stimulate brain

* The nucleolar size/lifespan correlation might have been somewhat inflated by binning the dependent variable (lifespan) *before* regressing against nucleolar size. Other markers found by Zachary Pincus may be in the same range as this nucleolar marker.

activities that will point you to food. The fact that nutrient restriction also extends health and lifespan is unlikely to be the main point of DR—except when reproduction is also slowed—but rather a fortunate by-product of these shifts in cellular metabolism.

One could ask why there are so many different pathways that mediate DR's effects on longevity. This may be related to the complexities of most animal diets, and the different needs animals each have. Perhaps the different components trigger different pathways in order to elicit the appropriate behavior or cellular response. For example, if you are starving for a particular dietary component, you might still be able to carry out some tasks, but not others. Lack of amino acids might preferentially slow development and growth, lack of nucleic acids might prevent rapid reproduction, lack of sugar might impact brain function, and lack of fat might shift metabolism. Temporal dynamics might also matter; long-term starvation is a different problem than bursts of intermittent fasting or even normal circadian feeding cycles. Each of these components has sensors and signaling pathways that are specific to that component, but some of their pathways converge downstream to regulate the whole system.

For example, the TOR pathway is sensitive to changes in amino acid levels. The branched-chain amino acid transferase enzyme BCAT1 is specific to branched-chain amino acids (BCAAs) and seems to regulate lifespan at least partially through TOR, while the insulin pathway is sensitive to carbohydrate/sugar levels. The downstream effects of these regulators include proteostasis, mitochondrial changes, and autophagy, all of which help the cell either conserve energy or "clean up" damage, which ultimately helps the organism live longer. Together, these results all suggest that nutrient sensing, TOR, AMPK signaling, and protein translation are shared aspects of DR-regulated longevity.

One of the exciting directions that we will cover in the last chapter is the possibility of tricking the body into thinking it's being dietarily restricted when it is not, by treating with a DR "mimetic," a compound that elicits the effects of DR without fasting. Such a drug might extend lifespan or increase healthspan, but without the painful and irritating process of actually experiencing DR. Some compounds for such development target AMPK or mimic exercise. From a pharmaceutical company's perspective, the perfect drug would be one that would trick your body into thinking that you are practicing DR (but in reality, you are eating normally) or exercising when you are not—but you'd have to keep taking the drug the rest of your life to reap the benefits.

Hugo Aguilaniu's team (then in France, now in Brazil) may have found exactly this: they carried out a clever screen where they started calorically restricting worms, and then started feeding the worms again—but their food was bacteria from the RNAi library. Of course, most of these clones would knock down a gene that was unimportant for DR, and the worms would resume their normal lifespan, since, as we've seen before, once DR stops, the beneficial effects, including lifespan extension, stop as well.[89] But at least one clone had a fascinating effect: although the worms were eating again, their fat storage and lifespan resembled those of worms that were being calorically restricted. Starting the DR and then knocking down a particular sphingosine kinase by RNAi essentially stuck the worms' DR program in the "on" position, even once they were happily eating again.* A drug that inhibits this sphingosine kinase should have the same effect. This would be like if I asked you to try caloric restriction for a few weeks, and then said, "OK, if you take this drug, now you can eat whatever you want to again, and still experience the benefits of DR." Sounds too good to be true—what's the catch? You have to take the drug the rest of your life, because if you are anything like Linda Partridge's flies, the benefits will end once you stop restricting.

While it's been clearly demonstrated in well-controlled model-organism research that there are likely benefits to be reaped from DR, there is also no doubt that such diets are extremely challenging for most people to follow, and so it's reasonable to ask whether DR is worth doing at all. Recently, some skepticism about translating the beneficial effects of periodic fasting to humans has been finally cropping up. For example, some versions of intermittent fasting in mice (single meal feeding) show less of a benefit in lifespan extension than simply calorically restricting without fasting periods (11% vs. 28% for IF vs. CR, respectively).[90] In flies, some forms of IF can even shorten lifespan (yikes!), and time-restricted feeding has to be done at *exactly the right time* to reap any benefits and worked only at a certain time in adulthood.[91] Paying better attention to our circadian rhythms and eating less (that is, being less

---

* As far as I know, this research was never published, but Aguilaniu gave a talk on it at Cold Spring Harbor Laboratory, and wrote a patent on it: "Methods for prolonging the health benefits triggered by a dietary restriction using a sphingosine kinase inhibitor," EP2166094A1, European Patent Office, GooglePatents, https://patents.google.com/patent/EP2166094A1/en.

obese) are likely to make us healthier, but clinical evidence for clear benefits of TRE in humans is lacking.[92] Most worrying, reducing caloric intake seems to reduce brain function[93]—not exactly a ringing endorsement of DR. That is why it is so important for us to understand the underlying molecular mechanisms; if we can pinpoint the beneficial and deleterious pathways downstream of reduced calories, we might be able avoid the negative effects that some diets can cause. Using all of this information to make a good "mimic" of DR sounds like a cop-out to avoid doing the hard work of starving ourselves, but if we can be smarter about which pathways we leave alone and which are good targets for intervention, we might find a drug that truly improves health without causing unwanted side effects. A mimetic of DR might be an effective way to keep you healthy, without the difficulty of DR or the potentially dangerous psychological effects of actually starving yourself. . . . In other words, my mom would be happy.

8

Taking out the Trash

MOLECULAR HOMEOSTASIS IN THE REGULATION OF LONGEVITY

Trash is something you get rid of—or disease. I'm not something you get rid of.

—LANA TURNER, *LANA: THE LADY, THE LEGEND, THE TRUTH*

SANITATION WORKERS ARE THE unsung heroes of our society. My family was in Washington, DC, a few years ago during a government shutdown. Everything, including trash pickup, had ground to a halt. Within days, garbage bags and piles of trash overflowed containers, crowded sidewalks, and spilled into streets. As the stench rose, the rats became bolder, standing on the tops of garbage receptacles in broad daylight. Normally we don't think about the work that sanitation workers do, but when there is an interruption in their work, you learn to appreciate them pretty quickly.

Cells are a lot like cities: there are thousands of proteins with different jobs to do, crowding, tons of traffic, and, of course, trash: damaged macromolecules, a result of wear and tear on the cell. When cellular damage accumulates to dangerous levels, everything stops working properly. In order to maintain order, cells need cleanup systems, like sanitation workers, to get rid of their trash. Without these important mechanisms to keep damage from accumulating, the aggregation of damaged macromolecules can gum up the works. Maintaining consistent cell function requires "homeostasis," or the ability to maintain a cellular status quo—and these homeostasis mechanisms are often the first things to go with age.

Proteostasis: How to Keep Proteins Happy and Functioning—or Get Rid of Them when They Aren't

We have established that organisms may need to extend their lifespans under particular circumstances, such as low nutrient conditions, but how does an organism *actually* slow aging? To get at this answer, first we must understand what aging is at the cellular level: damage to the many components of a cell (proteins, DNA, RNA, lipid membranes, and organelles) that accumulates with time, often as a result of ROS (reactive oxygen species) generated by the very organelles that power the cell, or a decline in quality control as these components are being made. Cells can deal with this damage on a number of levels. Some cells can simply die and replace themselves by way of stem cells and regeneration (chapter 10), but some cells cannot. Non-regenerating cells can overcome such damage for a time by sensing, eliminating, and replacing their damaged components. Longevity-regulating pathways such as insulin signaling and dietary restriction "turn on" several cellular repair mechanisms simultaneously, fixing components and recycling those parts that are too damaged to maintain, while also slowing production of proteins in order to limit the opportunity for damage. Cells have also developed elegant methods to prevent proteins from misfolding as they are produced, to detect defects in proteins and target them for degradation, and to recycle their amino acid components. Other processes detect, tag, and eliminate defective RNA and mitochondria, as well. These cellular mechanisms that repair damage are critical components in the slowing of aging.

If you recall the central dogma (DNA → RNA → protein), you'll remember that every cell's nucleus contains DNA, which encodes all of the proteins that the cell will make. DNA exists, not just for its own sake and to make copies of itself for propagation purposes, but to generate gene products: RNA and proteins. To keep a cell functioning, messenger RNA has to be made properly, and that mRNA must be translated into protein in order to make functional enzymes and other bits that the cell needs in order to keep working. Each step of this process can go awry, resulting in the accumulation of damaged products (figure 4). But, like Elizabeth Warren, the cell has a plan for that.

I'm going to go backward from protein to RNA (and DNA in chapter 10) because I have no doubt that you are already aware of some of the more famous cases of protein aggregation, such as those that cause neurodegenerative disorders like Alzheimer's disease. Protein homeostasis, or "proteostasis"

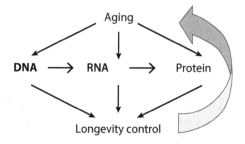

FIGURE 4. Each of the components of the central dogma—DNA, RNA, and protein—is regulated by processes that aging affects, and they in turn can each affect longevity.

(a major buzzword in the aging field), describes the maintenance of protein integrity and processes that prevent the accumulation of damaged proteins into toxic aggregates. A cell can reduce the amount of damaged protein it contains through several different proteostasis mechanisms: reduced translation to slow down making proteins in the first place, improved protein folding as they are being synthesized, proper trafficking of proteins through the correct organelles, tagging and degradation of damaged and aggregated proteins, and recycling of the parts to allow new proteins to be made. As we age, proteostasis mechanisms fail, leading to the accumulation of damaged and aggregated proteins.[1]

Long-lived animals can attribute much of their cellular health to improved proteostasis relative to their shorter-lived counterparts. *C. elegans* represents an extreme example of the need for constant protein maintenance and cellular repair, because it doesn't replace its adult somatic cells at all; it cannot rely on stem cells and regeneration to replace its damaged cells as it ages. The exact same cell must persist for the entire lifetime of the adult animal, so it must keep those cellular parts functioning the whole time. To do that, it takes a lot of protein homeostasis machinery. When we identified the genes that kept working in *daf-2* worms to keep them functioning twice as long as their wild-type counterparts,[2] it was quite obvious that they cared a *lot* about maintaining protein function, because they regulated a ton of proteostasis genes. We saw upregulation of expression of genes encoding proteins that help other proteins fold correctly (called "chaperones"), including heat shock proteins, others that repair and refold damaged proteins, and still others that help eliminate them altogether when they are too damaged (through autophagy). (By contrast, an upregulation in DNA repair genes was notably absent, but that is likely because

I was mostly looking at somatic (non-germline) cells, the kind of cells that don't divide or turn over anymore.) DAF-16/FOXO has a major influence on the parts of proteostasis that function in folding, trafficking, and recycling of proteins. Later, we also found that a reduction in expression of genes associated with the ribosome and protein translation is shared among all longevity mutants, suggesting that not only are proteostasis genes upregulated but reduction of translation is a common theme. Most long-lived animals have better proteostasis levels than shorter-lived species, as measured by protein degradation, protein flux, heat shock protein response, and proteasome activity assays performed on cells isolated from animals with different longevity quotients (that is, cells from naked mole rats and long-lived marsupial sugar gliders have better proteostasis function than cells from mice and opossums, respectively).[3]

A big part of proteostasis can be thought of as reducing the load of damaged protein the cell has to deal with, since repair is a tougher job than just making less damaged protein in the first place or getting rid of it altogether. Just as I can keep my house clean by taking my whole family on vacation, perhaps the most efficient way to prevent the accumulation of damaged proteins is to never make them in the first place. Low ribosomal protein activity and downregulation of the expression of ribosomal proteins is a hallmark of longevity conditions. Downregulation of translation—that is, reducing the rate of translation from the mRNA nucleotide code into a string of amino acids that will fold into a protein—is a major cellular preservation mechanism that is employed by all longevity pathways, but by dietary restriction (DR) in particular, as you learned in the last chapter. Cells can decrease total levels of protein synthesis by inhibiting the translation machinery, or by making less of the protein components that are necessary for the ribosome (the translation complex) to function. One of the major findings of the whole-genome RNAi screens done in the early 2000s in C. elegans was that knockdown of translation machinery, such as translation initiation and elongation factors and ribosomal components, extends lifespan.[4] Translation inhibition is a very efficient way to reduce protein damage, as long as the cells can survive with slightly less function, or with slower rates of protein turnover (replacement). (Some nuclear pore complex proteins, such as those in the brain, are extremely long-lived and essentially never turn over,[5] perhaps eventually leading to age-related problems.) Most of these components act in the DR pathway, as shown by genetic experiments. For example, the mTOR pathway is a major regulator of protein translation inhibition that occurs in DR.

Perhaps because protein translation is so critical to DR-mediated longevity regulation, the size of the nucleolus, where ribosomes do their job, correlates inversely with longevity: the nucleolus of young animals is small and gets bigger with age in worms, flies, and mice, and stays small in long-lived mutants.[6] In fact, biopsies after a few weeks of exercise in middle-aged people show that the nucleolus shrinks with exercise, a promising result. The inverse correlation between nucleolar size and longevity indicates how strong the connection is between reduced protein synthesis and increased lifespan and health. Thus, nucleolar size is a good cellular biomarker for aging, indirectly reporting on ribosomal function and protein translation.

After mRNAs are translated into strings of amino acids, or peptide chains, folding these chains correctly into their final, functional form is important; otherwise, their inside parts, which are gummy (owing to the packing of hydrophobic amino acids generally on the inside of the protein), can end up incorrectly on the outside. This may cause them to stick together and stop working or, worse yet, gum up the rest of the cell by making aggregates of these unfolded proteins. These aggregates are a problem not just because of the loss of the proteins' functions but also because the aggregates can be toxic to the cell. Protein aggregation at this nascent step is combatted by proteins known as "chaperones," which can detect those gummy hydrophobic parts and help guide proteins to fold properly in the first place and then be trafficked to the right cellular location (hence the name), or to disaggregate and refold if they are misfolded. Because undoing misfolding is an energetically costly process, some chaperones, such as "heat shock proteins" (HSPs) use the energy generated through ATP hydrolysis to undo improper protein-protein aggregation.[7] Heat shock factor 1 (HSF-1) is not a chaperone itself, but rather a transcription factor that regulates the expression of HSPs and other genes that help the cell cope with stressful conditions through a highly regulated activation process. HSF-1 works together with the DAF-16/FOXO transcription factor downstream of insulin signaling, and is critical for the *daf-2* mutant's longevity, as the two transcription factors both regulate expression of multiple chaperones.[8]

Protein Aggregates and Neurological Diseases

One of the more famous types of protein dysfunction is aggregation, which most of us are at least somewhat aware of because of its association with neurodegenerative diseases, like Alzheimer's disease (AD), Parkinson's disease, and ALS, and some of the really scary infectious prion diseases, like mad cow disease and Creutzfeldt-Jakob disease (CJD). These protein aggregates are particularly well known because of their disease associations, but undoubtedly other damaged proteins also accumulate in a disordered manner as cells age and become dysfunctional.

Two of the major characteristics associated with AD are A-beta peptide amyloid aggregates (also known as senile plaques) and neurofibrillary tangles (NFTs) made by the microtubule-binding protein tau. Other protein aggregates, such as SOD1 and α-synuclein, have been associated with the neurodegenerative disorders ALS and Parkinson's disease, respectively, but smaller oligomers might in fact be the toxic form, causing synaptic dysfunction through membrane and ion dysregulation, and inflammation. Huntington's disease is rarer but is known to be caused by a trinucleotide repeat in the gene for huntingtin protein, causing the protein to become longer, get cut into smaller fragments, and aggregate via its polyglutamine repeats.

The accumulation of beta-amyloid and tau protein aggregates, outside and inside the cell respectively, has been associated with AD pathology, and as a result, reduction of A-beta amyloid aggregates has been the intense focus of the AD research field for the past 30 years. However, the amount of senile plaques, or plaque burden, does not correlate well with levels of cognitive impairment, and people without dementia also have similar neuropathology,[9] meaning that A-beta plaques might not be the real cause of AD. Interestingly, some of the larger aggregates of A-beta may even be protective, perhaps functioning as a sink for the more toxic soluble forms of A-beta intermediate oligomers that might actually be the cause of the disease. This may be why AD drugs aimed at eliminating the large aggregate plaques have failed,[10] and in some cases might have even made the disease worse;[11] other therapies targeting toxic oligomers and different proteins altogether (i.e., tau instead of A-beta) may prevent spread between cells in AD and Parkinson's.

Prion-like proteins also aggregate but are particularly concerning because they act as infectious agents; one protein in a specific state (that is, the misfolded "prion"—protein infection form) can in turn cause other proteins to adopt that misfolded form, acting as a "seed" for aggregated protein forms, which may even spread to other cells. Prion proteins cause such diseases as chronic wasting disease, CJD, fatal familial insomnia, scrapie, kuru, and bovine spongiform encephalopathy (BSE—more commonly known as mad cow disease), which can lead to CJD when it spreads to humans. BSE from beef caused by the feeding of meat and bonemeal from infectious animals caused a CJD outbreak in the United Kingdom in the late 1980s that killed 177 people and led to the elimination of about 4 million cows. The prion aggregates are extremely stable, so they aren't eliminated when cooked—which is how it is thought that a sporadic case of CJD led to kuru, when the Fore people of Papua New Guinea cooked and ate the brain in cannibalistic funeral practices. Today most cases of CJD seem to be the result of accidents in labs that study prion-infected tissues, since prions are difficult to disinfect. Shockingly, in France, more than a dozen lab accidents involving prions have occurred in the last decade, including a cut and infection with a variant form of CJD that led to the onset of the disease about seven years after the accident. Since then, research on prions in France has been suspended.[12]

You might wonder why a protein that can adopt a prion-like structure would even be expressed in a cell—what would be the point? There may be a natural role for prion proteins in the formation of long-term memory (which might make sense, since prions themselves make a molecular form of memory), as has been suggested through experiments in *Aplysia* and flies,[13] and prions therefore might be used by other animals for memory as well. Other studies suggest that there might also be roles for prions in innate immunity, stem-cell function, and programmed cell death.[14] Whether prions are regulated in these natural situations is unknown, but an exciting extension of this possible regulation—that is, the reversal of prion formation—is that identifying these regulators might provide insights into the potential for "undoing" of pathogenic aggregates. Whether their formation is accelerated with age is unknown but would be interesting to investigate.

AGEs: A Dangerous Candy Coating

While you might love the caramelization on your crème brûlée, and a bit of a Maillard reaction is fine for searing your steaks, these sugary reactions are not so great for the proteins inside your cells, where metabolic (glycolysis) by-products can damage proteins by reacting with amino acids. These sugar modifications are a form of damage that can increase with age and in metabolic disorders. Reactive α-dicarbonyl compounds from sugars can accumulate with age and react with proteins, forming "advanced glycation end products," or AGEs, which can impair function.[15] High-sugar diets, conditions like diabetes in which there is chronic hyperglycemia, and the overreliance on dry-heat-cooked food can all increase the level of glucose catabolism by-products; they in turn may contribute to diabetic neuropathy. AGEs are particularly noted to appear in end-stage renal failure and may contribute to age-related neurode-generative disorders. Long-lived proteins are particularly susceptible to AGEs, since they are not replaced frequently. On moderately high levels of dietary glucose, worms live short;[16] loss of the enzyme that would normally detoxify AGEs, glyoxalase (GLOD-4), further shortens lifespan under high-sugar conditions.[17] Invertebrate models offer a system in which AGE-reducing chemicals could be screened to find therapeutic targets.[18] AGEs also contribute to "inflammaging," as a receptor for AGEs (RAGE) activates inflammatory signaling through the immune regulatory NF-κB, exacerbating aging effects in metabolic and neurodegenerative diseases.[19]

Besides removing aberrant glycation groups from proteins, cells can also sort damaged proteins to prevent the accumulation of sugar-coated proteins in new daughter cells to help retain their pristine state. Oxidatively damaged, carbonylated proteins accumulate with age in old yeast mothers, but are preferentially retained in the mother cell as daughter cells split off; this asymmetric inheritance is regulated by Sir2 in yeast.[20] However, the elimination of carbonylated proteins that refreshes the oocyte in C. elegans does not rely on SIR-2, instead using the proteasome (see below) in the germline to reset protein status.[21] Learning how to harness these sorting and refreshing processes to rejuvenate proteins could be a powerful approach to keeping our cells functioning longer and better.

How Protein Bits Are Recycled: The Ubiquitin-Proteasome System and the Unfolded Protein Response

Like the sticky notes for "recycle/donate/throw away" that you might use when you are cleaning out your closet, proteins can be tagged with small molecules called ubiquitin by "ubiquitin ligases." This process tags proteins with chains of ubiquitin, and these ubiquitin tags are recognized by other proteins that help guide the tagged proteins to their ultimate fate—degradation—by dragging them to the "proteasome," a barrel-like complex that the protein is fed into. The proteasome takes apart the damaged protein so that its parts (amino acids) can be recycled. The ubiquitin-proteasome system, or UPS, is a major protein recycling mechanism that is required for longevity and begins to fail with age, possibly contributing to the aggregation that we observe in so many age-related diseases. Long-lived daf-2 mutants regulate this pathway by increasing the expression of the proteins that function in the UPS, and the UPS is also regulated by FOXO in mammalian cells.[22]

We previously mentioned the unfolded protein response (UPR) that is activated in the mitochondria to help maintain cellular homeostasis.[23] Mito-UPR, or UPR[mt], requires the activity of the ATFS-1 signaling pathway,[24] a clever mechanism that activates a stress response when the "everything's OK" signal disappears upon mitochondrial dysfunction. Both mitochondrial and nuclear components are required for UPR[mt], including the CBP/p300 acetyl-transferase, Jumanji histone demethylases, and MET-2/SETDB1 histone methyltransferase, suggesting that chromatin regulation in the nucleus is important for repair of the mitochondria—which makes sense since mito-chondria make only a few of their own transcripts and proteins, and rely on nuclear-encoded genes for their function.[25] Since mitochondrial function is critical to the cell, and mitochondria fail with age, cellular responses to mito-chondrial protein dysfunction are critical for health—and finding ways to boost this response could extend lifespan.[26] Neurons might be particularly sensitive to mitochondrial stresses and appear to send signals not only to the nucleus but also to neighboring cells—and perhaps across generations[27]—to respond to neuronal mitochondrial stress.

Unfolded and misfolded proteins also accumulate in the endoplasmic re-ticulum (ER), because this organelle is where proteins that will be secreted or sorted get special tags after being translated. If the process of proper folding after translation is disrupted, then the protein can be degraded by ER-associated degradation.[28] Like the UPR[mt], the UPR[ER]—or, as Joaquin Navajas Acedo (MadScientist) likes to call it, "Panic! at the ER"[29]—involves signaling be-tween the organelle and the nucleus to control the activity of components that carry out the job, and is critical for dealing with cellular stresses and damage.

Proteostasis Mechanisms Fail with Age

Of course, if proteostasis mechanisms worked forever, then cells and organ-isms would never get old—but alas, all proteostasis mechanisms begin to fail with age. Long-lived mutants make more of these components than do their normal-lived, wild-type counterparts, partially explaining how animals like daf-2 mutants can stay healthy twice as long. Similarly, dietarily restricted ani-mals dial down protein synthesis to make fewer proteins to get damaged, and also activate autophagy to clean up the damage and reuse the parts. And some very long-lived species and species with exceptional longevity quotients have improved levels of proteostasis compared to their phylogenetically related, shorter-lived species. Thus, proteostasis mechanisms are critical for healthy cells and long lives.

Of course, old cells also beget new cells if they divide, and proteins get damaged in adults, but somehow they produce babies with pristine new cells. How does the "immortal germline" deal with potentially damaged proteins— especially if the oocytes are from fairly old mothers? As I mentioned before, Goudeau and Aguilaniu showed that damaged proteins accumulate in the unfertilized oocytes of C. elegans right up until they get to the spot where they can be fertilized, and then, *bam*, the damaged proteins disappear. This seemingly magical process of protein cleanup is driven by the activation of a V-ATPase in oocytes by the presence of sperm[30]—an elegant mechanism to flip the oocyte rejuvenation switch to "on" only when it is necessary; that is, when fertilization is finally a viable option because sperm are around.

The ability to co-opt a mechanism like the oocyte protein rejuvenation program might enable old somatic cells to similarly renew themselves, which could extend lifespan if done in the whole animal. Although the initial studies of proteostasis focused on cell-autonomous mechanisms, the systemic regulation of proteostasis is being appreciated more and more. In fact, some mechanisms of sensory regulation of longevity (chapter 13) use an outsourcing approach: instead of just rejuvenating the neurons involved, those neurons sense environmental cues (like caloric restriction, gases, or temperature) and then send out signals to other tissues, where proteostasis processes (like activation of the UPR) are then induced and coordinated across the animal.[31] This means that a cell doesn't have to experience damage itself to benefit from the renewal of proteostasis mechanisms, but instead can benefit from the sentinel cells, the canaries in the coal mine, to anticipate repair needs.

Autophagy and the Lysosome— the Cell's Recycling Center

The proteins you'd tag with the "recycling" Post-it go through a regulated process called "autophagy," or "self-eating." Proteins and cellular components that are damaged beyond repair need to be sequestered away from the rest of the cell and eliminated to avoid causing more damage. Autophagy allows the cell to degrade these damaged proteins and cellular components into their amino acid components, which can then be used again. Autophagy is critical in the regulation of longevity.[32] Slowed or impaired autophagy is a feature of aging and several age-related diseases, including cancer,[33] so understanding autophagic processes and searching for pharmacological activators of autophagy might help stave these off.

Autophagy comes in a few flavors: micro-, macro-, and chaperone-mediated autophagy (CMA). Some organelles have specialized processes of breakdown, as well (e.g., mitophagy, ER-phagy, and lysophagy). In macroautophagy, an "autophagic vacuole" is built around damaged proteins and sent to the lysosome, a specialized organelle that is specific for this breakdown process. A key feature of the lysosome is that it is a membrane-bound vesicle where large pH changes enable acidic enzymes to degrade proteins. By restricting the reactions to the inside of a membrane-bound vesicle, the rest of the cell is protected from potentially damaging reactions. The lysosome itself can also suck in proteins through "microautophagy." CMA is a more selective process in which a special receptor on the lysosome membrane called LAMP2A receives chaperone-bound complexes for delivery into the vesicle. Vacuolar ATPases are required for proper function of the autophagic lysosome because they enable the proton (H^+) pumping necessary to drive the vesicle to have an acidic pH. When budding yeast daughter cells—the immaculate, new cells—bud off from their mothers, the difference in cytosolic pH (more acidic in daughters) driven by an asymmetry in the distribution of a plasma membrane proton ATPase helps the daughter cells' proteins "stay young."

Autophagy also recycles other macromolecules, not just proteins. You might not be surprised to learn that autophagy is induced under low nutrient conditions—recycling of your amino acids, lipids, metabolites, and carbohydrates is a great way to simultaneously clean up and get some good out of your cells' parts when they're starving. In fact, every major longevity pathway induces autophagy, including dietary restriction (via TOR and AMPK signaling and TFEB nuclear localization) and insulin/IGF-1 signaling (via FOXO transcription).

Perhaps to prevent errant, willy-nilly degradation of proteins and organelles within the cell, autophagy is sequestered within specialized membranes and highly regulated. Autophagy involves a *huge* group of proteins, including beclin, ATG proteins, and p62/SQSTM-1, an autophagy receptor also known as "sequestosome 1" in human cells. Many of these were of course first identified in small model systems like yeast, worms, and flies, but are highly conserved in mammalian cells. Phosphatidylinositol 3-phosphate is important in autophagy because of its role in building cargo-bound vesicles that will fuse with the lysosome for degradation.[34] As you might have gathered from the number of different ways that autophagy can be achieved, it is a critical process in keeping cells healthy. Each different route to the lysosome involves many different protein components and is regulated in a different manner. While several drugs increase autophagy (e.g., metformin, rapamycin, urolithin A)[35] and improve health in model systems, negative regulators of autophagy are also necessary to prevent deleterious effects, so homeostasis of autophagy is an important feature of healthy cells.

RNA Homeostasis: Keeping the Message Alive

For many years, proteostasis and DNA repair grabbed all the attention in the aging field, but recently there has been newfound respect for that middle part of the central dogma, RNA. In retrospect, it seems a bit silly that RNA quality control was so ignored by the longevity field, given the focus on all other aspects of cellular and molecular quality maintenance. What if the RNA that is needed to make proteins gets damaged? There must be systems in place to maintain RNA quality as well, and they must be important for longevity. In fact, RNA is highly regulated through processes like splicing, editing, nonsense-mediated decay, and other post-transcriptional processes, so it would make sense for the cell to have "RNA surveillance" mechanisms to monitor changes with age.[36] It had already been recognized that production of mRNA, or transcription, becomes more variable and less well controlled with age, and that longevity mutants maintain this control better.[37] In fact, rising transcriptional heterogeneity is a biomarker of aging (albeit a bit too vague to use for a blood test). But *how* RNA quality control is affected by aging was not well studied until relatively recently.

Different kinds of RNA damage require different repair mechanisms. Sometimes mRNAs accidentally get truncated while they are being made, resulting in aberrant transcripts, which can be dangerous for the cell, particularly if they were to be translated into nonfunctional, partial, and unfolded proteins. "Nonsense-mediated decay," or NMD, specifically eliminates these transcripts, many of which have premature termination codons and other markings that indicate they are nonfunctional. Seung-Jae V. Lee's lab (Korea Advanced Institute of Science and Technology) had previously found that a *C. elegans* RNA helicase called HEL-1 regulates longevity through the IIS/FOXO pathway, and a component of the NMD pathway was isolated in his lab's RNAi screens.[38]

By using a green fluorescent protein reporter that includes a premature stop codon, Lee's group visualized the appearance of aberrant transcripts, and found that the generation of these mistakes increased with age, which is perhaps unsurprising. They also found long-lived *daf-2* mutants do a better job in suppressing the appearance of these aberrant transcripts, through upregulation of the expression of the proteins that carry out NMD surveillance.[39] NMD is required for the full longevity effect of not only *daf-2* but also other longevity pathways, including mutants in mitochondrial function (*isp-1*) and dietary restriction (*eat-2*), suggesting that RNA quality control through NMD is both necessary and shared across longevity mechanisms. Thus, RNA quality maintenance is an underappreciated aspect of cellular homeostasis.

Another level of RNA quality control is through proper splicing. In most eukaryotes, genes can be expressed in many different flavors, because the final mRNA products are made by gluing together pieces expressed from sub-gene parts called "exons," which have "introns" in between them that get cut out, a kind of mix-and-match process that allows variability within the context of a single type of gene. The process of cutting out the introns and using only the relevant exons is called "RNA splicing" and is a highly regulated process. You can imagine how being able to make slightly (or very) different versions of the same gene to carry out different jobs, or to be expressed in different places, would be useful in a complex organism, or over different ages. But, as with any complex mechanism, errors can arise. Using Masatoshi Hagiwara's fluorescent reporter system to visualize splicing in *C. elegans*, Heintz and colleagues found that splicing errors increased with age, but that the rate of these errors is suppressed under dietary restriction conditions.[40] A splicing factor called SFA-1 is required for this activity and is regulated by the AMPK/TORC1 pathway. Similarly, a splicing factor called RNP-6, or poly(u) binding splicing factor 60 (PUF60) in mammals, was identified in a *C. elegans* genetic screen for novel cold-stress-resistant mutants that Adam Antebi's lab (Max Planck Institute for Biology of Aging) carried out; this splicing factor acts upstream of SFA-1 and is involved in immunity in both worms and mammalian cells.[41] The role of RNA splicing in DR-induced longevity was further supported by the examination of transcription in the livers of calorically restricted rhesus monkeys. Rozalyn Anderson's group (University of Wisconsin) carried out an extensive "omics" analysis (global assessments of transcription, proteins, metabolites, lipids, and post-translational modifications) in primates that had been dietarily restricted, and in addition to the expected changes in ribosomes and metabolism, the "spliceosome" (components of the splicing machinery) scored highly in an analysis of enriched changes.[42] They went on to assess the results of splicing, which again reflected changes in metabolism. So in addition to maintaining quality in general, DR alters the products of RNA splicing to induce necessary changes in metabolism.

Other RNA functions also change with age. MicroRNAs, the small RNAs that play mostly regulatory roles, change in their expression with age, are biomarkers of aging, and are pivotal in regulating longevity.[43] Some of these miRNAs play important roles through their nonautonomous regulation of longevity. For example, C. elegans miR-71 changes in expression with age,[44] conveys the state of germline-less animals, and links food odor to longevity through its regulation of intestinal proteostasis.[45]

A more somewhat more esoteric function, RNA editing, has also been associated with longevity in both humans and in worms. ADAR (adenosine deaminase acting on RNA) proteins mediate an adenosine-to-inosine (A to I) post-transcriptional modification, which changes how miRNAs recognize their target RNAs.[46] A-to-I editing modifies susceptibility to RNA interference, thus altering the regulation of translational inhibition by miRNAs. This RNA editing process helps the cell's innate immune system distinguish between its own RNA and that of an invading, pathogenic organism.[47] Paola Sebastiani and colleagues found that 18 SNPs in the RNA genes ADARB2 and 5 in ADARB1 were associated with longevity in centenarians,[48] but since it's difficult to know what they are doing in humans, the authors turned to testing in C. elegans. Even though it's not clear whether the SNPs would increase or decrease activity in humans, mutations in adr-1 and adr-2 shortened the lifespan of both wild-type and daf-2 mutants. The loss of the RNA interference gene rde-1 rescued the lifespan of the double mutants back to wild-type, but the underlying mechanism is not clear, nor is the corresponding reason for the SNP associations in ADAR genes with human centenarians. Later it was found that at least ADARB2 SNPs are associated with metabolic disorders, likely through serum adiponectin levels (that is, ADARB2 affects fat metabolism, leading to high BMI and other visceral fat changes).[49]

Circular RNAs are an old and almost forgotten RNA species that has recently become investigated again (come full circle, so to speak) for its possible role in aging. Circular RNA has been found in plants, fungi, and animals, including humans. Perhaps fitting with its assumed character as a by-product of splicing, circRNAs have been shown to increase with age in almost every organism.[50] Drosophila insulin mutants accumulate circRNAs more slowly than wild type with age[51]—but at least one circRNA can increase lifespan. Some of these functional circRNAs might work through translation into a peptide, while others might act as miRNA antagonists[52]—offering another level of previously unappreciated regulation that needs further exploration.

Quality control is critical for maintenance of cellular and organismal longevity, and cellular cleanup is the means by which organisms keep their cells functioning as long as possible. Because RNA is such a critical component of the cell, it is likely that in coming years we will find that every step of RNA quality control is critical for healthy cells, declines with age, and is maintained in some type of longevity-extending conditions—as we seem to have already established with every step in the life of a protein. The question will then become *How can we manipulate these quality control mechanisms to remain "on" longer with age?* While none of the individual mechanisms of protein or RNA homeostasis are likely to be the one and only solution for aging, boosting some of the most critical of these processes late in life—particularly if it is done in combination with other macromolecular cleanup quality control and regenerative mechanisms—could be a powerful weapon in slowing aging.

Keeping the sanitation workers doing their jobs longer will help us all, and we should appreciate them.

9

Powering Longevity

MITOCHONDRIA'S ROLE IN AGING AND LONGEVITY

Slightly more than half of everything I am is thanks to you
—CADAMOLE, *A BIOLOGIST'S MOTHER'S DAY SONG**

IF YOU'RE OVER the age of 35, you may not sprint as fast as you used to. That certainly is true for me; I could never hope to run the 100-meter hurdles as fast as I did as a senior in high school. Some older athletes defy the odds, of course, and today's Masters athletes have even exceeded performances at the first Olympic games: in 1896, at the age of 21, Thomas Burke (USA) ran the Olympic 100 meters in 12.0 seconds, but in 2009, 57-year-old Oscar Peyton ran the same distance in only 11.56 seconds. And for the centenarian benchmarks: at 105, Hidekichi Miyazaki ran the men's 100 meters in 42.22 seconds in 2015, and 101-year-old Julia "Hurricane" Hawkins ran it in only 40.12 seconds. On April 30, 2022, Lester Wright, a 100-year-old New Jerseyan whose track career was interrupted by World War II, ran the fastest ever 100 meters for a centenarian in 26.34 seconds at the Penn Relays.† I'm personally inspired by Ida Keeling, the 105-year-old world-record-holding sprinter who is trained by her daughter, Shelley Keeling (also a world record holder), and Man Kaur,

---

* As much as I hate to explain jokes: this song will make more sense once you realize that we inherit all of our mitochondria from our mothers' oocytes.

† I was lucky enough to be there, since my 14 year old son was running his own race in the same meet.

the Indian woman who took up running at the age of 93, urged on by her 79-year-old son, Gurdev Singh.[1] These inspirational athletes show us that one can be active and competitive even at advanced ages. The exercise guru Jack LaLanne literally lived up to the hype, working out until the day before he died at the age of 96. Another hero of centenarian athletes is Robert Marchand, a French cyclist who set the 100–104 age group world track cycling record for covering 26.93 kilometers in one hour. While most centenarian athletes show a decline of 78% when world records are compared, Marchand's decline was *only 8% per decade over more than six decades.* This might still sound like a lot, until you consider the fact that Marchand's record is only 50% slower than Bradley Wiggins's (54.53 km)[2]—a feat that many of us could not achieve even in mid-life. While anaerobic (sprinting) events generally show greater changes with age—just like the worm's decline in maximum velocity (see chapter 5)—aerobic performance also declines, but at a later age. (This is why I do triathlons now instead of running 100-meter hurdles, although maybe I should start again?) This decline in performance is so well appreciated that marathon organizers set different qualifying times based on age: a woman could qualify for the 2019 Boston Marathon if she ran a previous race in 3:35, but if she was over the age of 35, she got an extra five minutes, and a woman over 70 could qualify with a 4:55.

Of course, most of us are not concerned with setting world records, or even qualifying for the Boston Marathon, but rather with staying healthy with age. Loss of skeletal muscle performance with age is a real issue, as sarcopenia (muscle degradation) develops and frailty sets in, leading to falls, broken bones, and a downward spiral of grave illness in the elderly. Declining muscle performance with age has been attributed to the inability of our muscles to efficiently consume oxygen and deal with the buildup of lactic acid in the blood. At least part of this decline originates with changes in our mitochondria, organelles within the cell that regulate energy production. Mitochondria are critical for maintaining homeostasis in the cells. They regulate energy production, carbohydrate metabolism (the TCA cycle), lipid metabolism, oxygen sensing, apoptosis (programmed cell death), and the influx of calcium and other ions. The function of mitochondria is perhaps best studied in skeletal muscle, but they are necessary in most cells (mature red blood cells are the rare exception)—including liver, fat, even in the synapses of neurons, to provide energy and regulate metabolism. While mitochondria have always been intimately associated with theories of aging, our view of the role of this organelle in aging has evolved significantly over the years, from one of decline with age and a source of damaging agents to the current view of mitochondria as sensors of stress and inducers of protective mechanisms, essentially acting as endocrine organelles.

So which came first, declining mitochondrial function with age or age-induced decline in mitochondrial function?

———

If there is one thing you remember from your high school biology class, it's probably that "mitochondria are the powerhouses of the cell." I'm not going to dispute that here, but these tiny organelles are even more interesting than that old description implies. Mitochondria (from the Greek *mito* [particle] and *chondria* [lines]) convert the breakdown products of nutrients into energy in a more efficient way than cells without mitochondria can, because of the packing of specialized enzymes into their many inner membrane folds ("cristae"), where energy molecules (ATP) are produced. It is most likely that the engulfment of an aerobic prokaryote (*which* prokaryote, exactly, is still a matter of some debate, but it seems to be an α-proteobacterium) into a nucleated cell about 1.8–1.45 billion years ago allowed the development of a mutually beneficial symbiotic relationship: the mitochondrion gained efficiency by transferring most of its genome to the nucleus, allowing rapid amplification and proliferation of its own small genome, and the cell gained the ability to perform aerobic respiration and, as a result, produce energy more efficiently. In fact, it was a huge win for the engulfing cell, as mitochondria generate up to 38 ATP per glucose molecule through aerobic respiration, while anaerobic processes make only 2 ATP/glucose. This efficiency is due to the mitochondria's use of a proton (H^+) gradient, which creates an imbalance in electrons, and an "electron transport chain" (ETC) to shuttle electrons from protein complex to protein complex. This process ultimately builds units of ATP from the energy of molecules (NADH and $FADH_2$) obtained from nutrients. ATP molecules essentially store energy like a battery does, releasing that energy when the bonds are broken and phosphate units are transferred. This whole process is known as "oxidative phosphorylation" or OXPHOS. OXPHOS uses oxygen to help drive energy production, which is why aerobic organisms need air. The difference in protons (H^+) on either side of the mitochondrial membrane is key to this process, creating the "pressure" to move electrons to equilibrate and thus relieve this difference.

Mitochondria have been intimately linked with the regulation of longevity for at least half a century. You'll remember that one of the earliest ideas about the causes of aging, Harman's free radical theory of aging (1956), involves mitochondria, because oxygen radicals (reactive oxygen species, or ROS) are produced as a by-product of the ATP-generating process.[3] For decades, ROS production was regarded as *the* major cause of age-related damage. As ATP production and efficiency decline in aged mitochondria, superoxide radicals are made as a by-product. These oxygen radicals were thought to induce mutations through damage to the nearby mitochondrial DNA (mtDNA), a circular genome encoding fewer than 40 genes (tRNAs, rRNAs, and some proteins of the ETC complexes); the idea that the circular mtDNA becomes damaged and doesn't properly replicate, causing aging, was a prevalent theory for many years. However, when the rates of errors induced by replication versus those induced by oxidative damage were quantified, only 15% of the mutations could be attributed to ROS-induced damage, putting a serious dent in the ROS-induced mitochondrial DNA damage theory, or at least the idea that it is solely responsible for aging. Moreover, cells contain hundreds to thousands of copies of mtDNA, since most mitochondria include several copies, and there are hundreds of mitochondria per cell; perhaps because of this redundancy, most cells can withstand a high level of new mtDNA mutations (up to 60%–90%) before exhibiting significant deleterious phenotypes.[4] It is true that mtDNA mutations are a cause of serious human disorders, particularly those that are present from birth, and large deletions of mtDNA in mice (e.g., via a deletion in polymerase gamma) do cause early-aging ("progeroid") phenotypes, like gray hair, osteoporosis, cardiac hypertrophy, and sarcopenia, but it is seeming less and less likely that ROS-induced mutations are a major cause of normal aging.

Nevertheless, the loss of mitochondrial function has been linked with aging and age-related neurodegenerative disorders, including Parkinson's disease, Alzheimer's disease, and ALS. One way to resolve this apparent discrepancy is to understand that many critical cellular operations are happening in the mitochondria all the time, so it shouldn't be surprising that many of them break down with age, and when they do, dysfunction of the cell soon follows. In theory, a vicious cycle could then take over: as the components of the ETC become damaged, the efficiency of the respiratory chain drops, and oxidative damage—not just to DNA, but also to proteins and lipids—from ROS increases. These damaged mitochondria then generate more oxidative radicals, which in turn cause more damage.

Of course, the free radical theory was also attractive because it meshed well with Pearl's 1928 "rate-of-living" hypothesis, as faster-living animals' cells should require more energy and therefore should produce more damaging ROS, resulting in faster aging. Indeed, specialized enzymes that battle ROS reside in both the cytoplasm and inside mitochondria: superoxide dismutase (SOD) converts oxygen radicals into hydrogen peroxide (H_2O_2), and catalase converts that H_2O_2 into water and oxygen (O_2). Thioredoxins, glutathione, peroxiredoxins, and glutathione peroxidases are enzymes that all further work to scavenge ROS, protecting the cell. George Martin, who was one of the leaders of the aging field for decades (and even at over 95 was still a force in the longevity field), proposed oxidative damage as one of the "public"—that is, evolutionarily shared—mechanisms of aging in 1996.[5]

There is substantial evidence supporting at least some aspects of the free radical theory of aging. Many long-lived animals express higher levels of SOD, which, along with catalase expression, is often used as a marker for the induction of longevity pathways. These high levels of expression of SOD and catalase have been reported in long-lived mutant worms and flies, and even in clams that live for decades or centuries. Furthermore, it seems that in some cases having more of these enzymes can help animals live longer. For example, fly lifespan was increased when SOD and catalase were both overexpressed.[6] SOD seems to offer a protective effect to fibroblasts as well, as cell senescence rates appear to be inverse to the level of SOD: decreased SOD levels are associated with some forms of senescence, and high SOD expression slowed telomere shortening and extended fibroblast lifespan. Mice overexpressing human mitochondrial catalase were also healthier, as they exhibited less insulin resistance.[7] And in *C. elegans*, the application of the antioxidants *N*-acetylcysteine and vitamin C rescued the premature aging phenotype of a mutant in *alh-6*, a mitochondrial proline catabolism enzyme.[8] The *isp-1* mutant, which reduces activity of the mitochondrial iron sulfur protein of complex III, was originally theorized to live longer because of its lower rate of oxygen consumption and subsequent lower production of ROS, but now is thought to also affect other longevity mechanisms, including autophagy.[9] Therefore, until relatively recently, it seemed that there was ample evidence that the free radical theory of aging was *the* theory of aging, and that the best thing for us to do was to follow in the footsteps of Linus Pauling, gobbling up vitamin C and other antioxidants in copious amounts.

But, eventually, cracks appeared in the simple model that fast living generates damaging ROS, which in turn cause aging. For example, long-lived bats present a conundrum, as they seem to produce low levels of ROS despite high metabolic rates, but do maintain proteostasis with age.[10] In *C. elegans*, long-lived insulin signaling mutants and dauer-stage animals express higher levels of *sod-3*, and for several years it was theorized that this was a key component of the longevity of these animals, fitting with the antioxidant theories. However, when testing the targets of my *daf-2/daf-16* microarray results, I found that RNAi knockdown of *sod-3* and catalase causes only a small (about 5%) decrease in lifespan of the long-lived *daf-2* mutant.[11] This was a bit controversial at the time, both because the oxidative damage theory of aging was so popular, and because a paper claiming that loss of *sod-3* completely abrogated *daf-2*'s long lifespan had been published a few months before—but it turned out that the strain in question didn't have mutations in either *daf-2* or *sod-3*, so it was just a wild-type worm (and the paper was eventually retracted). Later studies showed that deletion of *sod-2* and *sod-3*, or all five superoxide dismutase genes together, had no significant deleterious effects under normal conditions (although loss of *sod-3* and pairs of SODs were deleterious under hyperoxic and other stressful conditions), suggesting that they are not as indispensable as previously thought.[12] Furthermore, expressing MnSOD or catalase in the mitochondrial matrix significantly *shortened* the lifespan of flies.[13] Finally, Shelley Buffenstein's lab showed that everyone's favorite ugly methuselah, the naked mole rat, seems to *not* owe its longevity to lower rates of oxidative damage, prompting articles such as "Is the Oxidative Stress Theory of Aging Dead?"[14]

When Julie Ahringer's *C. elegans* RNAi library became available in the early 2000s,[15] the Kenyon and Ruvkun labs both carried out genome-wide screens, and both found that mitochondrial gene knockdown—rather than overexpression—actually extended lifespan.[16] Reduction of several different components of the mitochondrial electron transport chain and oxidative phosphorylation function (which work together to generate energy in the form of ATP) all extended lifespan. Similarly, blocking ETC function using antimycin A, a complex III inhibitor, increased lifespan. Later these results were replicated in *Drosophila*; RNAi of complex I and IV components extended the lifespan of flies, even when knocked down in neurons alone, suggesting that a similar mechanism is utilized in flies and worms.[17] And this was not just an invertebrate phenomenon: mice with reduced function of the cytochrome c oxidase assembly factor SURF1 or the p66shc electron transfer redox enzyme lived longer, as well.[18]

At first glance, these results appeared to mimic the *isp-1* results, supporting the model that slower living reduces ROS and extends lifespan. Not surprisingly, this knockdown of mitochondrial ETC function in the worms was accompanied by decreased ATP production, which again could be interpreted as supporting the ROS theory. However, Dillin and Kenyon found that extension of lifespan—an adult phenotype, obviously—was achieved *only* when the ETC components were reduced *during development*, and had no effect when they were decreased during adulthood, even though ATP levels were still reduced. This result uncoupled impaired mitochondrial function—and presumably lower ROS production—from longevity extension. Furthermore, restoration of expression of the components during adulthood by blocking the RNAi machinery also restored ATP levels, and still allowed long lifespan. This result meant that the ATP levels were uncoupled from the longevity effect of mitochondrial gene knockdown, suggesting that the mechanism of lifespan extension could not be a direct result of low ATP and ROS. Instead, it pointed to the existence of a "critical period" of mitochondrial function during development that somehow was sensed and then set the future lifespan of the animal, a radical (*see what I did there?*) notion of temporal uncoupling of mitochondrial function and its regulation of longevity. The idea that ROS might not be the reason that these worms live long was almost heretical in the aging field, which had embraced the free radical model for so long that it raises the question: *If ROS-induced damage is not the key to longevity caused by reduced mitochondrial function, then what is?*

————

In 1943, Evelyn Witkin, a bacterial geneticist working at Cold Spring Harbor, discovered the SOS response when studying radiation-resistant mutants and a checkpoint response to low doses of UV damage.[19] (She is now retired and lives in Princeton, after having been at the Waksman Institute at Rutgers University for many years—and as of March 2021, she is now a centenarian!) It's hard to even imagine now what was not known at that time, but she made this discovery back when there was still debate about the nature of hereditary material, which was still thought to be protein, since DNA seemed too simple to carry the code of life. In the late 1960s and early 1970s, Witkin and Susan Gottesman (National Institutes of Health, NIH) discovered the Lon protease/Sfi cell division inhibitory mechanism behind the UV-induced DNA damage response. In the SOS response, stress does not act as a damaging agent itself but rather acts a *signal*; a low level of stress induces a response, and the components of the stress response do the job of protecting the cell.

Dillin and Kenyon's mitochondrial RNAi results suggested that loss of mitochondrial function might induce a signal that would ultimately affect the organism's lifespan, but only if the signal happens during a critical window during development. Instead of a *direct* mechanism in which ROS induce damage, it seemed that mitochondria must instead produce some sort of signal when distressed that in turn activates a protective response that increases longevity. Such a mitochondria-to-nucleus effect was first recognized in yeast and mammalian cells, where the "retrograde response" is turned on when ETC components are not functioning well. The knockdown of ETC components or inhibition of activity, rather than simply decreasing rates of oxidative damage, could induce a similar type of protective activity. To get such activity, the organisms wouldn't need to send a signal from every cell—just from important ones. For example, reducing mitochondrial function specifically in the worm's neurons and intestine extended lifespan both separately and nonadditively, and there were also cell-nonautonomous effects.[20] Furthermore, these manipulations uncoupled lifespan from other mitochondrial phenotypes, such as small size, reduced fertility, and movement defects, suggesting that a regulatory, rather than direct, mechanism was at work. Similarly, Sylvia Lee's lab (Cornell) showed that overexpression of a transcription factor called CEH-23, which is present in *C. elegans* neurons and intestine, extends lifespan, and is required for the longevity effect but for none of the other phenotypes of ETC knockdowns.[21] Mild mitochondrial stress of worm neurons—or neuronal "mitohormesis"—extended lifespan, as well.[22] Therefore, there appears to be a signal that travels from neuronal mitochondria to other tissues to regulate lifespan.

The mitochondrial unfolded stress response, or UPR^{mt}, was a good candidate for this trans-tissue longevity effect.[23] When the UPR^{mt} is activated in response to mitochondrial damage, hundreds of genes that function to repair damaged mitochondria, resist pathogens, and protect proteins are expressed. The UPR^{mt} induces the stress-dependent shift from mitochondrial to nuclear import upon mitochondrial dysfunction, utilizing an elegant mechanism. But before I reveal how the UPR^{mt} works, let me first tell you what it reminds me of: There's an episode of *Mr. Robot** in which the main character, a hacker named Elliot, is being threatened by a drug dealer he has ratted out; it's clear

---

* Aired on USA Network (NBCUniversal, 2015–19).

that the dealer is going to kill Elliot as soon as he has the chance. But Elliot's smart: using the drug dealer's brother's phone, he's captured the entire drug operation's data, then set up an automatic system to send this information to the police in a message *that only Elliot can disable*. Every day, Elliot needs to respond to this message to turn off the alert. The day that he doesn't, the message isn't inactivated, and all the dealer's data gets sent to the police. This simple system ensures Elliot's safety.

That's basically how the UPR[mt] works, and a transcription factor called ATFS-1 is key to the whole thing. Like Elliot's automatic message, ATFS-1 disappears when things are fine, but in an emergency, like dysfunction of mitochondria (like Elliot's death) ATFS-1 doesn't receive the "turn off" message, and springs into action.[24] As long as the mitochondria are functional, ATFS-1 is imported into the mitochondria and is subsequently degraded there by Lon protease, the same protease involved in the bacterial SOS response (and remember, mitochondria originated from a bacteria-like prokaryote). But if the mitochondria are too sick and dysfunctional to import ATFS-1, it moves into the nucleus instead, because ATFS-1 also has an NLS (nuclear localization signal) tag. Once in the nucleus, ATFS-1 induces transcription of stress response genes. This elegant mechanism ensures that mitochondria do not have to be functional in order to set a stress response into motion—breaking the system, like killing Elliot, sets the chain of events in motion. It's interesting to think about how such a system may have evolved, considering it utilizes an ancient bacterial stress system (the Lon protease of the SOS response) but relies on a transcription factor that functions in the nucleus but is degraded in the mitochondrion.

When the UPR[mt] was tested for its role in the systemic effect of neuronal ETC knockdown, it was found that reduction of mitochondrial ETC components indeed induced expression of a marker of UPR[mt] activity.[25] Other components, such as the homeobox transcription factor DVE-1, the ubiquitin-like protein UBL-5, the mitochondrial protease ClpP, and the mitochondrial peptide transporter HAF-1 are also required for UPR[mt]. Loss of the UPR[mt] ubiquitin-like cofactor UBL-5 blunted the longevity induced by loss of ETC components. Moreover, knocking down the ETC genes in neurons induced the UPR[mt] response in intestinal cells, suggesting that a trans-tissue signaling mechanism was at work. Andy Dillin's lab (University of California, Berkeley) has also identified a pair of histone demethylases, *jmdj-1.2* and *3.1*, that may confer a more long-lasting epigenetic imprint of the stress signal,[26] possibly explaining the temporal delay between mitochondrial gene knockdown in

L3/4 larvae and their subsequent adult longevity through an epigenetic mechanism. Because *jmdj-1.2* and *3.1* have mammalian counterparts that are also associated with UPR[mt] genes, and mammalian ATF5 acts homologously to ATFS-1, it is possible that this epigenetic mechanism of UPR[mt] response is evolutionarily conserved from worms to mammals.[27]

The term "mitokine," analogous to the immune system's "cytokine" signal, was coined to describe the (unidentified) signal that communicates mitochondrial stress from one cell to the rest of the animal.[28] Mitochondrial dysfunction on its own is not responsible for the worms' long lifespan, but, rather, the resulting stress signal and activation of stress responses allow the animal to live long. Signaling from *C. elegans* neurons to its intestine (where ATFS-1 activates the UPR[mt]) involves the FLP-2 neuropeptide, which appears to be a signal that conveys the message from neurons to the intestine.[29] (However, one might argue whether FLP-2 meets the definition of a "mitokine," since it doesn't originate in the cells where mitochondria were damaged, but rather acts in a relay of cell–cell signaling from neurons to intestine.) Similarly, stress in glia and neurons controls a signal that is interpreted by a neuropeptide receptor (NPR-28) which in turn regulates the UPR[mt].[30] Together, these findings suggest that mitochondrial dysfunction on its own is not entirely responsible for worm lifespan but, rather, that neuronal stress signals nonautonomously activate the UPR[mt] to regulate longevity.

As was found in worms, muscle-specific ETC disruption of mitochondrial function in fly muscles prevented age-related muscle deterioration (a cell-autonomous effect, that is, happens in the cell itself) and increased lifespan (a cell-nonautonomous or systemic effect). The UPR[mt] was induced and insulin signaling was systemically downregulated by induction of the insulin-antagonizing peptide ImpL2.[31] The fact that increased catalase or glutathione peroxidase prevented this effect suggested that, as in worms, elevated ROS are necessary for the beneficial systemic effects. Intriguingly, these results link two of the major longevity pathways, mitochondrial downregulation and insulin signaling, which had been thought to be independent. The authors also noted that the mammalian ortholog of ImpL2, IGFBP7, might affect signaling in response to mitochondrial perturbations as well. (Later work showed that the TGF-beta pathway functions in muscle-to-fat communication via mitochondria,[32] which also has links to insulin signaling.)

Mammals also seem to use mitokines to signal between tissues. For example, the mammalian hormone fibroblast growth factor 21 (Fgf21) links autophagy, mitochondrial dysfunction, and activation of the transcription factor Atf4.[33] Mice lacking an autophagy gene, Atg7, specifically in skeletal muscle were surprisingly healthy at the whole-animal level, despite their impaired skeletal muscle mitochondrial function: they were leaner and were protected from diet-induced obesity and insulin resistance.[34] These health effects seem to be due to the induction of Fgf21 by the Atf4 transcription factor in response to the impaired skeletal muscle, resulting in increased mitochondrial fatty oxidation and "browning" (induction of mitochondria) of white adipose tissue. That is, autophagy deficiency led to mitochondrial dysfunction in muscle, which caused the Atf4-dependent increase in Fgf21 expression, which in turn signaled to fat, leading to browning of white adipose tissue and fat loss. Thus, Fgf21 appears to act as a mammalian mitokine, and the general mechanism of nonautonomous induction of mitochondrial activity in response to mitochondrial dysfunction elsewhere seems to be conserved from worms to flies to mice. If Fgf21 sounds familiar, you might remember that it has also been called "the starvation hormone," because it is produced in the liver of intermittently fasting animals and prevents the production of insulin-like growth factor (see chapter 7). Mice that express high levels of Fgf21 live longer, which is consistent both with IF and with the effects of mitochondrial dysfunction in worms. These results further underscore a theme of communication between mitochondria and peripheral tissues and the linking of mitochondrial and IIS longevity pathways.[35]

An additional link between mitochondria and insulin signaling was established with the discovery of the "mitochondrial-derived peptides" (MDPs) humanin and MOTS-c. These small peptides are transcribed from small open reading frames (that is, where genes are expressed) within the genes for ribosomal RNAs in the mitochondrial genome. MDPs were only recently discovered, as they are both tiny and kind of hidden within other genes of the very small mitochondrial genome. The MDP humanin was found through a screen for factors that bind to insulin-like growth factor binding protein 3.[36] IGFBP3 appears to be protective, as it normally declines with age, is higher in individuals of long-lived families, and was found to prevent in vitro Alzheimer's disease–induced neuronal cell death. Binding of the mitochondrially derived peptide humanin to IGFBP3 is another connection between insulin signaling and mitochondria-derived signals. Another MDP, MOTS-c, comprising only 16 amino acids, is encoded in a short open reading frame within the mitochondrial 12S rRNA.[37] MOTS-c appears to regulate metabolism, particularly folate levels, glucose metabolism, the glycolytic response, and fatty acids in skeletal muscle. AMPK may regulate the activity of MOTS-c, linking mitochondria and nutrient metabolism. Treatment of high-fat-diet-fed mice with MOTS-c peptide prevented obesity and insulin resistance; thus, MOTS-c may act essentially as an endocrine signal, and has promise as a possible clinical therapeutic.

In addition to MDPs and mitokine activation of the UPR[mt], other signals from mitochondria may activate protective responses. For example, "mitochondrial-derived damage-associated molecular patterns," or DAMPs, can act as nonautonomous signals of mitochondrial stress and activate an immune response.[38] Excitingly, the existence of mitokines, MDPs, and DAMPs in mammals implies that if you could provide such a molecule systemically, you wouldn't have to knock down mitochondrial function to elicit healthy responses, or maybe even promote longevity. These mitochondrial distress signals may represent a new class of longevity treatments to be explored.

How can we resolve the seemingly conflicting notions that mutations in mitochondria can cause serious disabilities in humans—like deafness, early-onset neurodegeneration, and other severe defects—but their reduction can *extend* lifespan in flies and worms and even mice? This paradox may be resolved, not by assuming that the difference is because of the organism being tested (and, frankly, that's usually a cop-out), but rather by considering it may be because of the dosage. Knockdown by RNAi and hypomorphic (loss-of-function) mutations are milder perturbations than null (total deletion) mutations. Whereas complete mitochondrial gene knockouts render worm and fly progeny inviable, mild knockdown simply reduces activity. Shane Rea and Tom Johnson (University of Colorado) carried out a careful RNAi dilution analysis of mitochondrial gene knockdown in C. *elegans*, and found that there are three phases of inhibition, resulting in an inverted-U-shaped curve for lifespan; they found no lifespan effect at low inhibition of ETC components, an extension of lifespan at medium levels of knockdown, and, finally, decreasing lifespan at the highest levels of RNAi knockdown.[39] Thus, reduction of function of the same gene can cause either lifespan extension or be deleterious, depending on how much the gene's activity is decreased—a little bit of reduction is a good thing, but a lot can be lethal.

The idea that low-level damage could induce a stress response that might ultimately lead to longer life is the general concept behind hormesis. By turning on a stress response, the organism might not only survive a stressful condition but even have its lifespan extended as a result—a kind of Nietz-schean response to temporarily tough times. Together with the UPR[mt] data, a new paradigm emerges, that of mitochondrial hormesis, or "mitohormesis": mild mitochondrial stress and ROS generation induces a systemic stress response, and this stress response helps the animal live longer.[40] Thus, inducing *low levels* of oxidative stress might increase lifespan. In this model, mild mitochondrial disruption might increase ROS, and ROS act as a signal rather than a damaging agent. Cynthia Kenyon's lab showed that a burst of ROS and hydrogen sulfide production upon germline elimination leads to longer lifespan through induction of both the UPR[mt] and a transcriptional response,[41] suggesting that longevity pathways previously thought to be separate, such as the germline pathway and mitochondrial gene knockdown, might ultimately utilize common and parallel signals and responses to tune longevity responses to stress.

A somewhat alarming prediction of the mitohormesis model is that blocking the damage and the ensuing oxidative stress signal will prevent the beneficial effects of stress conditions. Indeed, treating healthy young men with vitamins C and E before a four-week exercise treatment *prevented* improvements in insulin sensitivity and other health benefits induced by exercise.[42] Yes, you read that right—it's not just that vitamin C and E did nothing, they actually *blocked the beneficial effects of exercise.* This finding is particularly shocking given the advice that we've been told for years, to ingest antioxidants because they will help us live longer; instead, they may prevent the very beneficial effects we seek.

————

So what makes mitochondria "high quality," and how can they stay that way with age? Given the fact that mitochondria are so critical to cellular and organismal homeostasis, but operate at high metabolic rates, thus incurring damage, they must repair and renew themselves constantly. They do so through several different mechanisms, regulating the amount and quality of DNA, proteins, and the organelle itself.

One major problem that mitochondria must continually solve is proper stoichiometry (ratios of proteins) of its enzyme complexes. Since the mitochondrial complexes that make ATP through oxidative phosphorylation require both nuclearly and mitochondrially encoded subunits, there must be a "mitonuclear balance" to coordinate protein levels; if the components made in the nucleus and in the mitochondria are no longer coordinated to build functional complexes, as in the case of defective mtDNA or defective aminoacyl-tRNAs in the mitochondria, mitochondrial function also declines. "Matching" of mitochondrial and nuclear DNA was shown in both worms and mice to be required for healthy aging.[43]

Mitochondria are also in communication with and dependent on nuclear DNA and the nucleus. Mitochondria do not encode their own DNA polymerase or repair enzymes, so they must rely on proteins that are encoded by nuclear chromosomes. Mitochondrial DNA is made by a DNA polymerase called pol gamma (pol γ), which replicates mtDNA and also has "exonuclease" (DNA-cutting) activity that is important for preventing mutations that may arise during the replication process, a kind of proofreading function. In fact, deleting the exonuclease activity from pol gamma causes the accumulation of mtDNA mutations. There is also communication between telomeric (chromosome end) damage and mitochondrial biogenesis, underscoring the importance of communication between mitochondria and the nucleus.[44]

At the level of protein damage, mitochondria maintain quality through mitochondria-associated degradation, a ubiquitin-proteasome mechanism to move damaged or excess proteins to the outer mitochondrial membrane (via Vms1), ubiquitin tag them (via E3 ubiquitin ligases), extract them from the membrane (via Cdc48p), and send them to the cytosolic proteasome for degradation.[45] This system is analogous to the ER-associated degradation pathway.

Another way to get rid of your trash is to dump it into someone else's bin. Mitochondria can become so damaged over time, at both the DNA and protein levels, that they cease to function properly, and must be eliminated somehow, either through mitophagy (destruction) or just moving them elsewhere. Much like the ability of yeast to apportion damaged proteins to the mother cell rather than the daughter cell,[46] yeast cells actively sort better-functioning mitochondria (as defined by redox potential) to the developing daughter cell;[47] disruption of this sorting shortens replicative lifespan. A pH gradient and lysosomal function are important for this process.[48] Similarly, mammalian stem cells preferentially keep their aged mitochondria in the old progenitor cell, allowing the newer daughter cell to carry younger mitochondria.[49] Like restricting Dorian Gray's signs of aging to his portrait in the attic, mother cells keep the old mitochondria, while pristine daughter cells maintain "stemness" better than cells that cannot carry out asymmetric mitochondrial inheritance.

Mitochondria are built and expanded through the process of biogenesis, which relies primarily on the activity of two proteins, PGC-1α and PPAR-γ, which regulate transcription within the mitochondrion with the help of the mitochondrial transcription factor TFAM, and nuclear respiratory factors NRF1 and NRF2.[50] Endurance exercise may stimulate repair prior to mitochondrial biogenesis. Once the mitochondria are made, the process of maintaining mitochondrial quality relies on their ability to fuse together, split into smaller parts (fission), and subsequently eliminate poor-quality mitochondria. Fusion is likely the reason that very high levels of damaged mtDNA, proteins, and lipids can be tolerated; each fused mitochondrion can contain many copies of mtDNA, with the intact copies complementing the activity that damaged mtDNA lacks—sort of a communist system of maintaining functional mitochondria by sharing. Healthy, young skeletal muscle cells in C. elegans usually have long tubules of mitochondria, which promoted the viewpoint that this network-like form is healthier than the smaller, punctate (often called "fragmented") mitochondria that appear in muscle cells with age. Mitochondria clearly exhibit changes in their morphology with age: they become fractured and fragmented, and they are more easily damaged. They even exhibit "swirled" shapes in aged Drosophila cells.[51] Clearly these aged mitochondria are at an irreparable stage, but mitochondria have several layers of quality control mechanisms to combat damage as they age.

Once proteins in a mitochondrion are damaged beyond repair, they can become sequestered to one portion of the organelle, and then when the mitochondrion divides through fission, the damaged proteins are sequestered to one mitochondrion, which can later be eliminated. Fission, the splitting of long (fused) mitochondria into smaller bits, is necessary for the cleaning up and getting rid of damaged mitochondria, a process known as mitochondrial autophagy, or "mitophagy." When mitochondria are so damaged that they no longer properly maintain their inner membrane potential, they are destroyed through a specialized system. Like the disruption of Lon-mediated degradation of ATFS-1 that signals the UPR^{mt}, mitophagy is signaled by the failure to import and degrade PINK1, a mitochondrially targeted serine/threonine kinase; this failure to import PINK1 results in the accumulation of PINK1 on the mitochondrial surface. PINK1 recruits the ubiquitin ligase parkin, which tags outer membrane proteins for autophagic elimination. Subsequent lysosomal fusion degrades the mitochondrion. In order to do all of this, the mitochondrion must be small (punctate, or fragmented). (As the names imply, parkin and PINK1 have been associated with Parkinson's disease, and mtDNA has been associated with other forms of neurodegeneration.) By tagging and degrading damaged mitochondria after preferentially sequestering damaged proteins and fission, the cell can eliminate damaged mitochondria and recycle their parts. The process of mitophagy is critical for healthy longevity; blocking this pathway prevents germline-mediated longevity in *C. elegans*.[52]

Are tubular, networked mitochondria "better" than punctate ("fragmented") mitochondria? That was certainly the viewpoint for a long time, but the picture has been getting more complicated lately. In general, disruption of normal mitochondrial morphology and dynamics alters lifespan, but different studies report results in opposite directions.[53] Excitingly, the induction of fission in middle-aged flies through the expression of a mitochondrial fission protein, Drp-1, increases their lifespan and healthspan,[54] and similarly, the promotion of fission increased lifespan in worms. This was surprising, given that the fused state of the mitochondrial network is often observed in young, healthy animals, while fragmentation of mitochondria increases with age,[55] and insulin signaling mutants may require the fused state.[56] The fly results seem to suggest that the fissioned (punctate) state of mitochondria might be healthier, perhaps because of an induction of mitophagy that may remove damaged mitochondria. However, fused mitochondrial networks have been associated with youth and longevity in multiple systems, and inhibition of both fission and fusion extended lifespan of worms but shortened that of yeast.[57] Moreover, both fission and fusion are required for longevity mediated by dietary restriction (DR) and intermittent fasting (IF),[58] and both seem partially required for mitochondrial RNAi (*cco-1*) longevity, but neither appears to be required for IIS-mediated longevity.

For the most part, these previous studies, particularly in *C. elegans*, focused on fission and fusion mutants, or imaging of skeletal muscle mitochondria; however, what might not be obvious as one reads these papers on skeletal muscle mitochondria is that most (more than 90%) of the mitochondria in *C. elegans* are actually packed into its germline[59]—but the morphology and role of these mitochondria in longevity regulation have been largely ignored. (And one might rightfully wonder whether the longevity effects seen in mutants are actually due to the mitochondria in the germline, rather than those that were imaged in somatic cells.) By examining the morphology of these germline mitochondria, Vanessa Cota, a graduate student in my lab, recently discovered yet another role for mitochondrial dynamics: maintaining oocyte quality.[60] While young wild-type worms have tubular mitochondria in their oocytes, the morphology of these mitochondria doesn't change with age or with declining quality. By contrast, the oocytes of long-reproducing *daf-2* worms *always* have punctate mitochondria—exactly the opposite what we had expected, since their skeletal muscle mitochondria are tubular—and they remain punctate whether they are young or old. In fact, this punctate morphology seems to be how *daf-2* maintains healthy mitochondria in its oocytes: the punctate mitochondria are *necessary* for high-quality oocytes. Forcing *daf-2* oocyte mitochondria to become fused decreases oocyte quality and reproductive span, and the fission-promoting DRP-1 is required for *daf-2*'s long reproductive span.

The punctate morphology of *daf-2* oocyte mitochondria primes them for repair, it seems, by enabling faster mitophagy, as the mitophagy protein PINK1 is also necessary for *daf-2*'s high oocyte quality and long reproductive span. Based on her *daf-2* data, Vanessa hypothesized that promotion of fission and mitophagy might improve normal aging oocytes; in fact, she found that treatment of middle-aged wild-type worms with urolithin A, a compound that promotes mitophagy, extends their reproductive span, suggesting that pharmacological manipulation of mitochondria might improve reproduction. Therefore, the common characterization of mitochondria as "fragmented" is a biased term suggesting that they are damaged; this may be true if the mitochondria were previously tubular or elongated and became punctate, but it seems that *daf-2*'s oocyte mitochondria were never elongated, and instead are simultaneously "good" and punctate.

Don't worry, you are not the only one who is confused—there is a lack of consistency in the mitochondrial morphology results I'm conveying to you, suggesting that we don't yet know the whole picture. How can all of these seemingly conflicting data be resolved? One possibility is that the demand for different mitochondrial dynamics or morphologies in different conditions— or cells—may be the key. Dietary fluctuations may require mitochondrial remodeling, concomitant with metabolic shifts in specific tissues and cells, as well as interactions with other organelles, such as peroxisomes and lysosomes,[61] to extend lifespan, while reduced mitochondrial function induces protective mechanisms through the UPR^mt. Insulin/IGF-1-signaling-regulated longevity may employ many fewer mitochondria-dependent mechanisms, since FOXO also regulates many other cell-maintenance systems, so IIS may need only to use the last-ditch process of mitophagy, which is also used in germline longevity regulation. Further analyses of the metabolic states of mitochondria in different cells and under different conditions will be necessary to clear up some of these mysteries, even in the relatively simple system of

C. elegans. Understanding how mitochondria regulate lifespan and change with age from a tissue-specific and systemic viewpoint in invertebrates will be very helpful to understanding how they function in mammals, including us.

How can we leverage what we now know about mitochondria into effective therapeutics that will help us live longer? One thing to remember is that simply blocking mitochondrial function or inducing the kind of mitochondrial stresses that induce long life in invertebrates is not wise. As Cole Haynes (who uncovered the ATFS-1 mechanism) pointed out to me, "mitochondrial dysfunction is bad. This gets lost in some of the worm and fly longevity studies. For example, taking cyanide tablets is not a good longevity strategy." That is, being deficient for mitochondrial activity is not a good thing in itself, even for worms, despite their long lifespan: mitochondrial mutants and mitochondrial gene RNAi knockdown animals suffer from poor fertility and poor learning and memory, and, frankly, they just look horrible—scrawny and miserable. Even worse, these mitochondrial genes must be knocked down during the juvenile stage to achieve long lifespan. This would be the equivalent of stunting the growth of teenagers just to allow them to become centenarians in the future. Overall, mitochondrial dysfunction mutants are a poor model for healthy longevity extension.

Likewise, mitochondrial dysfunction can lead to severe human diseases. In fact, mitochondrial mutants are an example of the discrepancy between healthspan and lifespan that Heidi Tissenbaum's work (chapter 5) warned against. (This is why Tissenbaum's revelation that not all long-lived animals are healthy was no surprise to most worm biologists, at least those of us who have ever looked at a mitochondrial mutant.) But the identification of these mutants helped identify mechanisms that are required for mitochondrially associated homeostasis mechanisms, such as the UPS, UPR^{mt}, and mitophagy, so their study was important in order to uncover potentially healthy mechanisms of extending lifespan and health. In fact, the results of mitochondrial gene knockdown in flies and mammals are more encouraging, since mild mitochondrial stress (mitohormesis) can increase systemic stress resistance and overall health, and this can be achieved even if knocked down in adulthood. (And exercise, of course, is a form of mild mitohormesis.) The mechanisms we would like to harness are those that increase longevity through their increase in healthy mitochondrial function.

In some cases, however, reducing mitochondrial function might be just the ticket. We found this somewhat by accident through a collaborative project between Rachel Kaletsky in my lab and Vicky Yao in Olga Troyanskaya's lab

(Princeton) that resulted in our identification of potential Parkinson's disease genes in humans, which we tested in worms.[62] Vicky and Olga developed a method called *diseaseQUEST* to combine tissue-specific gene networks (in this case, neurons) and GWAS disease data (here, Parkinson's disease) to identify new candidate disease genes. Since we didn't know what Parkinson's would look like in worms, we used a simple, unbiased method to test these genes—we just asked, if we knocked down these candidate genes in adult worms, what behaviors might we see as they age? For several of the top candidates, we found that they had a bizarre curling and freezing movement, a spasm of sorts. The top candidate from both the *diseaseQUEST* list and in our behavioral assay was BCAT-1, a mitochondrial enzyme involved in the metabolism of branched-chain amino acids (that is, leucine, isoleucine, and valine). BCAT-1's reduction by RNAi caused such a severe phenotype that it was obvious even in still pictures because the worms were all curled up into little C shapes. When we checked in human disease samples, we found that the *BCAT1* gene is reduced in sporadic cases of Parkinson's in the substantia nigra, which is the site of much of Parkinson's-induced dysfunction. In the worms, we could see that *bcat-1* knockdown caused neurodegeneration in the cells that regulate motor function, which is why we saw the spasm-like phenotype.

But we didn't know two things: *why* knockdown of the *bcat-1* gene caused neurodegeneration, and how we could fix it. Danielle Mor and Salman Sohrabi, both postdocs in my lab, carried out a drug screen to identify new candidate therapies. By treating worms first with *bcat-1* RNAi bacteria for most of their adulthood and then treating them with drugs in mid-adulthood after we could already see the curling, they found that a few drugs were able to stop the neurodegeneration in it its tracks.[63] One of them was metformin (which we will hear much more about in chapter 17), the diabetes drug that may be a complex I inhibitor. (Metformin's exact activity is both unknown and ascribed to several different functions.) The biggest surprise came when Danielle looked at the function of the mitochondria: *bcat-1* knockdown actually *increased* mitochondrial activity and subsequent ROS generation, and metformin treatment reduced this mitochondrial activity and decreased ROS. In fact, treatment with another mitochondrial activity blocker, azide, also "fixed" *bcat-1*-knockdown-induced Parkinson's symptoms (but you wouldn't want to take this as a drug!).[64] So at least in some cases it might be the case that runaway mitochondrial activity can cause ROS damage that kills neurons, and reducing this activity helps fix the problem. Whether metformin could be a viable treatment for at least some types of Parkinson's disease is not yet known, but perhaps for cases where increased mitochondrial activity is the root of the problem, it could be a good candidate.

———

Mitochondria have been central to aging theories for more than half a century. However, the view of the mitochondrion's role in aging has evolved from one of incidentally generating ROS and causing oxidative damage that accumulates with age to a view of mitochondria as stress sensors, integrators, and signal secretors that dynamically regulate rates of aging across many tissues. Regulatory processes that signal, repair, and eliminate mitochondria are critical to mitochondrial regulation of longevity. Since we can't all be blessed with the genetics of a supercentenarian, and longevity drugs are not yet in the pipeline, can we use what we know about mitochondria to help us live longer?

Dietary restriction and pharmacological activators may stimulate mitochondrial processes, but one of the easiest ways to extend our lifespan through mitochondrial mechanisms is to exercise more. The fact that exercise might mimic some of the aspects of mitohormesis suggests that, as long as our mitochondria can continue to respond to stresses, exercise might help us to maintain better health with age. Exercise helps prevent sarcopenia and prevents age-related declines in cognitive activity and cardiac function, at least in part through the stimulation of mitochondrial biogenesis and the maintenance of healthy mitochondria in brain and stem cells.[65] Exercise has clear, positive effects on muscle: 60-year-old triathletes show obvious slowing of aging in comparison to nonathletes, and, as we discussed in chapter 5, fitness is a good predictor of lifespan in the elderly. Eline Slagboom and Adam Antebi found that even short-term exercise induces a hallmark of improved cellular function, a smaller nucleolus.[66] Monica Driscoll's group figured out how to make worms exercise by forcing them to swim several times a day, and found that not only do they live longer, but their memory is maintained longer, as well.[67]

Not surprisingly, drugs that promote the mitochondrial mechanisms we've just covered, pharmacological exercise mimetics, are being developed (see chapter 17). For example, NAD$^+$ levels, which are necessary for proper mitochondrial function, decline with age, and there is the hope that supplementing with the NAD$^+$ precursor nicotinamide riboside (NR) will help stave off age-related decline. (In fact, NR, in combination with pterostilbene, a resveratrol-like molecule, is in the Basis pills from Guarente's company, Elysium.) The ketoacid α-ketoglutarate, the AMPK activator compound 14, urolithin A, and ellagitannins have been proposed to act as exercise mimetics, since they stimulate the same pathways that exercise affects. The now discontinued GlaxoSmithKline drug GW501516, an activator of PPAR-δ, is being studied by Ron Evans's lab for its possible therapeutic use for muscular dystrophies, arguably a better and more immediate use of such a drug.[68] The problem is that the long-term effects of many of these exercise mimetics are not yet known, so as attractive as it might be to get your exercise in a pill, it's probably better for now, at least, to follow Nike's advice and *Just Do It*. We may not all be Robert Marchand, our super athletic centenarian cyclist, but it's never too late to try.

10

Dracula and Wolverine

HOW DNA REPAIR AND CELL REPLACEMENT CAN HELP US LIVE LONG

You know, when you make a copy of a copy, it's not as sharp as the original.

—MICHAEL KEATON (DOUG KINNEY), *MULTIPLICITY*

ANYONE WHO'S WATCHED A Marvel film can tell you that science—often science gone wrong—is at the heart of most superhero and supervillain back-stories, from Hulk to Spider-Man to Captain America. (In fact, most of Spider-Man's foes arose through some sort of scientific or engineering misuse, and an absurd number of them have PhDs.) Mary Shelley, arguably the first writer of science fiction, imagined how a scientist might animate nonliving tissue in her 1818 novel *Frankenstein; or, The Modern Prometheus*. From Gothic villains to modern-era heroes, the ability to escape death and regenerate tissues after what should be a mortal wound is the stuff of legends, comic books, and block-busters. Blood, DNA, mutation, and regeneration all figure heavily—if largely incorrectly—in these works of science fiction, and in some ways presaged a few of the current directions in longevity research. And longevity-wise, Stan Lee himself, the creator of many of these works, was no slouch—he lived until he was almost 96, and made cameos in his films into his 90s (and even one posthumously).

What comic book and science fiction writers realized long ago was that to have some sort of invincibility, one must overcome the normal dangers of injury and death, but to do so, some pretty superhuman things have to happen at the cellular level. This is because with time (and battles) our cells become damaged, including the genetic material inside them. Because DNA encodes the instructions for cellular function, it is critical that it remain undamaged. While our cells try to repair damage, at some point with increasing age they are not able to keep up, losing homeostasis. Our bodies constantly replace our old cells with new ones, but as we age we lose this ability, as well.

What limits how long cells can divide and make new, perfect daughters? As with anything that needs to be done with perfect fidelity, the process of cell replacement has several points of vulnerability, and those are the steps that go awry with age. Sometimes these problems not only kill cells but even cause them to become toxic to other cells, accelerating the aging process and causing neurodegeneration and cancer. In order to develop therapies to slow these processes in regular humans, we must first understand what happens in all cells, including stem cells, at the level of DNA.

DNA Replication, Damage, and Repair—and How It Can Go Wrong

Before we launch into understanding cell division's problems with age, let's go back to high school biology and review the central dogma:

$$DNA \rightarrow RNA \rightarrow protein$$

That is, DNA acts as the blueprint for the cell, providing the template for messenger RNA (mRNA), which is then "translated" into protein. Each of the components of the central dogma has special attributes that make it perfect for its job. DNA needs to reliably store information and be replicable in a highly error-free manner; RNA must be able to dynamically convey specific parts of this information at the right time; and proteins are, for the most part, the workhorses of the cell.

DNA and RNA are similar, as they are both made up of strings of nucleic acid, but DNA is more stable than RNA because it is double stranded and is less chemically reactive than RNA. In our cells, DNA is wrapped around proteins called histones, and those histones are packaged into structures called nucleosomes, which are further folded into chromosomes. Nucleosome packaging helps keep DNA protected and compacted, allowing incredible lengths of DNA to fit into a tiny space until genes need to be "transcribed" into mRNA.

mRNA is more dynamic, a transient and degradable message whose job is to be translated into protein at the right time and in the right place. (You probably remember this from getting your Pfizer or Moderna Covid vaccine— mRNA is injected into your cell, makes spike protein, and then is rapidly degraded.) Proteins made from those mRNAs are necessary to do the actual cell biological work, such as enzymatic reactions and making up the structure of the cell and its subparts.

DNA's most important jobs are to encode genes and to limit the amount of change—mutation—that happens to its sequence over time. The effects of sequence changes that happen to a gene in a "somatic" (non-germline) and non-stem cell are largely restricted to that tissue, while changes to DNA in stem cells can affect dividing cell tissues, and mutations to DNA in germline cells can lead to permanent changes, since those can be passed on to progeny. Mutagenesis cuts both ways. These changes in an individual can cause major problems, including genetic diseases, cancer, and even death. But mutagenesis is also the mechanism by which evolution and adaptation happen. Mutations led to humans being what we are as a species, and even differences that lead individuals to have specific traits that keep us from just being clones of one another. (But, no, DNA mutations won't give us web-slinging powers or make us X-Men, that was always a stretch). The single-nucleotide polymorphisms (SNPs) that we track in genome-wide association studies (GWAS) are exactly these kinds of mutations, and a few of these SNPs are associated with exceptionally long lifespans.

In order to make their replacements, cells must copy their DNA, and, importantly, that copy needs to be pretty much *exactly* the same as the original. That is, each base (A, T, C, and G) must be laid down in the same order as its original sequence, which is the beautiful thing about base pairing: since A pairs with T and C pairs with G, the double strandedness of DNA, coupled with the pairing of opposite nucleotides, allows the replication complex to "read" the opposite strand and instantly know how to make a perfect copy of the original sequence. This double-stranded complementarity of DNA is the major defining feature that enables this molecule to be used as a copy-able template, leading to the "*A-ha!*" moment of understanding how this could happen once its structure was determined from Rosalind Franklin's crystallography data. DNA already has the instructions on how to copy itself—it just needs to be unzipped into single strands, and the right enzymes have to attach the complementary nucleotide (As with Ts, Cs with Gs) and then link them together into a new strand.

There are some changes to the DNA sequence that might not matter as much. For example, a change could occur in a region of DNA that doesn't encode or regulate a gene, and a mutation might not affect the outcome significantly if it happens to be in a stretch of nucleotides that don't affect the sequence of a protein or the regulation of any important sequence-specific processes—like transcription-factor binding or RNA splicing, for example. Often SNPs are in areas with generally low sequence-specific function, because they don't change the outcome, so there is no evolutionary selection for or against them. Also, since there are 20 amino acids that are specified by 61 three-letter codons, several codons are associated with each single amino acid (3 of the 64 are "stop" codons), allowing redundancy; if the change to the DNA doesn't affect which amino acid will get incorporated at that codon (a "synonymous" mutation), then the protein output will be the same.

But in general, cells care a *lot* about the fidelity of DNA replication. There are as many different mechanisms dedicated to fixing DNA as there are ways to damage it. DNA can be affected by what goes on inside the cell (endogenous) and outside it (exogenous). Endogenous damage includes errors during DNA replication, which can cause mismatch errors and double-stranded breaks. The first category also includes our old friend ROS-induced damage, the kind we've heard about from reactions that go on in mitochondria, which causes oxidation of the DNA. Radiation, including UV, X-rays, and gamma radiation, and toxins originate from exogenous sources, and they cause all kinds of damaging chemical reactions on the DNA itself: oxidation, alkylation, changes to the bases themselves (deamination, depurination), mismatches, bulky adduct formation, and both double- and single-stranded breaks. If that sounds like a lot, well, it is—even for a fairly stable molecule, each nucleotide of DNA can be altered in multiple ways, and the strand itself can be broken. And since we are not superheroes, avoiding UV-induced DNA damage might be the best way to stay young looking; as the Pulitzer Prize–winning *Chicago Tribune* columnist Mary Schmich famously recommended to the Class of '97: "Wear sunscreen."[1]

To fix what's been damaged, there are complexes and proteins that are dedicated to fix each type of damage: mismatch repair, base excision repair, and nucleotide excision repair to fix damaged parts to the DNA, while nonhomologous end joining and homologous recombination use the DNA sequence itself to make repairs.[2] Moreover, cells will pause at specific times, called "cell cycle checkpoints," to slow cell division and allow time for DNA repair. I'm not going to explain here all the differences in each mechanism of DNA repair, but you can tell how important perfect DNA replication is by the fact that cells have dedicated so much energy and evolved so many mechanisms to repair virtually any type of error it encounters.

DNA Repair, Aging, and Progerias

For a long time, it was thought that the *real* problem in aging was the damage that ROS caused to DNA. The "DNA damage theory of aging" posits that unrepaired DNA damage is *the* cause of aging, and of course there is some experimental evidence to support this viewpoint, particularly in the examination of aging diseases that have their roots in DNA or nuclear functions.[3] Several rapid-aging diseases (progerias) are caused by defects in DNA repair mechanisms,[4] while other progerias are associated with changes in the nuclear envelope and chromatin accessibility. While loss of DNA repair mechanisms can cause variable phenotypes, many of these defects result in phenotypes that at least partially resemble rapid aging.[5] Whether one can equate progerias with normal processes in aging is debatable, but the study of these diseases does shed light on processes that are necessary to keep a cell functioning.

DNA repair is required constantly to maintain normal health and function, and a few progerias are due to severe defects in particular aspects of DNA repair. Werner's syndrome is one example of a rapid-aging disease that affects adults, causing prematurely gray hair, skin aging, atrophy, and atherosclerosis, usually starting in the individual's 20s. The WRN protein is a RecQ helicase (DNA unwinder) that is required for repair of double-stranded DNA breaks and possibly oxidative DNA damage. Bloom syndrome is another disease of the RecQ helicase family (RECQL3). Similarly, Cockayne syndrome (aka Neill-Dingwall syndrome) is caused by a defect in an important DNA repair gene, ERCC8, which is necessary to fix damaged induced by UV or free radicals. (ERCC8 has also appeared in some GWAS of exceptional human longevity and exceptionally long-lived animals, you might remember from previous chapters.) UV rays are so damaging to patients with xeroderma pigmentosum (XP) that they cannot go out into sunlight; XP causes defects in the ability to repair nucleotide excision, another type of DNA damage. Several of these disorders also have links to mitochondrial dysfunction and NAD^+ depletion.[6]

Perhaps the most severe progeric disease is associated with problems other than DNA repair, but likely affects nuclear integrity, cell division, and gene expression. Hutchinson-Gilford (HG) progeria causes rapid aging very early in life, and the kids who have it usually do not live beyond their early teens, primarily owing to problems with their cardiovascular systems. A change of a GGC to GGT in in the LMNA (lamin A) gene results in the production of a truncated form called "progerin."[7] This truncated protein causes morphological abnormalities of the nucleus and problems with properly packing DNA into heterochromatin. Progerin expression has also been found to be a biomarker of aging in normal skin cells,[8] suggesting that its role in aging is not only evident in HG progeria. Recently, there have been two breakthroughs in HG progeria research: David Liu's group (Broad Institute/MIT) used an adenine base editor to correct the dominant negative (C-T) mutation in LMNA in cultured patient fibroblasts and in transgenic mutant model mice, which improved lifespan and vitality in the mice.[9] Another group used CRISPR/Cas9 to both express progerin to model HG progeria and then to restore expression of lamin A in cardiomyocytes later in life; this restoration of expression appeared to benefit the animals, perhaps offering a treatment for this severe disease.[10] The possibility that there could finally be a treatment for this devastating disease is heartening.

Each of these rapid-aging diseases is known as a "segmental progeria" because patients display some, but not all, characteristics of normal aging. Strangely, even though progerias do have a general characteristic—defective DNA repair—in common, each progeria is distinct, affecting specific tissues and at different times in life. While there is no doubt that DNA repair and cell regeneration are important in cell quality maintenance with age, the relationship between progerias and normal aging is less clear.

Stem Cells: A Renewable Resource—at Least for a While

DNA repair is only one of the cell's quality control maintenance mechanisms. Another strategy to never get old would be to simply replace damaged cells with brand new ones. With this approach, you could not only avoid getting older but also quickly heal wounds and regenerate tissues, just like Wolverine (but hopefully without all the military experimentation). Some animals already do this brilliantly. For example, you might remember the planaria flatworm from your high school biology class; you can cut these little critters into pieces, and the pieces will grow into whole new animals. The reason they can do this is that they have specialized cells called "stem cells"—cells from which all other cells can develop—sprinkled throughout their bodies, so they can continuously regenerate their cells, building new tissues.

Extremophiles: Nature's Superheroes

It turns out we already have a few superheroes in our midst—they're just not the size we're used to seeing on the silver screen. Extremophiles are organisms that can live in the harshest conditions on earth—deep ocean thermal vents, highly alkaline and saline lakes, acidic and sulfuric volcanic vents, metal-laden mine shafts, the high pressure of the deep Mariana Trench, extremely desiccating desert conditions, and even hypergravity. One such extremophile, *Thermus aquaticus*, revolutionized molecular biology: its heat-stable polymerase, called Taq, allowed the first cycling PCR reactions to be done without intervention (because the cycles of high temperatures to denature the DNA don't kill Taq polymerase, unlike polymerases from normal bacteria)[a]—we rely on variations of this process for almost all modern cloning and sequencing approaches today, including the PCR test for Covid you might have taken. You are also familiar with bacteria and archaea that are salt tolerant, since halophilic archaebacteria produce the pink color of expensive salt crystals.

These extremophiles are inspirational in a superhero sense: they have adapted to withstand all kinds of conditions that would kill other organisms. For example, the bacterium *Deinococcus radiodurans* has figured out not only how to survive desiccation but, most impressively, how to avoid mutations from UV and ionizing radiation it might be exposed to in space or at high altitudes.[b] It protects its DNA similarly to how your backup hard drive works, by making multiple copies of its genome (4 when stationary, and up to 10 times that many when dividing) and recombining it constantly so that the information is never lost, no matter how many DNA breaks it endures. (Ironically, *D. radiodurans* may have evolved the ability to not evolve.) After extensive testing on earth under a vacuum and with radiation, the ability of *D. radiodurans* to survive in space was tested outside the International Space Station, and it was found that even after three years, it could still be revived! This appears to be thanks to the combination of its DNA recombination and building outer layers of cells that protect inner cells. Some species of cyanobacteria use a similar survival mechanism, building protective layers of cells, to survive on mountains exposed to intense solar radiation. DNA transfer via vesicles

Continued

Extremophiles: Nature's Superheroes (*continued*)

between individuals and subsequent recombination is likely one mechanism for DNA repair in some thermophilic archaea species,[c–e] yet another way to solve the problem.

But my favorite extremophile—besides *C. elegans* dauers, of course—are the tiny, multicellular, eukaryotic tardigrades (aka "water bears"), perhaps the cutest microscopic stress-resistant superheroes. You might remember them as the pudgy swimmers Ant-Man saw as he was shrinking down to the "quantum realm." In real life, these microscopic animals live on moss and have four pairs of legs that allow them to walk slowly (hence their Latin name) and can survive an amazing range of conditions: ice, boiling water, desiccation, high salt, UV, and X-ray irradiation. They do this by expelling their water and forming a metabolically inactive, highly protected cryptobiotic state, the "tun" (kind of like a *C. elegans* dauer, but even tougher), which can last years. Like *Deinococcus*, tardigrades have also survived in space—after being exposed to outer space and direct sunlight, cryptobiotic tuns revived when they were rehydrated. It's not great that in 2019 an Israeli lunar lander that was carrying a digital archive of human knowledge layered with DNA and tuns crashed on the moon; it may have contaminated the surface with tardigrades—but the tuns would need water to revive, one assumes.

Some extremophiles can survive decades or more in their protected states. The record holder is a spore-forming, salt-tolerant *Bacillus* bacterium extracted from a *250-million-year-old* salt crystal from the Permian Salado Formation, which is below a deep mine shaft in Carlsbad, New Mexico. Amazingly, this ancient *Bacillus* recovered from its cryptobiotic spore state.[f] The survival and recovery of such an ancient organism evokes images of possible interplanetary survival. Might extremophiles have led to our presence here on Earth? At least some of them could survive a harsh meteoric trip between planets.[g–i] Even if this is not the case, we can learn some molecular lessons from these extremophiles that might help us better survive aging, or at least climate change.[j]

Differences in regeneration rates may contribute to differences in lifespan, and some species with extreme lifespans might live long by increasing regeneration. For example, many species of the freshwater polyp *Hydra* use this strategy, essentially "escaping aging" by continually shedding their cells and renewing themselves through their three separate stem-cell types[11]—although other sea creatures use other mechanisms to maintain tissue homeostasis with age.[12] Of course, it is legitimate to ask when this individual ceases to be itself, but rather becomes a clone of itself, if all of its cells have been replaced over time, like the planks of Theseus's ship. The absolute Benjamin Button of the animal world, the "immortal jellyfish" *Turritopsis dohrnii*, takes this concept to the extreme by becoming adult and then returning to a juvenile stage, a cycle that could go on indefinitely. While most of us would prefer not to return to middle school in order to live longer, or to have to generate a whole clone of ourselves,* cell regeneration is a great strategy to keep tissues youthful.

Stem cells are not just for weird marine life and superheroes; they are critical for healthy function in normal adults throughout our lives. Understanding the molecular mechanisms these organisms use to reverse their development, and then figuring out how to harness the regulators of this process in key cells could help us repair some of our most age-susceptible tissues. In particular, our skin, liver, gut lining, and blood absolutely require regeneration for proper function—and once they stop functioning, there are no more replacements. You have probably already heard about stem cells and the promise that they may help us live longer (and, ridiculously, are falsely invoked in "anti-aging" skin creams and other dubious products), but what do they normally do?

First, there are several different kinds of stem cells. Germline stem cells, as we have already discussed, can give rise to the whole animal, and are the precursors of sperm and oocytes; they must remain pristine to avoid having mutated progeny. Embryonic stem cells (ESCs) are similar, but usually we are referring to cells taken from human embryos specifically. ESCs are important because they "pluripotent"—that is, they can develop into any type of tissue, and therefore are critical for research into many different diseases, which is why they are often in the news. Peripheral "somatic" (non-germline) adult stem cells can divide and differentiate into certain specific cell and tissue types. Stem cells divide in two ways, asymmetrically (like yeast cells) to make the daughter cells that will go on to maintain and repair tissues, and symmetrically to make their own future stem

---

* Or maybe not, judging from the ongoing success of Tom Brady in *Living with Yourself* (Netflix, 2019).

cells. Serious diseases arise when any of these stem cells have difficulty functioning. We need them constantly in our bodies to help heal wounds, regrow liver after damage, and replenish our blood cells. Neural stem cells (NSCs) are thought to be critical for maintenance of neuronal and cognitive function. With age, NSCs become less able to form the precursors of neurons, called neuroblasts, and instead shift to making more astrocytes, non-neuronal support cells. As rates of neurodegenerative diseases rise, so has the interest in maintaining pools of NSCs.

A super-fancy type of stem cell has been engineered, called "induced pluripotent stem cells" (iPSCs), which start out as regular, post-mitotic, differentiated cells and are then genetically manipulated to turn them back into stem cells. In 2006, Shinya Yamanaka's lab showed that introducing just four transcription factors—Oct4, Sox2, Klf4, and c-Myc, or OSKM, now known as the "Yamanaka factors"—into mouse and then human adult or embryonic fibroblasts was sufficient to cause the cells to become pluripotent; that is, they could then be differentiated (developed into) to other cells.[13] These iPSCs can then be re-differentiated into a cell type we want to study. In addition to obvious clinical applications, another benefit is that we can take, for example, skin cells from a patient with Parkinson's or another disease, and then differentiate those cells into neurons for experiments, without having to take the patient's neuronal cells directly.[14] These types of cells are good for personalized medicine because they should have the same genetic issues as the individual whose cells they were created from, and can then be studied and possible therapeutics explored, while also avoiding ethical issues surrounding ESCs.

Excitingly, there is now evidence that one might be able to "reprogram" cells to maintain youthfulness by expressing the Yamanaka factors, essentially resetting the aging clock, at least for that specific cell type.[15] Juan Carlos Izpisua Belmonte's lab showed in 2016 that cyclic induction of the expression of these Yamanaka factors in a mouse model of premature aging (lamin A) reduced aging phenotypes and extended lifespan, and partial "reprogramming" with these factors improved aging hallmarks in both mice and in human cells.[16] This study was particularly exciting, as it suggested that even older animals could benefit from reprogramming. Since this work in 2016, partial reprogramming by cyclic or short-term expression of OSKM or OSK has demonstrated "rejuvenation" in chondrocytes and muscle stem cells. Reprogramming might also be possible in neurons, at least peripheral neurons; expressing three of the four Yamanaka factors in peripheral retinal ganglion cells reversed blindness in a glaucoma model.[17] And at least one part of the brain, neurons of the mouse dentate gyrus, is amenable to reprogramming: expression of Yamanaka factors in mature mice resulted in a slowing of neurogenesis decline and epigenetic changes with age, and appeared to somewhat improve object recognition, a type of memory, in middle-aged mice (10 months old).[18] Whether these improvements will last through aging (at least 24 months in mice) or apply to other cognition tests must still be shown. In any case, the likelihood that stem cells might offer some sort of realistic treatments has moved from science fiction a bit closer to reality in recent years, although the possibility that tumors might be induced must also always be carefully monitored.* This is a fast-moving field, in both academic and biotech settings, so its applications to aging may be seen soon.

The Hayflick Limit and Telomeres

Stem cells don't last forever. For example, we are born with about 20,000 blood stem cells ("hematopoietic stem cells," or HSCs), and we're using about 1000 of them at any point to renew the supply, with a self-renewal rate of every 25–50 weeks. But those clones will accumulate somatic mutations; once they have too many mutations that render the DNA damaged, or their telomeres have been shortened too much, those cells will die, reducing the usable pool of stem cells. When a 115-year-old Dutch woman named Hendrikje van Andel-Schipper died in 2005 (the oldest human at the time of her death), she donated her body to science, so that we could ask what helped her live so long. A study of mutation allele frequency revealed that, at the time of her death, about two-thirds of her white blood cells could be traced back to *only two stem cells!*[19] Since we start with about 20,000 of these HSCs, it's clear that we can exhaust this supply.

For many years, it was thought that cells could be cultured in dishes (in vitro) indefinitely, that one could just keep splitting the cells, give them more nutrients (usually in the form of a serum derived from an animal), and they would continue to thrive. But in 1961, a researcher at the Wistar Institute in Philadelphia challenged this notion. Leonard Hayflick tested the theory that vertebrate cells are immortal and can divide forever: by carefully tracking cell divisions, he discovered that mammalian cells can divide only a certain number of generations (40–60 times) before they cannot divide any further, known as "replicative senescence." This was a controversial notion at the time, as cells

---

* Somatic stem cell therapy for the treatment of joint pain is already a thriving industry, albeit not yet FDA approved. The most common approach takes adipose tissue from the patient, extracts stem cells and serum factors, and injects this mixture into the patient's joints. More controversial is the use of placental stem cells, which is also offered at some clinics.

were thought to be able to divide indefinitely, but Hayflick showed that the misperceived immortality was likely due to contamination from the serum used to feed the cells—more cells were accidentally being added from time to time. (Oops.) In reality, if not contaminated with new cells, the original cells will lose their ability to grow and divide, become "senescent," and die.[20]

This "Hayflick limit" was later found to be due mostly to the shortening of telomeres, the very ends of chromosomes. It's easy to imagine how you can replicate DNA faithfully for a long strand of DNA, until you get to the ends of the chromosomes, then you might need to use a different approach. This is because DNA replication requires a complicated machinery that relies on information from both strands beyond the site of replication; with each round of replication, the replicated portion gets shorter, known as the "end-replication problem." If you can't copy out until the very end, then your chromosome will get shorter with every round. Each time a cell divides, it must replicate its chromosomes, but this gets difficult at the tips, so it loses a bit more of the chromosome end with each division, resulting in an unstable chromosome when the ends become too short. If the shortening ends up eating into a sequence that is important, such as a gene or a regulatory region, then it could cause big problems for the cell. To avoid this fate, chromosomes have special ends called "telomeres"—like those little caps on the ends of your shoelaces, "aglets" (as you crossword puzzle fanatics know)—protecting the ends of the chromosome. This cap is a long stretch of sequences (e.g., CCCCAA) that are repeated thousands of times to create a long (11 kilobases!) region at the end of the chromosome. Elizabeth Blackburn, Carol Greider and Jack Szostak were jointly awarded the Nobel Prize for their discovery of a special enzyme called "telomerase" that is required to renew the repeated sequence cap to chromosome ends, allowing them to maintain their length in dividing cells.[21] Despite telomerase's best efforts, though, the telomeres still get shorter with age; as telomerase levels decline, this protective process slows, eventually resulting in cell senescence. The Hayflick limit is rooted in the concept that as telomerase levels decline with age, this protective process slows, eventually resulting in cell senescence.

Because telomerase activity and telomere length decline with age, telomeres have been posited to be one of the determinants of longevity—but it is not at all clear whether just adding more telomerase or longer telomeres will increase lifespan.[22] In fact, a mutant of *clk-2*, the *C. elegans* Tel2p telomere binding protein homolog, causes short telomeres but is long-lived—but this might also be related to the fact that DNA damage in germ cells might induce systemic stress resistance.[23] Additionally, the variation in the length of human telomeres far exceeds the expected attrition of telomeric sequences during all cell divisions. Furthermore, the levels of expression and activity of telomerase are inversely proportional to lifespan when comparing various mammalian models and humans, making it difficult to draw a straight line from telomere length and telomerase function to long life.

Certainly, however, having dysfunctional telomerase or short telomeres is detrimental for cells. Interestingly, telomeres also shorten with increased stress; Blackburn, Elissa Epel, and colleagues (UCSF) showed that financial strain, socioeconomic status stress, and urban stressors correlate with shorter telomere length in the immune cells of African American men in mid-life, suggesting the possibility of accelerated aging in these individuals.[24] By contrast, leukocytes from individuals in the Nicoya region of Costa Rica, one of the Blue Zones we mentioned earlier, have longer telomeres than the leukocytes from individuals from other regions of Costa Rica, even when controlled for age.[25]

Bigger Is Not Better

Telomere length is not the only thing limiting the lifespan of an old dividing stem cell. Like the lifespans of small dogs and cells with small nucleoli, smaller seems better when it comes to cells that need to keep dividing indefinitely.[26] Both budding yeast and mammalian stem cells show a correlation between larger cell size and decreased proliferative potential—that is, the bigger (and older) cells get, the less proliferative potential they have. While this might seem counterintuitive, stem cells are generally small, while senescent cells are large.[27] Angelika Amon's lab (MIT) showed that HSCs grow larger when there is DNA damage, causing cell cycle arrest, but protein production and growth continue.[28] One result of this increased cell volume is that, as RNA and protein don't scale, the cytoplasm is essentially diluted. Not surprisingly, the size of the nucleolus—the membraneless organelle where proteins are made, and whose size also is inversely correlated with lifespan[29]—also increases in stem cells with age and with DNA damage. Of course, the question is: If HSCs function better when they are small, will forcing them to be small make them function longer? Surprisingly, keeping them small does indeed maintain HSC function; rapamycin treatment to inhibit mTOR (thus blocking protein synthesis and growth) prevented radiation-induced HSC growth (without fixing the DNA damage) and partially restored proliferative function—but rapamycin could not reverse HSC enlargement in old cells, suggesting there is a limit to this effect.[30] Inhibiting the entry into the DNA synthesis phase of the cell cycle (S-phase) pharmacologically also improves HSC function, so perhaps some combination of these treatments might slow stem-cell exhaustion.

Why would cells not simply have evolved a mechanism to make them functional forever? First, if the most severe deficits occur post-reproductively, or if there are benefits for development that don't have costs until late in life (remember antagonistic pleiotropy?), there may be no evolutionary pressure to make cells survive much longer, as we have previously discussed. The energy necessary to perpetually combat DNA damage could be used for other functions, and is necessary only until the animal has properly reproduced. Another possibility is that loss of stem cells is coupled to levels of unrepaired DNA damage. Despite the function of many different mechanisms of DNA repair, a certain level of DNA damage may be unavoidable after many cell divisions. If DNA is not repaired, but cells continue to divide, those damaged cells might wreak havoc—in particular, they might cause cancer. (Turns out that copying copies over and over can lead to some problems.) Although we think of senescence as a bad thing, it may in fact be necessary to kill cells that have accumulated damage. Therefore, cellular senescence may be thought of as a major tumor suppressor mechanism, preventing cancerous cells from running amok. Boosting levels of telomerase may hold the key to the maintenance of regenerating tissues, but since telomerase would not help with DNA repair, it must be carefully balanced against the risk of promoting cancer. Perhaps doing so while slowing the cell cycle to allow repair and slowing protein production would be a winning recipe.

SASP: The Senescence-Associated Secretory Phenotype

Normally, cells are able to repair their DNA because they go through several "cell cycle checkpoints"—if DNA damage is detected, cell division is slowed to allow time for the DNA to be repaired. But the accumulation of unrepaired damage can cause a shift into a different state, called cell senescence. The presence of different kinds of DNA damage, such as short and dysfunctional telomeres and chromatin instability, as well as stress signals, like high levels of the tumor suppressors p53 and p16^{INK4a}, induce cell cycle arrest that can lead to cell senescence. In addition to being damaged and dysfunctional themselves, senescent cells also begin to secrete factors that are not produced by normal, healthy cells. Judith Campisi (Buck Institute) first recognized this "senescence-associated secretory phenotype," or SASP.[31] A pathway involving proteins called cGAS and STING has been found to be critical for the activation of

SASP, through the detection of cytoplasmic chromatin.[32] This makes sense—DNA shouldn't be floating around outside of the nucleus, and seeing it in the cytoplasm is a pretty good signal that something is very wrong. The activation of LINE1 retrotransposons, which are related to ancient viruses and are normally silenced by the proteins RB1 and FOXA1, is also an indication that things have gone awry.[33] (I'll note that while most of this chapter is about mammalian cells, DNA damage can also trigger systemic, cell-nonautonomous effects in *C. elegans*, even though the cells are not dividing; the response to this DNA damage can increase general stress responses and increase lifespan as well.)[34]

Senescent cells not only lose their ability to divide, they also take on specific characteristics that are different from healthy cells. Senescent cells are marked by changes in their morphology, increased "ploidy" (they have lost control over regulation of chromosome number, which is disastrous), and they are less able to resist stresses. Senescent cells are metabolically active but dysfunctional, changing the structure of their chromatin, how their mitochondria function, the patterns of gene expression, and the set of proteins that they secrete. Moreover, the shift into senescence appears irreversible. If senescent cells just died, that would be one thing, but they damage other cells on their way out (like Crossbones trying to kill Captain America with his own bomb vest). Senescent cells can be toxic to other cells because of what they secrete to their surrounding tissues, exacerbating the effect of aging.[35] Extracellular vesicles can carry miRNAs and other cargo that are part of the SASP.[36] Several biomarkers are associated with senescent cells; these include inflammatory regulators, such as IL6, as well as β-gal, metalloproteinases, and a decrease in lamin B expression.[37] The SASP response induces the secretion of growth factors, interleukins, inflammatory cytokines, chemokines, IGF-binding proteins, matrix metalloproteinases, extracellular matrix proteins, and even nitric oxide (NO) and ROS—it's a whole nasty milieu. The increase in inflammation and inflammatory factors by senescent cells has led to the concept of "inflammaging." While we normally think of our immune system as helping us survive, inflammaging describes the runaway inflammatory responses that rise with age and eventually damage tissue. Senescent cells can also promote tumors by increasing proliferation, cell motility, and differentiation of neighboring cells (epithelial)—undermining the whole point of dying in senescence to prevent tumors.

Since senescent cells are so terrible, it might be a good idea to get rid of them. Using markers of senescent cells to pick them out of a crowd and then delivering drugs to eliminate them is a strategy that has emerged.[38] Promising work from the labs of Judith Campisi, James Kirkland, and others on drugs that either kill (senolytic) or modify (senomorphic) senescent cells suggests that aging could be slowed, or some of its more debilitating side effects be blocked, by using such drugs (see chapter 17).

Stem Cell Renewal via Parabiosis:
Vampires and Blood Boys

Perhaps the strangest mechanism to extend lifespan and maintain youthfulness truly reeks of science fiction: the process of connecting animals so that they share a circulatory system, known as "parabiosis." Paul Bert, a French physiologist (and later diplomat in French colonial Indochina) was interested in how gases were dissolved in blood, for example, at high altitudes and in deep seas—and for this work, he is regarded as one of the fathers of space and deep-sea exploration and hyperbaric treatment.[39] In 1864, Bert stitched two rats together, and observed that their circulatory systems merged—the first parabiosis experiment.[40]

The fascination with using blood to stay young has a dark history that is reflected in fiction. Although Bram Stoker's *Dracula* (Archibald Constable, 1897) drank blood to rejuvenate himself, it's probably even more relevant that blood transfusions had just been coming into use when Stoker wrote his book, as the method played a prominent role in the narrative. (Blood groups were not discovered until 1901 by Karl Landsteiner, which is probably why the character Lucy was able to receive several transfusions from her suitors without dying.) Blood treatments caught the attention of a Russian Bolshevik, Alexander Bogdnanov, who was such a fan that he included the technique in a sci-fi novel about a communist society on Mars called *Red Star* (Saint Petersburg, 1908), and he even founded his own institute on transfusions. (Perhaps unsurprisingly, he himself died from a botched transfusion.) The modern reinvention of parabiosis is no less popular and has caught the public's eye—so much so that when rumors got out about investments in Ambrosia, a now shut-down company that transfused young donor blood into patients for a hefty fee, *Silicon Valley* soon aired an episode depicting a CEO in his living room transfusing blood from a young athlete that he called his "blood boy."[41]

About a century after Bert's parabiosis experiments, Clive McCay (you remember him from his pioneering work on dietary restriction, chapter 7) carried out the first parabiosis-based experiments to study aging, joining young and old mice, known as "heterochronic parabiosis." Isochronic parabiosis means joining animals of the same age, and is a good control for the effects of the joining procedure itself. McCay found that the old mouse appeared younger—and, sadly, vice versa. In 1972, Ludwig and Elashoff (University of California, Berkeley) showed that heterochronic parabiosis increased the life-span of older rats by more than four months.[42] This Frankenstein-like approach was abandoned for many decades because the animals died of "parabiotic disease," but was recently revived to study stem-cell regeneration, with the hypothesis that the blood of young animals might help simulate regeneration.

The parabiosis renaissance in research can be traced to the early 2000s at Stanford University, where Amy Wagers had started to use the technique as a postdoc, taught by her advisor, Irv Weissman.[43] This might not have happened if Weissman hadn't been taught the technique as a teenager the 1950s while working with a pathologist in his hometown. Wagers studied the effect of heterochronic parabiosis on rejuvenation of blood stem cells. Later, postdoctoral colleagues Irina and Michael Conboy in Tom Rando's lab nearby started to use the technique to study muscle and liver regeneration.[44] They already knew that aged muscle could be repaired when grafted onto young muscle, but not vice versa, suggesting that there must be a systemic factor that allowed the repair, but that it changed with age. Sure enough, heterochronic parabiosis enabled muscle repair after wounding, through the activity of resident progenitor cells. They went on to show that serum from the young cells was sufficient to recapitulate the effect. Liver could be rejuvenated through a similar approach. Since then, the field has mushroomed, with the discovery of circulating factors that appear to maintain stem-cell divisions, muscle function, and even brain activity.

In all of the studies, it became clear that systemic circulating factors were responsible for these rejuvenating—and aging—effects. While heterochronic parabiosis was how blood factors were originally found to help tissues regenerate or stay young, now there is less of a need to actually stitch two animals together. (Still, a 2021 study used long-term [three month] heterochronic parabiosis, then measured various parameters two months post-detachment, and compared the effects with longevity interventions.)[45] Instead, transfusion of plasma from a young animal to an old animal mimics the effects of joining the animals; experiments can be carried out that are less taxing on the animals than actually stitching them together and forcing them to share a circulatory system, although obtaining the plasma can be difficult.

Studying plasma allowed investigators to start to drill down on the factors that might cause rejuvenating effects. Mass spectrometry of the plasma factors can reveal the differences in proteins and peptides between young and old animals. Notch signaling and TGF-beta signaling, both already known to be involved in growth regulation, were identified as circulating factors that could affect stem-cell maintenance in different tissues with age. Work from Amy Wagers and Richard Lee suggested that a factor called GDF11 could help regenerate aging heart muscle, reversing age-related cardiomyopathy.[46] The work, although somewhat controversial (you don't see the word "reverses" in the title of many papers), had obvious implications for treating cardiovascular disease, a major factor in age-dependent mortality. Other factors that were already known from previous aging research also appeared and were identified as blood-borne factors that affect stem-cell regeneration—these included oxytocin, the hormone that is known for its role in maternal attachment, and Klotho,* an activator of FGF21 that was implicated in mammalian aging several decades earlier.[47] Part of the controversy in the field was likely rooted in the excitement about the possibilities for treating aging and neurodegenerative diseases, and several companies, such as Alkahest, have been formed based on discoveries from heterochronic parabiosis research.

Neural stem cells are another hot area of study, with the reasoning that regeneration of this cell type might be critical for the maintenance of cognitive function with age. Saul Villeda, then a graduate student with Tony Wyss-Coray and Tom Rando (Stanford), used heterochronic parabiosis, and then blood plasma, to show that adult neurogenesis could be affected by blood-borne factors. This is a two-way street—when young mice are exposed to plasma from old mice, it reduces their synaptic plasticity and causes deficits in several forms of learning and memory.[48] Using proteomics and subsequent testing, the team identified six factors that were up in old unpaired and young heterochronic mice. One of these factors, CCL11, also rises with age in the plasma and cerebrospinal fluid of humans, making it a great candidate to test. When they injected CCL11 alone into young mice, it caused them to have poor memory, and anti-CCL11 blocked the effect. The results strongly suggest that circulating factors in blood affect NSCs and memory, and may even damage the memory of a recipient, a fascinating idea. More excitingly, later work

---

* Klotho was named after the Greek goddess of fate who spins the thread of life. I didn't expect to hear the word "klotho" on TV ever, but in a Netflix film about a band of immortal warriors—more superheroes, or I guess antiheroes—called *The Old Guard* (2020), the antagonist was a young, evil CEO who was anxious to get the warriors' blood for his longevity company.

from Villeda and Wyss-Coray suggested that factors from young mice could *reverse* age-related cognitive impairment in old animals, improving memory through activation of CREB, one of the major transcription factors that function in memory (see chapter 14).[49] Wyss-Coray's lab also identified proteins present in human umbilical cord plasma that can improve brain function in aging mice. They found that injecting a protein called TIMP2 (for tissue inhibitor of metalloproteinases 2) increased synaptic plasticity and improved hippocampal function and memory in old mice, and specifically depleting TIMP2 blocked this improvement.[50] Villeda's lab at UCSF later showed that exercising mice produce blood factors that can be given to couch-potato mice to improve their memory.[51] These studies of course have huge implications for what we might be able to learn from blood-borne factors, and how we might improve our chances of maintaining our health and cognitive function with age—not to mention giving hope to the sedentary public, at least to not suffer from our laziness.

A few years ago, Saul Villeda and I were both invited to present our work at the Sun Valley Conference, aka "Summer Camp for Billionaires." (You'd recognize the pictures: every time one of these rich tech guys is in the news, he is wearing a vest and a Sun Valley ID badge from this meeting, because as you walk through the entrance, photographers are snapping pictures of the CEOs—not so much of us scientists.) Saul gave a great presentation about his work to preserve cognitive function in older mice by giving them the blood of younger animals. He explained to the rapt audience that by providing the older animals with factors isolated from young blood, he was able to keep the older mice functioning better with age in several different learning and memory tasks. The implication is, of course, that there are factors in blood that change with age, and if we could determine what these are, we might be able to give them to people to help them retain their cognitive abilities—and this was not lost on the audience. Later that evening, I was sitting next to a tech CEO at dinner and we were chatting about the day's talks. He looked around the room of billionaires and commented about Saul's talk, "We all watch *Silicon Valley*. . . . Why didn't he just say, 'Blood Boy'?"

———

As crazy and sci-fi as some of the newest longevity research directions may seem, from killing our own dying cells to transferring blood factors to old animals, they offer new potential directions to treat various aspects of aging and other diseases. CRISPR may someday be used to correct inborn errors that affect metabolism and DNA repair so that progerias might become a thing of the past; renewable sources of stem cells could help keep our tissues working longer; factors first identified in blood may help slow aging and neurodegenerative diseases; and drugs to block senescent cells from becoming toxic to their neighbors, or eliminate them altogether, might slow aging. But remember that cell senescence likely arose to prevent runaway cell division—so at least some caution is necessary as we move forward, to make sure that we don't cause cancer in our zeal to avoid mortality. Regardless of the risks, the huge developments in the areas of stem cells and blood factors are making it more and more likely that we will all benefit from some of this tech in our lifetimes, even if we can't all become superheroes.

11

Use It or Lose It

REPRODUCTIVE AGING, THE GERMLINE, AND LONGEVITY

Listen, the best advice on aging is this: What's the alternative? The alternative, of course, is death. And that's a lot of shit to deal with. So I'm happy to deal with menopause. I'll take it.

—WHOOPI GOLDBERG, *NEW JERSEY MONTHLY*, MAY 2013

WHEN I DECIDED TO DO my postdoctoral work on aging, it occurred to me that even if I could live to be 150, it wouldn't change the most pressing decision in my life: when I would have kids. Women start to have reproductive problems in their mid-to-late 30s, with rates of infertility, miscarriage, and birth defects rising with age. This "biological clock" and its accompanying stress is a recurring, unavoidable pop-culture theme, woven throughout movies, chick lit, and panic-inducing media stories. The biological clock can be a harsh governor of young women's lives; while everyone knows that our time to have children is limited—and if they don't, their mothers, friends, and *Newsweek* are happy to remind them—they have no easy way of knowing how fast their own eggs have aged.

It may seem odd to discuss reproduction—a characteristic that declines well before middle age—in a book about aging and longevity, since when we think of aging, we usually think of problems that affect people—and, let's face it, mostly men—in their 70s and beyond. Likely for this reason, at least when I started in the field, the aging research community rarely discussed reproductive aging, focusing instead on the more obvious signs of late-life aging, such as cognitive decline, obesity, metabolic disorders, and cardiovascular disease. (This problem has recently started to change, but reproductive aging has been largely ignored as an aging phenotype.) In fact, female reproductive aging in the mid-30s is the first significant sign of aging in humans, and I would argue that we should pay closer attention to it.

The increase in infertility, miscarriage, and birth defects that women experience beginning in their mid-30s comes well before any other obvious signs of aging are present. While menopause has been studied in some detail, it is not generally appreciated that reproductive aging starts about 15 years before the onset of menopause, suggesting that the two processes may be uncoupled. Reproduction itself is a key output of the nutrient-sensing pathways that I described in previous chapters, such as insulin signaling and dietary restriction, and thus is intimately connected to longevity regulation. Additionally, historic data show that women who can reproduce later in life are most likely to live exceptionally long, suggesting that longevity mechanisms may be set in motion far earlier than we have appreciated. Therefore, studying reproductive aging might not only shed light on ways to extend healthy reproduction but also provide biomarkers of longevity.

And before you ask, "What about men?" let's be clear that women's reproductive window is smaller than men's. As Jeffrey Kluger noted in *Time* magazine in 2013, "There are a lot of downsides to being male. We age faster and die younger. But give us this: we're lifetime baby-making machines. Women's reproductive abilities start to wane when they're as young as 35. Men? We're good to go pretty much till we're dead."[1] Not entirely true, but the slope of reproductive aging is far steeper for women.

In this chapter I will discuss the surprising connections between reproductive aging and the regulation of longevity, and what that might mean both for modern motherhood and total lifespan.

———

When Boston University gerontologist Tom Perls and his colleagues published a study of a mother's age of last childbirth and her eventual longevity, it sparked headlines such as, "To Live Long, Have Your Kids Late!" Perls was quick to disabuse readers of the mistaken notion that having children late will somehow prolong one's life—the cause-and-effect correlation was misinterpreted by most media outlets. Of course, these data do *not* suggest what the newspaper summaries kept writing, that "having kids late makes you live longer!" In fact, to avoid such an erroneous conclusion, one study even warned, "our findings do not imply that intentionally delaying childbearing will increase the likelihood of living to age 90 years and do not support delaying childbearing, given the complications associated with older maternal age."[2] So don't wait to have kids, or start having kids, in order to live long. The real result is no less interesting, though: using historical data from the Long Life Family Study (see chapter 3), Perls's group found that women who had—that is, were *able* to have—children after the age of 33 were more likely to live to 95 than women whose last child was born before the age of 30.[3] Similarly, by examining records of women in the New England Centenarian Study cohort (Boston suburban centenarians) born in 1896 and comparing them to women born in the same year who had died by age 73, Perls's group found that those women who gave birth after the age of 40 were four times more likely to live to 100.[4] Obviously, women back in the early 1900s didn't have access to assisted reproductive technologies or great birth control methods, so these births were not the result of artificial insemination or in vitro fertility treatments; because of this, maternal age at last birth in that era is a reasonable measure of reproductive ability at that age. That is, if we view successful childbirth as a healthspan biomarker, we can see that women who are able to have kids (naturally) later in life are more likely to live longer; being able to have kids late, therefore, might be correlated with other healthspan metrics, as well as long life. To be clear, rather than being caused by *having children*, being *able to* give birth late in reproductive life may be a sign of slower aging. In short, late-life childbearing may be a biomarker—in fact the earliest biomarker—of future extended longevity. The first sign of what might eventually end up as a really long life is the ability to have kids later than normal.

If you are a parent, it might not be a great surprise to you to hear that beyond two or three kids, having more children (known as increased "parity") steadily *decreases* a woman's lifespan. Having no more than two children seems to be optimal for future lifespan, an observation that Perls made, as did Tom Kirkwood's group later, in an analysis of the lifespans of British aristocracy (and that you might agree with if you drive your kids to a lot of afterschool activities).

Molecularly speaking, why might there be an association between long life and late motherhood at all? The correlation between successful late childbirth and longevity suggests not that one causes the other but rather that there are underlying systemic reasons that allow both. Following up on their earlier findings, Perls and colleagues examined the length of telomeres in leukocytes (immune blood cells) of women over the age of 70 who had children after the age of 33, and found that their telomeres were longer than those of similarly aged women who had had their last child before age 29.[5] But this was not an unbiased candidate analysis, it simply asked whether one particular marker, telomere length, is significantly different in the two sets of women. Telomere length might not be the only difference and might simply be correlated with or a downstream result of a systemic difference between women who are destined to live long (and reproduce late) and their normal-lived, normal-reproducing counterparts. Figuring out how women with long reproductive spans are different from their peers might be a great way of predicting who will become centenarians.

———

Several years ago, I was at a Gordon conference on fertility to present some of my lab's work on reproductive aging, and I found myself at a table full of IVF (in vitro fertilization) doctors. Because I am interested in the issue of reproductive aging biomarkers and predictors, I took the opportunity to ask them, "Right now, how far in advance can a woman know how long she can still have children?" They all cheerily responded, "Months!" I was shocked that they were so satisfied with this answer, since "months" is a fairly useless timescale if you are trying to plan when to have a family. To be fair, the IVF doctors noted that they usually see patients only when fertility has become a problem, so in many cases it's already too late for such long-term predictions to be useful. But for most women, knowing how many childbearing years they have left would be extremely helpful in planning their futures; right now all they can do is worry.

A cynical view of the lack of focus on reproduction in aging research is that the problem of human reproductive aging has been primarily considered to be a concern only to women. In general, I think it is fair to say medicine and science have not focused equally on issues of men's and women's health, a problem that the NIH is now trying to correct when it funds clinical trials and studies. When I pressed the IVF doctors further, instead of talking about other methods to address this question, one of the older, male (Who am I kidding? They were all male) physicians told me about his own wife having their kids when she was in her 20s—as if that had anything to do with my question—and he asked me why one would want to delay having children; clearly, he could not even fathom it. This attitude by some experts in fertility may illustrate one reason why effective diagnostics for reproductive aging do not exist yet and have been slow to be developed; if the people who can solve a problem don't recognize it as a problem, then it is unlikely to be solved. The pervasive notion that women have "chosen" to delay having children in order to pursue educational and career goals—so it's their own fault—has certainly not been helpful in pushing reproductive aging research forward. In reality, of course, female reproductive aging affects both women and men, as any couple who have endured months of IVF treatments will tell you, and I think that more people are starting to realize that reproductive aging is a legitimate research focus. In 2019 the NIH put out a call for proposals to address reproductive aging, the first time I have seen such a call since I have run my own lab. Simultaneously, the Silicon Valley venture capitalist Nicole Shanahan, who has been open about her own struggles with fertility, also started a new grant program through the Buck Institute for Research on Aging, called the Global Consortium for Reproductive Longevity and Equality, which focuses primarily on reproductive aging. These funding initiatives are an exciting indication that the issue of female reproductive aging and women's aging issues are finally gaining traction as important research problems.

The big, practical problem for women who want to have children is that at this point there is no effective long-term diagnostic for reproductive aging. We cannot tell a woman how many more years she will be able to have kids; we can only point to a chart of *average* fertility decline and monitor very short-term indicators (think *months* at best) through measurements of specific hormones, such as follicle stimulating hormone and anti-Mullerian hormone (AMH). AMH is a transforming growth factor beta (TGF-beta) ligand, and signals to its receptor, AMHR, to regulate the follicle pool. AMH and imaging methods report on the egg count, or "ovarian reserve," which can tell us if a woman has too few eggs but provides no information about the *quality* of those remaining eggs. Depressingly, though, despite years of AMH being promoted and used as a clinical indicator of fertility and their ubiquity in IVF clinics, a 2017 *JAMA* study reported that in fact those fertility hormones have *no predictive power* for the time it would take to conceive at all.[6] Other studies showed that AMH could not predict fertilized embryo quality or pregnancy outcome, or time to menopause.[7] Together it seems that the reliance on these hormonal indicators in the artificial reproductive technology (ART) field is overly optimistic, even at the months scale; they do not have the power to predict the reproductive span of any individual woman, much less how many years of reproduction she might have left.

So how can we solve the problem of reproductive aging? The first step might be to better understand what causes reproduction to decline with age in the first place. To do that, we should first have a solid understanding of what we are studying when we talk about "reproductive aging" and "menopause." These terms are often used interchangeably in the mammalian aging literature, but of course most women use them to refer to different biological processes that happen at different ages. I will use "reproductive aging" to refer to the decline in oocyte quality that starts in the mid-30s and leads to infertility, and the term "menopause" specifically to refer to the hormonal changes and oocyte depletion that occur as a result of ovarian depletion much later in life. Menopause occurs when ovarian hormone production decreases and menstruation ceases. The average age of menopause in the United States and northern Europe is 51. You'll notice that that is about 15 years later than the onset of loss of fertility that has many women concerned about running out of time to have children. When I first started giving talks about my own lab's work on reproductive aging in C. elegans, I was very confused when a famous male scientist framed his question all around "menopause," which in my mind is so separate in time from reproductive aging that I couldn't even figure out what he was talking about, much less what that meant for a worm. That's when I started adding arrows to my slides to point to "Menopause" at 51 years and "Reproductive Aging starts in the mid-30s" so the audience could appreciate the 15-year difference in these processes. (I have a feeling that most women are already pretty clear about these differences, since they know they will have trouble having kids well before their 40s, and menopause mostly becomes an issue for them much later.)

While it is *possible* that the mechanisms that regulate menopause and oocyte quality are shared or linked, there is no formal evidence yet to explain how the processes might be related. Recent genome-wide association studies (GWAS; see chapter 3) of the age of onset of menarche and menopause have identified many genes associated with time of menopause onset, including genes involved in hypothalamic signaling, DNA repair, and mitochondrial maintenance,[8] but of course such studies are not able to determine *where* the genes might act—in the eggs themselves, or in the surrounding, supporting tissues, such as ovaries, or in another tissue altogether—or *when* they will act. (Remember that GWAS examine static genomic DNA from all cells, rather than examining gene expression—that is, the mRNAs that are made dynamically, or the proteins made from those mRNAs—in reproductive tissues.) Therefore, such genes most likely affect hormone synthesis, but could also have some effects on the production of oocytes in the ovary, or other processes that might determine the time of menopause onset. Like late reproduction, there is some good news on late menopause, despite the increased risk of some cancers: menopause after age 55 is correlated with a 2-year longer life expectancy and lower risk of cardiovascular disease.[9]

There does seem to be a correlation between *early* menopause (onset between 40 and 45 years) and mortality risk—that is, the risk of dying. A 1989 study of Seventh Day Adventists in California included a "Lifestyle Questionnaire," and the authors found that early natural menopause (occurring before the age of 40) is associated with a high rate of mortality.[10] Unfortunately, estrogen replacement use, at least in the medical treatments of that era (1970s and 1980s), did not seem to reduce this mortality. This type of observational study was not able to distinguish correlations from causation, of course, so it is not clear whether early menopause caused high mortality or was a neutral biomarker of short lifespan. In any case, GWAS of early menopause seem to implicate a different set of genes than those associated with normal menopause.

But there is evidence that menopause might be more than just a biomarker of longevity; in fact—terrifyingly—the onset of menopause may actually *accelerate* the rate of aging.[11] Steve Horvath, a geneticist and biostatistician at UCLA, developed an "epigenetic clock" to measure the age of a person using just a small set of methyl modifications on DNA, called the DNA methylation signature.[12] This DNA methylation clock (which I will discuss in more detail in chapter 15), can be used to distinguish biological from chronological age; in brief, Horvath found that biological age is correlated with a set of a 353 DNA methylation events that can be tracked using microarrays or sequencing. Morgan Levine and Horvath asked the question, *Does aging accelerate after menopause, and can we distinguish cause and effect?* They showed that in various non-reproductive tissues, such as blood, saliva, and cheek cells, the DNA methylation marks associated with aging increased after menopause.[13] It would be difficult to distinguish cause and effect with just this observation, since aging effectively causes menopause. However, they did further analyses, looking at the DNA methylation patterns caused by surgically induced menopause, independent of age, and with menopausal hormone therapy (MHT). They found that aging marks accelerated whether the menopause was natural or surgically induced, and reassuringly, the aging clock was slowed by hormone therapy.[14] Regardless of the time of onset of menopause, the time since menopause was associated with DNA methylation age acceleration. The fact that MHT decelerated epigenetic changes suggests that menopause itself is responsible for the acceleration of aging in other tissues. Together, these results argue that reproductive system hormones regulate the rate of aging of non-reproductive tissues—the opposite of what you might expect if aging sets the rate of all tissues independently. Thus, the age of onset of menopause is likely genetically controlled, as GWAS results suggest, *and* the process of menopause itself seems to in turn regulate the rate of aging of non-reproductive tissues.

The concept of reproductive hormonal control of aging and longevity already had support from a set of *C. elegans* experiments done in the mid-1990s. Honor Hsin was a home-schooled student who was interested in aging, and her advisor, Cynthia Kenyon (of *daf-2*/insulin signaling fame, chapter 6), taught Honor how to use a laser to kill cells in developing *C. elegans*. By killing a single pair of precursor stem cells (the "parent" cells that eventually divide to make the rest of the germline), they prevented the development of all of the future germline cells in the adult, including oocytes, sperm, and mitotically proliferating nuclei. Hsin and Kenyon showed that killing these germline precursor cells extended lifespan.[15] If they had stopped there, this experiment would have looked like more support for a "trade-off" theory of longevity— that is, the idea that animals can either invest in reproduction *or* longevity, but not both, aka "resource allocation"—since these animals were both sterile and lived longer. But they did the next killer experiment (pun intended): laser ablating the stem cells that would make the supporting somatic gonad tissue, which also results in sterility because no germline development is supported, restored the lifespan to normal. So even though the animals were sterile in both cases, the effects on longevity were opposite, ruling out a simple trade-off theory at work. It is important to note this refutation of a simple resource allocation trade-off model, which much of the field still seems to not be aware of, even more than 20 years later. Instead of a trade-off, their results suggested that the animals normally produce a "pro-longevity" signal from the somatic gonad that balances an "anti-longevity" signal from the germline.

Kenyon's lab went on to show that the production of the "anti-longevity" signal must come from the mitotically dividing nuclei at the very tips of the germline (as opposed to the fully developed sperm or oocytes), and that it regulates the activity canonical longevity transcription factor DAF-16/FOXO (see chapter 6), as well as new, germline-specific factors.[16] This includes a transcriptional elongation and splicing factor called TCER-1, or TCERG1 in humans, that Arjumand Ghazi's lab (University of Pittsburgh) has found to be important in the balance between pathogen resistance, fat metabolism, and reproduction.[17] The germline anti-longevity and gonadal pro-longevity signals have not yet been identified, but several insulin-like peptides that can act as agonists (activators) and antagonists (inhibitors) of the insulin/IGF-1 receptor are good candidates, as is piRNA/Hedgehog signaling, and David Vilchez's lab (CECAD) identified prostaglandin signals as an important communication mechanism between adult germline stem cells and somatic tissues.[18] Together, these pro- and anti-longevity factors create a balance of "go" versus "slow" signals—the "go" signal tells the rest of the animal that there are enough nutrients, so the germline cells should keep dividing and the animal should reproduce and have a normal lifespan, while the "slow" signal tells it that there are not enough nutrients, so the animal should slow cell division, put reproduction on hold, and slow somatic aging until conditions are better.

For a while, the discovery that the germline and gonad together control longevity seemed to be yet another weird worm thing, and in fact one study in *Drosophila* using sterile mutants did not support the idea. However, this sterile mutant affected not just the germline stem cells but also the supporting somatic gonad tissue. By a decade later, results in both flies and mice supported Kenyon's view of the germline's regulation of somatic longevity. Work in *Drosophila* by Marc Tatar's group (Brown University) showed that germ-cell loss induced by a mutation that does not also affect the somatic gonad (unlike the previous fly study) extends lifespan by 30%–50%, similar to the effect in *C. elegans*.[19] Moreover, germ-cell ablation in both flies and worms requires insulin-like peptides and the FOXO transcription factor to achieve extensions in lifespan, suggesting that the basic mechanisms of germline-mediated longevity are evolutionarily conserved.

These results parallel findings in mice, as well: mice who had their ovaries removed (had been "ovariectomized") prior to sexual maturity and then received transplanted ovaries from mice of different ages had lifespans that corresponded to the remaining lifespan of the ovaries they received—that is, older mice who received young ovaries had the greatest extension of lifespan, while those mice who received age-matched ovaries saw the smallest effects.[20] Thus, young ovaries seem to protect the soma of the recipient and help her live long. The fact that, in women, menopause increased age-related changes and hormone therapy decelerated them fits with these observed lifespan results in mice, suggesting that ovarian hormones may slow aging, or that lack of these hormones causes acceleration of aging. Like the mouse ovarian transplant result and the hormone therapy slowing of menopause-induced aging, Kenyon's germline ablation results suggested that the reproductive system *actively* regulates the rates of aging in the rest of the body, rather than simply responding passively to choose between reproduction and longevity. Instead of a simple resource allocation model in which an animal just *reacts* to current nutrient levels, this model suggests that animals have a plastic response and modulate their longevity based on what the lack of nutrients will mean for their future reproduction: the germline of the animal *anticipates* the effects of nutrient deprivation, and responds accordingly, with signals that then affect how quickly the animal should age. Together, these results argue that reproductive system hormones regulate the rate of aging of non-reproductive tissues.

Do we have evidence that the reproductive tissues of humans might respond in a similar way? One story that frequently comes up when we talk about the fact that loss of germline signaling in worms, flies, and mice extends lifespan is the Korean eunuch paper (but there is some dispute about the numbers).[21] Eunuchs (castrated males) historically served as guards and servants in various cultures and were adopted into royal Korean families to preserve their lineage. A genealogical history of Korean eunuchs called the *Yang-Se-Gye-Bo* (1805) tracked 385 eunuchs, and from these data, Min, Lee, and Park were able to find the lifespans of 81 eunuchs. By comparing their lifespans to those of families of the time (Mok, Shin, and Seo) with similar socioeconomic status, they found that the average lifespan of the eunuchs was significantly longer, at 70 ± 1.8 years vs. an average of $51-56$ years for the non-eunuch males. Moreover, of the 81 eunuchs, 3 were centenarians, a huge enrichment not only for the time (when even male royalty lived only an average of 45–47 years) but even compared with the current Korean population. These data suggest that lacking reproductive tissues and male hormones might extend lifespan.

Use It or Lose It: It All Comes Down to Quality

If menopause is not the culprit in reproductive aging, what really limits how long women can reproduce? Despite the well-known fact that the number of eggs decreases with age, which is often mentioned when reproductive aging is discussed, this may be akin to only looking for one's keys under the streetlamp—we have relatively straightforward ways to measure egg number, but that doesn't mean that egg number is the first or major limitation in reproductive aging. Rather than quantity, it is the *quality* of the eggs that declines, causing problems with chromosome division, cell cycle arrest,

What "Reproductive Success" Means in Evolutionary vs. Popular Terms

Traditionally, "reproductive success" has been framed in terms of reproductive "fitness," which focuses on the total number of progeny produced, rather than on how late progeny are produced, which is the question we are interested in. (Here, "fitness" refers to evolutionary success in getting more of their own genomes out into the world, not being in good shape.) But the total number of progeny produced in the lab might not be a good reflection of optimal reproductive decisions in the wild, where conditions may fluctuate. For example, we can imagine conditions under which an animal might want to delay reproduction, such as when nutrients are severely limited. The animal might benefit from being able to slow reproduction temporarily, but then resume it once nutrients are available once more. In fact, Jonathan Tilly's group showed that caloric restriction, which is well known to extend the lifespan of most organisms (see chapter 7), also slows reproductive aging in mice, by maintaining mitochondria and chromosomal integrity. Such flexibility, or "plasticity," would make them more evolutionarily "fit" than animals that cannot adjust their reproductive schedules in response to low nutrients—even if that second animal produces more progeny under optimal conditions. In fact, having more progeny earlier would make the animals fitter, because they can beat out their competition by reproducing more and faster, thus spreading their DNA more effectively than competitors who delay progeny production, even if it causes them to have shorter lifespans. But this strategy is only useful if the body (soma) is actually healthy enough to support reproduction once it is restarted—and those systems that preserve the somatic health of the animal might also work to maintain late-life health and extend lifespan.

implantation, and fertilization. Decreasing egg number might correlate with declining fertility, but may not be the cause of the problem; in fact, decreasing egg quality most likely precedes the decline in egg number—it's just harder to measure in situ.

At the cellular level, oocyte quality decline is marked by the failure of chromosomes to properly segregate, which can lead to chromosomal abnormalities (e.g., Down syndrome and other aneuploidies [a type of abnormality of chromosome number] that rise with age), infertility, and increasing frequency of miscarriage, which itself might be due to problems with the fertilized oocyte. One of the issues is that oocytes are present in the germline of every baby girl, and those cells need to remain arrested in a particular state ("cell cycle arrest") for several decades in order to allow the proper resumption of cell development and division after fertilization. If the brakes are taken off before the cell is fertilized, the cell will not be able to properly function once fertilized.

Another mechanism thought to underlie decreased oocyte quality is the reduction in mitochondrial number and activity with age. Mitochondria, as we learned in chapter 9, are critical for continued cellular function, and their proper maintenance with age is key to longevity in all systems. Maintaining oocyte mitochondria in the highest-quality state, both in their morphology and activity, is critical for reproduction; this has been appreciated in human reproduction for many years, and can be assessed by simply looking at the structure of the mitochondria under a microscope. It's one of the few cellular markers that can be used to tell whether an oocyte is high-quality enough for IVF. In fact, one idea is to rescue old oocytes with new mitochondria (mitochondrial replacement therapy), either from the IVF patient herself (autologous transfer), or from a donor—particularly in the case of treating mitochondrially derived diseases.[22] This of course raises other questions, since mitochondria have their own DNA, but is an interesting approach. We recently found that insulin signaling mutants also use several clever mechanisms to keep mitochondria healthy in their oocytes with age and extend reproductive span, at both the morphological and metabolic levels.[23]

Other mechanisms that accompany the decline in oocyte quality with age have been revealed through genomic analyses of old and young oocytes. For example, comparing the genes that are "on" or "off" (that is, their mRNA is present or not) in oocytes from young and old mice pointed to declining mitochondrial function, stress resistance, cell cycle arrest maintenance, DNA repair, and chromosomal stability with age, results that were echoed in comparisons of oocytes (discarded from IVF procedures) from women of different ages.[24]

We wanted to use worms to ask the same questions, so that we could do genetic experiments that would identify regulators of reproductive aging. Before I describe worm oocyte quality decline in detail, I should first mention a big question that we had to address before we could use C. elegans as a model for reproductive aging. Because there are thousands of germline stem cells in the hermaphrodite that never get used up, even when mated with males, there had been an assumption in the C. *elegans* field that worms could reproduce their whole lives. But, in fact, the ability to produce progeny when mated does decline with age, a point illustrated nicely by Kerry Kornfeld's group. They mated spermless females with young males at successively later ages and examined the number of progeny the mothers had each day; the progeny production's peak does not shift to the right, as one would expect if the worms could mate indefinitely, but rather simply decreases.[25] That means that there is a limit to oocyte quality with age—just as in mammals, especially humans. We also found that egg quality declines, because the eggs from older mothers were less likely to hatch, and more likely to have lost one of their chromosomes.[26] (In C. *elegans*, hermaphrodites have two X chromosomes, and loss of an X chromosome—an XO animal—results in a male. So males are basically a "mistake" that mother worms produce with increasing age and under some stresses—but can be useful for increasing outcrossing under those stressful conditions.) Our and Kornfeld's results tell us that like women, worms have an unexpected "use it or lose it" mechanism that limits their reproductive capacity—they can't just hold off having progeny and hope to still be able to reproduce as well later in life.

Having established that worms do indeed undergo reproductive aging, Shijing Luo in my group carried out transcriptomic experiments comparing oocytes from young and old mothers, and found that the exact same processes—in fact, even the *exact same genes*—declined in the oocytes of "aged" (day 8) worms as in aged mammalian eggs.[27] At first, we were surprised at the incredible similarity in the gene lists in worms and mammals, not only because they are different species but also because worms, mice, and humans have vastly different reproductive spans—worms are reproductive for a week, mice for about a year, and of course women are reproductive for several decades. Upon further reflection, though, we realized that every egg, no matter the source, has the same challenge: to protect its genome and to be competent for fertilization when needed. Our data showed that the safeguard mechanisms, like chromosomal maintenance, DNA repair, and cell cycle arrest, are the processes that are lost with age, and the energy to restart once fertilized requires ATP generated by mitochondria.

Remarkably, a woman's reproductive ability is maintained for decades, meaning that the quality of the oocytes is maintained over that entire period of time, until late in reproduction—how is it possible to maintain good eggs for so long, and what changes when they go bad? We were interested in figuring how an animal might maintain oocyte quality with age, and while all the genes we have mentioned already (cell cycle arrest, mitochondrial regulators, chromosomal maintenance, DNA repair) went down with age, we had also noticed that there was a class of genes that went *up* in old mouse oocytes.[28] Genes that go up with age can fall into a couple of categories: they might be genes that are bad for the oocyte, or they might be genes that are "compensatory"—that is, they are turned on in order to fight against decline. Without testing the effects of knocking these genes out, it's impossible to know which category such genes belong to. There were several genes in one biological group that all went up in old oocytes, so it caught our eye. The genes in question had been previously associated with growth and development and are in the transforming growth factor beta (TGF-beta) signaling pathway.

To test the role of the TGF-beta pathway in reproductive aging, we turned to *C. elegans*, which has a mutant for every occasion. (Seriously—you can just look up the gene and often there is a mutant that you can order from the *C. elegans* Genetics Center at the University of Minnesota for a whopping $7.) Worms actually have two well-studied TGF-beta pathways: one that affects dauer formation (as we discussed in chapter 6) and another that is involved in the development of the male tail and in body size—that is, the mutants are small (Sma) and have a screwed-up tail (Mab, or male abnormal).[29] We then measured how long mothers can reproduce—the "reproductive span"—and the lifespan of these TGF-beta mutants. Wild-type worms can reproduce for an average of five to seven days. Our initial guess was that the TGF-beta dauer mutants would be the ones that affected reproductive aging, but we were wrong; their reproductive spans were the same as wild type's. To our surprise, though, the TGF-beta Sma/Mab mutants had extremely long reproductive spans, twice that of wild type. This was quite remarkable, because their lifespans were completely normal.[30] In human terms, this would be the equivalent of a woman having a baby—naturally—at the age of 75 or 80, while still looking like a 75-year-old. (Meanwhile, we discovered that the TGF-beta dauer mutants do have an extended lifespan, which was a canon-challenging surprise, but that is another story [see chapter 6].)[31] Our guess, based on the mouse oocyte data, that the reduction of TGF-beta signaling might actually improve reproductive aging ended up paying off.

Until we found the long reproductive spans of the TGF-beta Sma/Mab mutants, one could have argued that an extended reproductive span was just a by-product of everything being stretched out in longevity mutants; our favorite *C. elegans* mutant, the insulin/IGF-1 receptor mutant *daf-2* (for review, see chapter 6), not only has a very long lifespan but also doubles reproductive span,[32] which fits with this idea. But the TGF-beta mutant data showed that the two systems, reproduction and somatic longevity, are in fact independently regulated, even if they are at some level connected to one another.[33] The fact that the non-germline cells of TGF-beta mutants age normally while their germlines remain youthful leads to a somewhat horrible outcome: while they are happily reproducing at day 12 or 13, their bodies are breaking down with age, with the result that occasionally the poor mother worm will explode, essentially dying in childbirth ("matricide"). This gives us another clue about the importance of signaling between the germline and soma: when this communication is broken, the animal will be unable to properly coordinate the two systems, with disastrous results.

Finding out how mutants with long reproductive spans keep their oocytes healthy with age can give us clues about how we might slow the aging of women's oocytes, as well. Previously we had paid attention to the genes that decline with age, because that gives us a sense of what fails in the cell when oocyte quality declines. The daf-2 and TGF-beta Sma/Mab mutants keep their germlines and oocytes healthier with age by maintaining high levels of the regulators of chromosomal and DNA integrity, mitochondria, and cell cycle arrest,[34] as I described above, and we proved that these genes are important by knocking them down and measuring decreases in oocyte quality and reproductive span. But we also noticed that only a few genes (about 30) *increased* in expression in old wild-type oocytes, and of that small set, a notable fraction encoded a type of protein that degrades other proteins, called *cathepsin B proteases*. Moreover, these cathepsin B protease genes were not expressed in high-quality daf-2 oocytes. Thus, the presence of cathepsin B gene expression in an oocyte correlated with low quality.

There were two possibilities: (1) cathepsin Bs are protective, and the cathepsin expression rose in response to aging in a compensatory manner; or (2) the cathepsins are deleterious for oocyte quality. Only the latter case would cause *improvement* in aging wild-type oocytes if we blocked cathepsin B activity. Luckily for us, inhibitors of cathepsin B activity are widely available. When we fed the worms a cathepsin B inhibitor, we observed a significant improvement in oocyte morphology. Even more remarkably, we saw this improvement even when the drug was given to the worms more than halfway through their reproductive spans, suggesting that there may be ways to stop or slow reproductive aging even if initiated relatively late, by keeping the remaining oocytes from degrading.[35] This was all the more exciting because the role seems to be conserved with mammals, at least cows—it turns out the cattle breeding industry had already made the observation that low-quality oocytes have more cathepsin B, and so they test cathepsin B levels in their in vitro fertility procedures![36] If translated to women, this could extend reproductive span by several years, which could make all the difference for some patients. The fact that we could administer the inhibitor even in mid-reproduction—the equivalent to a woman in her mid-30s—is particularly promising.

————

When it comes to reproductive aging, some women may have hit the genetic jackpot. A few years ago (2007), a group of IVF specialists at an obstetrics-gynecology clinic in Israel noted that despite the fact that it was assumed by the field that women "cannot" have children after their mid-40s without ART (artificial reproductive technology), in a few years more than 200 women over the age of 45 had given birth in their clinic, all without ART. The researchers collected blood from a few of these patients and compared it to blood samples from women who stopped having children before the age of 30. (It should be noted that this may be on the early side, so the "controls" might have in fact been somewhat infertile.) Using microarrays, the researchers identified several gene expression differences between the two sets of samples.* It's difficult to tell what is going on in this analysis, as the terms are all super vague (e.g., "signal transduction," "cell metabolism") and the genes that are up- and downregulated were placed in the same lists, making it difficult for a reader to determine direction and causality. Nevertheless, buried in a category they called "apoptosis," they list the insulin/IGF-1 receptor and several other genes in the insulin signaling pathway, and all of them are downregulated in the blood of the women with exceptionally long reproductive spans. Sounds familiar, right? I think their data suggest that the women who live long and reproduce late are basically insulin signaling mutants—and our *daf-2* worms have the same phenotypes! That is, systemically low insulin/IGF-1 signaling might extend not only human lifespan but reproductive span as well. One

---

* Frustratingly, the researchers refused to send us their expression data when I requested it, so we can't do the analysis ourselves. This was shocking to me; I gather that it's common in the human genetics field, but I had never encountered this problem before, as sharing of data is expected in invertebrate research fields (and in fact is a stipulation of publication in most journals, whether they back it up or not).

of the signs of this long life might be that they can have children later than normal, without any reproductive interventions. Perhaps this is the connection between the long lifespans of some women, like the Boston women from the New England Centenarian study, and their ability to have children later.[37]

Post-reproductive Lifespan and the Grandmother Hypothesis: An Evolutionary Conundrum

When we think of aging, we usually think of an old person, well beyond the age of reproduction. But if we think in evolutionary terms, there can be no selection for post-reproductive long life, only for traits that occur before or during reproduction. That is because traits need to be possible to pass on to progeny to have any effect in future generations. So how could a signal from the germline, or reproductive hormones in a mammal, have anything to do with the regulation of post-reproductive longevity? If any benefits can be conferred to progeny by improving the odds of late-life reproduction, then those same changes may also extend lifespan.

The "grandmother hypothesis" was first proposed by the experimental theorists Williams, Medawar, and Hamilton to explain why long post-reproductive lifespans exist, and has been further studied particularly by the anthropologist Kristen Hawkes.[38] This theory posits that grandmothers contribute to the fitness of their grandchildren, so that is a case where the extended longevity of older women could benefit later generations, and thus be evolutionarily selected. The idea is that if women live well beyond their child-bearing years, they can help with the care of their grandchildren, so any genes that might help a grandmother live longer could be selected for. The problem with this theory is that it would be hard to extrapolate it from humans, who clearly invest a lot of energy into caring for their young, to C. elegans, which lays a bunch of eggs and hopes for the best ... yet women and worms have eerily similar reproductive and lifespan curves (figure 5), with their reproductive spans lasting about half their lifespans.

Both women and C. elegans females have significant post-reproductive lifespans, and similar genetic pathways appear to regulate reproductive span and lifespan, yet their long post-reproductive lifespans could be attributed to the grandmother hypothesis in only one of the species (unless there is more to C. elegans childcare than we are aware of—and there might be, see chapter 15). Not every animal has a long post-reproductive lifespan, either. This suggested

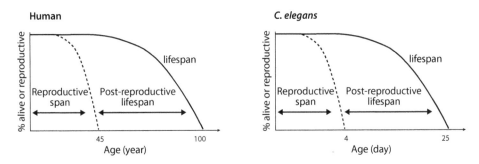

FIGURE 5. Women (left) and worms (right) each have reproductive spans that last about half of their lifespans, and long post-reproductive spans.

to us that something other than social interactions might be more likely to regulate the length of post-reproductive lifespan.

To understand what determines the length of an animal's post-reproductive lifespan, we needed to gather data on more than just worms and humans, to see if there is a factor that could explain why some animals have a long post-reproductive lifespan while others have almost none at all. Is there a unifying principle? After thinking about my own pregnancies, and watching plenty of worms suffer from matricide, I surmised that the more stressful it is to produce progeny, the healthier the soma of the animal would have to be in order to support reproduction; after reproduction stops, the healthy soma, even if not maintained, would lead to a long post-reproductive lifespan. By contrast, animals that produced small progeny might be able to reproduce much longer, and thus have a short post-reproductive lifespan. I was thinking about this problem when an undergraduate, George Maliha, approached me about writing a senior thesis. I explained that I had a crazy idea about the relationship between mother/ progeny size and post-reproductive lifespan that I would like him to test.

George proceeded to collect an extensive set of data on animals from all types of sources: historical records, agricultural data, and, best of all, zoos. He recorded average mother size, average progeny size, lifespan, reproductive span, gestational time, time to weaning, and all kinds of other information, whatever he could get his hands on. It was fairly easy for him to collect extensive data on mammals, and when he did careful analyses of the data, post-reproductive lifespan in mammals as a group best correlated with time of gestation—that is, how long an animal is pregnant. But, of course, mammals, by definition, care for their young, so this group might not be representative of *all* animals, which are what we are interested in. We needed four pieces of data (lifespan, reproductive span, mother size, progeny size) to test our hypothesis, but outside of mammals, it was hard to find all the data—particularly reproductive span in the wild. As a result, George ended up with only a handful of nonmammalian species to analyze; this group included alligators, chickens, birds, flies, worms, and sea urchins. The nice thing was that the group covered a wide range of lifespans and body sizes. With these data, George found that the ratio of progeny size to mother size ("offspring ratio") correlated fairly well with the length of the post-reproductive lifespan ($r^2 = 0.8$).[39] This supported the hypothesis that the more difficult the childbirth (which we estimated from the ratio of the progeny to mother size), the longer the post-reproductive lifespan. That is, somatic health correlated to some degree with the remaining time of life post-reproductively. Mothers who need to be healthiest to successfully reproduce might have the highest somatic health, thus the furthest to fall post-reproductively, resulting in a longer lifespan.

To illustrate my point, let's consider sea urchins, which are at one extreme. They can live for decades, some species even a century, some species can grow to be more than a foot in diameter, some with 12-inch-long spines, and often produce microscopic progeny until the day they die—so they have both an extremely short post-reproductive lifespan and a small progeny-to-mother size ratio.[40] We would hypothesize that their ability to reproduce almost until they die is because they don't need very high somatic health to reproduce, since they just have to squeeze out some tiny eggs. Women and C. elegans are at the other end of the graph, with progeny of very large size (relative to the mother's size) that can be difficult to birth. Humans' heads have evolved to be as large as they can be without killing all mothers during childbirth; worms, too, have this problem, as their eggs are fairly large, and in times of stress worms can explode or die when the eggs hatch internally (matricide, or, more colloquially, "bagging"). In both cases, the post-reproductive lifespan is very long, indicating that the level of somatic health and integrity required to successfully birth their offspring is very high.

With this in mind, we could revisit those C. elegans TGF-beta Sma/Mab mutants that I described earlier in the chapter, the worms with long reproductive spans but short lifespans. These include sma-2 (named for its small size) with its remarkable ability to extend reproductive span and maintain beautiful oocytes with age. While wild-type worms normally are done reproducing by day 5 or 6 of adulthood—about a third of their lifespan—sma-2 mutants could still have progeny way out at day 12 or 13, unheard of for wild-type worms. The problem: most of these worms explode (yes, really—their guts leak out), dying in childbirth, while still 100% fertile. We had tested them under sterility-inducing conditions to prevent progeny production to measure their lifespans, and found that sma-2's lifespan was normal.[41] sma-2 mutants' germlines and oocytes still look beautiful and healthy, while their bodies are falling apart around them. This makes them different from daf-2 mutants, which extend both lifespan and reproductive span. My suspicion was that the reason they all died in childbirth was because this was the equivalent of a 75-year-old woman giving birth; even if the baby was healthy, the process of being pregnant and giving birth is fairly traumatic, and might be difficult for an aged animal. Day 12 worms are no spring chickens, and sma-2's beautiful, youthful germlines did nothing to help them prevent the aging in the rest of their bodies. sma-2 mutant worms are small, but their eggs are still large and their lifespan is not extended; perhaps for this reason they often explode and die when their bodies have started to age, even while they are fully fertile. These poor mutants have an imbalance in the natural relationship between reproduction and longevity, and ultimately they pay the price.

———

It makes sense to regulate lifespan only if this regulation has a beneficial effect on reproduction, or is a by-product of reproductive choices. Thus, we can view the systems (regulatory pathways) that are usually thought to exist in order to "regulate aging" as in fact being present in order to regulate somatic maintenance under conditions where reproduction has been slowed (e.g., dietary restriction, or decreased insulin signaling). That is, these are not longevity pathways, they are reproductive timing pathways. This slowing of somatic aging then ensures successful progeny production once conditions are good again. Post-reproductive lifespan is just the result of the remaining health of the organism once reproduction is done. The more stressful it is to produce progeny, the better shape the soma has to be in, and thus after reproduction, the longer it takes to decline. So the pathways that regulate longevity are there just to adjust how long the soma needs to be healthy in order for the mother to still reproduce successfully. In fact, if we think in terms of a trade-off and reproductive fitness, every animal should have evolved to reproduce as quickly as possible and to have as short a lifespan as possible. But instead, we see that over evolutionary time animal lifespans grew longer and longer, with longer post-reproductive lifespans. That means that there must be some advantage in some cases for slowing down reproduction, which may often be the case when nutrients are not always readily available. The mechanisms that regulate longevity are intimately linked with the regulation of reproduction. In fact, I will go so far as to say that longevity is simply a by-product of the need to reproduce successfully later in life, and aging itself is not regulated. Without the need to regulate reproductive timing in response to nutrients, there would be no mechanisms that regulate longevity. To study longevity without carefully considering the impacts on reproduction is to miss half of the equation.

Given what we know now about the roles of TGF-beta and IGF-1 signaling, as well as dietary restriction, in slowing reproductive aging, it might be time to reframe the work in the aging field as a question of the evolutionary relationship between reproduction and longevity. Perls's data supporting a connection between a woman's ability to have children late in life and subsequent extended longevity counter the idea of a trade-off between longevity and reproduction, as they are in the same direction. Reproductive span extension—a goal for a modern problem—and the evolutionary definition of "reproductive fitness" are two very different, almost opposite ways of viewing reproductive success. While having 12 kids would get more of my genome out into the world, for me and most other modern women, being able to have a child late enough in life to complete my education was the goal. Because the focus has been primarily on the relationship between longevity and reproductive fitness, reproductive aging hasn't been a focus of the aging field, but I hope that the field is now shifting to look at the problems from a different perspective.

12

Sex, Flies (and Worms), and Videotape

THE BATTLE OF THE SEXES

[My] secret to a long life has been staying away from men. They're just more trouble than they're worth.

—JESSIE GALLAN, SCOTLAND, AGED 109

NOW, THIS IS PROBABLY the chapter you've been waiting for—sex and lifespan!

In many species, lifespan is a sexually dimorphic trait—that is, males and females of the same species can have different lifespans. Some of these differences are due to differential utilization of resources for reproduction, but other differences are caused by interactions with the opposite sex.* Sexual conflict can also influence lifespan and behavior in both sexes. For example, males of many species transfer peptides in their seminal fluid during mating that affect the receiving female's behavior and lifespan, and pheromones can play a role in behavior and lifespan determination as well. Some of these lifespan mechanisms are likely to be shared by many species, and others are species specific. What we would like to know is how many of these will apply to us humans.

---

* Note that I will use "sex" to indicate male and female in model system studies, since that is the biological definition, while "gender" is a social construct that is relevant for humans and is not the topic of studies using model organisms for aging. But I will talk about mating, too!

But before we get to the fun stuff, like how worms are a bit like Ming dynasty emperors, let's first talk about lifespan differences of men and women. In most countries today, women outlive men. Life expectancy has risen for everyone, but for women it has risen more (despite the obvious risks of maternal mortality). The sex difference in life expectancy is larger in some countries than others; for example, in most countries this gap is 3–5 years, but there is more than a 10-year difference in lifespan between Russian men (65) and women (76). Even Japan's very long-lived men are shorter-lived than their female counterparts (80 vs. 86 years). Some of this difference might be unsurprising, as rates of smoking have traditionally been higher in men (but are rising in women), men practice other unhealthy and risky behaviors at higher rates (particularly as adolescents—think motorcycles, guns, and alcohol), and men die at disproportionate rates in war—although the biggest wartime cause of death was the 1918 flu pandemic, which affected both men and women. The ongoing Covid pandemic is also bringing down life expectancy, and in 2020 increased the lifespan disparity in men and women in the United States from 5.1 (2019) to 5.4 years.[1] Other factors are biological, such as differences in how fat is stored (visceral vs. subcutaneous), sex-specific rates of survival of childhood diseases, and rates of infectious diseases. Paradoxically, older women suffer from illnesses at a higher rate, including in age-related degenerative diseases.[2] Women also suffer postmenopausal health problems due to changes in hormones, such as the loss of protective estrogen. While both men and women die of cardiovascular disease and cancer, women are far more likely to suffer from dementia, including Alzheimer's disease, even after adjusting for survival (although men with Alzheimer's die earlier and show more cognitive defects than women, suggesting an extra X chromosome might be beneficial).[3]

For many women, these extra years are associated with increased disease and disability—although women with exceptional longevity escape this grim fate. Regardless of the circumstances that led them there, the overwhelming majority (four of every five) of extremely long-lived individuals—that is, centenarians and supercentenarians—are women. In fact, all of the top 10 recorded longest-lived women lived longer than the oldest man. In an effort to better understand some of the possible genetic differences that might lead to differences in longevity, the Chinese Longitudinal Healthy Longevity Study compared the genomes of male and female Han Chinese centenarians to identify potential genes (genetic loci) that might be associated with long life in men versus women.[4] For men, several genes involved in inflammation and immunity emerged, while the PPAR-γ coactivator PGC-1α (which we discussed in chapter 9, about mitochondrial function in longevity), tryptophan metabolism, and estrogen metabolism emerged as loci associated with extreme female longevity. How these genes might contribute to longevity and whether they can account for the differences in male and female lifespan is still unknown. Our old friend FOXO3A, which you might remember is responsible for longevity in worms, flies, and mice, was primarily associated with long lifespan in male centenarians.[5]

Social Circumstances Contribute
to Differences in Aging

Before we discuss the underlying molecular mechanisms that evolved to help determine longevity, it's interesting to speculate on some of the evolutionary theories about these differences.[6] For example, it has been suggested that social structures that encouraged men to be polygamous might have both extended male reproductive span and shortened women's, while promoting a longer female post-reproductive lifespan. Coupled with this idea is the grandmother hypothesis (see chapter 11), which postulates that an extended female post-reproductive lifespan would have benefitted their grandprogeny.[7] A few nonhuman animals also have longer post-reproductive lifespans, and it is notable that these are primarily mammals, such as whales, that have some social structure and that care for their young. (However, as discussed in the previous chapter, worms have a long post-(self)-reproductive period as well, despite doing no child or grandchild care, throwing this hypothesis into some doubt.)

Some of the sex differences we see in human lifespan might be due to social circumstances. Depending on your perspective, you may or may not be surprised to discover that marriage increases the lifespan of men by almost two years, but decreases women's lifespan by more than a year (+1.7, −1.4, respectively),[8] while women with female companions live longer. This all seems to boil down to whether one's stress is increased or decreased by one's companions and social interactions. The negative effect of marriage on women increases with a greater age gap with their husbands. This same age gap increases the husband's lifespan, like vampires sucking away the years from their wives. The underlying reasons for this can only be speculated, but might include the fact that women provide more social interactions to men, which may not be a reciprocal function. This is also reflected in the worse health status of widowers than widows. The lifespan effect of women benefitting men even stretches across generations: a study of rural Polish families found that having more daughters specifically increased the lifespan of fathers, while sons didn't affect their lifespan one way or the other.[9] For mothers, the news is worse; having either sons or daughters decreased their lifespan.[10] And finally, women are more often the victim of intimate partner violence.[11] All of these social interactions seem to come at a cost to women, yet on average they live longer than men.

Biological Bases for Sex Differences in Aging

The underlying bases of sexual differences in aging are difficult to address experimentally in people and most animals, and ultimately we would like to understand the underlying biological changes that result in altered longevity in humans. To answer these questions, it would be good to be able to separate social effects from biological effects. Do other animals show differences in lifespan between males and females? If we consider animals in the wild, in most species the females live longer, with only a few notable exceptions, such as Brandt's bats. Female nonhuman primates, including orangutans, gorillas, and chimpanzees, live longer than males, but some species don't show sex-specific differences in longevity; work in rhesus macaques suggests there is no difference, but these results on captive primates might not completely reflect their lifespans in the wild, particularly in social groups.

It's perhaps unsurprising that animals with social castes, such as eusocial species of bees, ants, and termites, exhibit sex-specific differences in lifespan, since the jobs of drones, workers, and queens vary considerably. The queens also defy the notion of individual trade-off in reproduction versus lifespan, probably at least in part because the costs—finding nutrition, mating—are borne by workers and by drones who die soon after mating. The lifespans of male ants (*Cardiocondyla obscurior*) who were mated more frequently were much shorter than those of virgin or infrequently mated males (17 vs. 26 and 27 days), suggesting that these males experience a fecundity/longevity trade-off that queens do not experience.[12]

The mammalian workhorse of biology is the mouse, of course. Coming from my uber-feminist research field of C. *elegans*, where we ignore males unless we absolutely have to include them (because they are rare, pretty dumb, and a pain to work with, as I will explain in a moment), I was shocked to learn that most rodent experiments are done only in males. *What?* Rodents are expensive to maintain, and there was a now debunked but pervasive idea that female rodents were more variable than males, so most studies about anything but female biology used males exclusively.[13] The excuse for the male bias in mouse research has been "*something variability something something hormones*," but, in fact, recent data show that male mice are just as variable and hormonal as female mice.[14] (In fact, Smarr and Kriegsfeld showed that "in no instance do females exceed male variance, and in most instances, male variance exceeds female variance" [2022], countering the usual opposition to using female mice for research—yet the use of exclusively males remains pervasive.)[15] As a result, when you hear about a mouse or rat study, we might know only the result in males. This is not great, of course, as it might cause several types of error, possibly missing important biology that affects only females, or assuming that effects found in males will also be true in females. Yet much of the work in the aging field has used male mice and has assumed that these results will affect both sexes the same way; we have likely missed out on cures for diseases that affect women differently, and we may have overestimated the effect that a candidate drug would have on women. Recently the NIH recognized this problem and is now requiring researchers to include both male and female mice in their studies. The hope is that by using both male and female mice, sex-specific differences in disease and treatments will become apparent. Some scientists grumble about this because it can double the cost of a study, but the inclusion of females is undeniably important for our proper understanding of all biology.

The metabolisms of male and female animals might differ, and some of those differences will affect lifespan. In fact, longevity treatments might differently affect men and women, but we have not paid enough attention to these differences yet. However, Richard Miller's group (University of Michigan) has been running the Intervention Testing Program, which studies the effects of different candidate drugs on lifespan, and the group does compare effects in male and female mice. First, it's notable that the female mice controls in this study lived longer than the males, as we see in humans. (But this is not true of all mouse studies, and may not reflect lifespans of wild mice, as these lab studies are mostly done on inbred or F1 hybrid mice.) By separating the groups by sex, Miller's team could see whether a drug has an effect exclusively on one sex or another, or on both. They reported that some of these compounds, including acarbose (a dietary restriction mimetic), 17α-estradiol, and the anti-inflammatory nordihydroguaiaretic acid (NDGA) extend the lifespan only of male mice.[16] As another example, male mice that overexpress Sirt6 (see chapter 7) live longer than wild-type males, but that same genetic manipulation of Sirt6 does not increase female mouse lifespan.[17] (Would you be surprised to hear that those long-lived male mice have lower circulating levels of IGF-1 and other changes to the insulin/IGF-1 signaling pathway?)[18] Similarly, male and female mice have differential longevity responses to dietary restriction levels (20% vs. 40%), circadian rhythms, dietary composition, and timing of feeding, suggesting that that there are important underlying metabolic differences that affect how DR functions in males and females.[19] These differences will be critical to understand before we move to human trials.

Dena Dubal's group at UCSF has flipped the question on its head, asking if the extra X chromosome that women have is the reason they live longer, and if so, can we find out how and use the information to extend everyone's healthspan.[20] (Whether mitochondrial DNA, which is entirely maternally inherited, can affect lifespan is also being examined.)[21] Dubal is focusing on whether the X chromosome is the cause of women's longer lifespans and other differences. Normally that extra X is "silenced"—that is, its transcription is shut down, to prevent problems that would be caused by having too much of the genes that are made by the X chromosome. But Dubal's work suggests that in older females, some genes on the extra X chromosome may no longer be silenced ("X inactivation escape"), and these activated genes may improve resilience, helping older women live longer and retain cognitive function longer. The study of X inactivation escape genes may give us a clue to longevity that isn't from studying older men or male animals, yet such clues will be missed as long as we treat women's health and longevity as less important than men's health. Understanding what the molecular cause of this difference is might help everyone function better with age.

———

Model organisms can help us not only to observe sex differences but also to unravel exactly what is going on at the cellular level by using genetic and molecular approaches. But to do this, we need to find genetically manipulable species that exhibit longevity differences between the sexes. Whether there are sex differences in the lifespan of *Drosophila* appears complicated, as sex differences seem to depend on the particular strain being used, rather than being a characteristic of the species (although it still could be true in wild strains). Modulators of growth hormone (GH) and IGF-1 pathways have been reported to have greater effects on females than males in mice and in flies.

mTOR signaling and dietary restriction also appear to have a greater effect on females than males in several species[22]—this might be related to the connections between nutrient intake and sensing and reproductive needs, which are higher in females than males for the most part.

In the *C. elegans* field, the hermaphrodite has been used almost exclusively for most work because it is the predominant sex and it is so much easier to work with—hermaphrodites are essentially clones of one another, since they have their own sperm and oocytes and can self-fertilize, making them genetically homogeneous ("isogenic"). Males are rare in *C. elegans* populations, about 1 in every 1000 worms under normal conditions, which is great if you want a pure population of hermaphrodites. In fact, males are the result of the failure to properly segregate the X chromosome during very early development, creating an animal with only one X (an "XO" worm, since *C. elegans* has no Y chromosome)—a mistake that occurs more frequently as the worms get older, similar to aneuploidy errors that rise with human maternal age. Because males are rare, it can be difficult to get a lot of them to work with unless you use some genetic tricks to increase the frequency (called "high incidence of males," or "*him*" mutations). Also, males are kind of dumb, frankly—they have one-track minds. They ignore food and constantly crawl off the plates, or they clump together, attempting to mate with one another (or even with their own excretory canals). Douglas Portman's lab (University of Rochester) has shown the molecular reason this: males express much lower levels of a food-sensing receptor, ODR-10, than do hermaphrodites.[23] This low expression of ODR-10 causes the males to ignore food and instead leave the plates in search of mates, which of course is their only purpose in life. For these reasons, most of the studies I've told you about so far in this book used primarily unmated hermaphrodite worms.

What about male lifespan? In 2000, David Gems and Don Riddle reported that male *C. elegans*, at least when the worms are measured on plates with other males (as we do when measuring hermaphrodite lifespan) live shorter than hermaphrodites do when raised similarly (17 vs. 22 days).[24] They also showed that at least part of this lifespan shortening in grouped conditions could be avoided if the worms were not allowed to clump into groups, but instead kept one per plate; when the animals were kept on plates by themselves, the solitary males lived longer than solitary hermaphrodites. This greater male lifespan effect appears to be independent of insulin signaling, because *daf-2* males lived even longer than wild-type males (so it's not as though the males are already *daf-2*-like), and they do not seem calorically restricted.[25] Gems and Riddle interpreted the fact that solitary males live longer than grouped males to mean that the worms were causing physical damage to one another—male worms try to mate with whatever they are next to (it's shocking to see their behavior for the first time). However, later Cheng Shi in my lab discovered that if males have no germline, they still clump and presumably cause damage, but this has no effect on their lifespan, eliminating the "beating each other up" model.[26] Instead, it suggests that the presence of the germline is the most important factor—male worms just have to *think* they are mating, which increases the proliferation of their germ cells and sends a lifespan-shortening signal, much as Hsin and Kenyon had found in hermaphrodites.[27] Gems's group found that male lifespan is longer in most *Caenorhabditis* species they tested (9 of 12), whether androdioecious (hermaphrodites and males) or gonochoristic (males and female), possibly suggesting that there might be an evolutionary reason for this male longevity bias, at least in the lab under fully fed conditions.[28]

Reminiscent of Rich Miller's findings about acarbose and mouse lifespan, there is evidence for sexual dimorphism of *C. elegans* lifespan in the effects of diet and DR on longevity. For example, while in Cynthia Kenyon's lab, Seung-Jae Lee and I showed that glucose shortens the lifespan of *C. elegans* hermaphrodites, but Michelle Mondoux (College of the Holy Cross) later made the intriguing discovery that the same treatment on males slightly *increased* their lifespan and motility.[29] Similarly, Eisuke Nishida's lab (RIKEN, Japan) had previously found that intermittent fasting extends lifespan of *C. elegans* hermaphrodites in a DAF-16/FOXO-dependent manner, but males essentially have no longevity or FOXO response to DR or to the genetic mimic of DR, *eat-2* mutation.[30] While males do have gene expression differences in response to DR, they are still able to reproduce for several days, which is very different from the complete arrest of reproduction seen in females. Nishida's group tracked this sexually dimorphic DR response down to the X chromosome, the TRA-1 transcriptional repressor, and the nuclear hormone receptor DAF-12, which had already been linked to both DR and germline signaling.[31] The ancestral homolog of oxytocin (the love hormone), nematocin, has sexually dimorphic expression and effects on behavior;[32] as you might guess, this hormone is involved in mating. (Its ant homolog, inotocin, is involved in social foraging behavior.)[33] Finally, germline ablation in males doesn't cause the lifespan extension seen in hermaphrodites, suggesting that males may lack that somatic gonad lifespan-extending signal.[34] Altogether, it seems that what it takes to keep a male worm functioning is different from what hermaphrodites use. It's likely that this is not just a weird set of coincidences but, rather, that the somatic gonad longevity signaling is connected to DR effects on longevity. If males don't need to worry about their dietary intake in order to be reproductive—which makes sense when the nutrient needs of oocytes and sperm are compared—then it might make sense that they also lack a longevity signal from the somatic gonad.

This brings me back to the connection between DR and sex that I mentioned in chapter 7. Beyond whether DR works the same way in male and female animals, one thing that I've always wondered about is whether men and women are equally likely to stick to a DR regimen, which can be quite tough.[35] (This is different from "dieting," which women have been pressured to do since they were adolescents, with the media obsession with thinness, while the recently emerging commercial field of DR seems to be very male-personality driven.) This could be answered by examining the rates of and reasons for DR study dropout by men and women, which are collected but not necessarily reported broken down by sex.[36] My gut tells me that DR might be harder for women to carry out, perhaps owing to hormonal changes, but I am not aware of a formal study that asks this question—and I could be wrong, as all of my information here is anecdotal, taken from conversations with people who run DR studies and have mentioned higher drop-out rates of women. I know only a few women who have undertaken an extreme feeding regimen, while many men seem to embrace DR and especially intermittent fasting (IF). Even IF, despite its relative ease compared with DR, can be difficult to manage on a daily schedule, and may affect women unequally, particularly if the burden of providing family meals still falls on the mother. Our society even treats the same behavior by men and women differently, with substantially different weighing of their intrinsic values: "dieting" is considered superficial and vain (because it's what women do), while the Jack Dorseys of the world "disrupt" and "biohack" their daily food intake for the higher purpose of living long or enhancing their brain function. Gender and sex differences not only in the underlying biology but also in psychological effects and social roles ultimately will determine how effective a dietary strategy will be in extending the lifespan of men and women.

Sex and the Single Worm: How Mating Affects Longevity

None of the studies I have described so far actually take into account mating, but of course the "life goal" of an animal like the worm is not to live long but to pass on its genes to future generations. Although we've talked about sex and lifespan,

I haven't yet touched upon SEX—that is, mating and lifespan. Like Carrie Brad-shaw, I couldn't help but wonder, *How much does sex matter for longevity?*

Let's veer away from humans for a while, since they are so hard to experi-ment with, so that we can start looking more at molecular mechanisms. First let's discuss sexual conflict, or sexual antagonism. This is the concept that males and females are essentially in conflict with each other—like Prince's "World Series of Love"—to get what they want (passing on their genomes) at the expense of one another. Males often benefit from multiple matings, while females invest in the development of their offspring, and these differences leave the two sexes at odds with each other. To increase their chances of win-ning the competition, the males of many species have developed mechanisms to disable the female's future matings, or to change the behavior of females toward other males after they have reproduced. All of these terrible effects on females enable the males to "win" because they propagate their own genomes and discourage the female's future suitors, thus reducing competition with other males. The biological mechanisms that have evolved due to sexual con-flict are a house of horrors: spiky genitals in weevils, "penis fencing" in pla-naria, traumatic insemination by bedbugs, and "love darts" in snails (no, I did not make that up).* These are all ways that males cause injury to females dur-ing copulation in order to better compete with other males. And it could be worse—like female praying mantises, some species of nudibranchs actually eat their partners after mating. Billie Jean King and Bobby Riggs notwithstand-ing, the *true* battle of the sexes plays out in the animal kingdom every day. Other behavioral changes, like decreasing receptivity to future mates in flies and worms, or the transfer of anti-aphrodisiacs to female butterflies, are per-haps more subtle but have the same effect. With all this, you can understand why *Caenorhabditis* species evolved hermaphroditism three separate times—to get away from males! Just kidding—it's more likely that evolving to have their own sperm allowed *C. elegans* hermaphrodites to be able to wander off and survive as lone individuals—but same difference, right?

Breakthrough studies on the effects of sex and mating on lifespan were done in the labs of Linda Partridge (University College London) and Mari-anna Wolfner (Cornell).[37] In an analysis of the lifespans of *Drosophila* under high- and low-mating conditions, mating with males substantially decreased

* Seriously, check out the Wikipedia page on sexual conflict (Wikipedia contributors, 2022, "Sexual Conflict," *Wikipedia, the Free Encyclopedia*, accessed December 20, 2022, https://en .wikipedia.org/wiki/Sexual_conflict)—so much weird stuff!

the lifespan of female flies. The surprise here was that this was not all because of resource allocation, as one might predict from the prevailing evolutionary theory, since female flies with different levels of fertility had similarly shortened lifespans. Instead, the lifespan decrease appeared to be due to a deleterious effect caused by the males. If the male is able to make the female avoid mating again later with a different male, that makes it more likely for the first male to have his, and *only his* progeny produced, which helps him win in the race to get his genome propagated. The males do this in several ways. First, mating with a male causes the female to increase the number of eggs that are available for his sperm to fertilize. Secondly, and perhaps more remarkably, something that the female receives from the male causes her to change her behavior, so that she is less likely to mate with other males she encounters in the future. This remarkable effect benefits bachelor #1 at the expense of the female, all in his effort to compete against other males.

This deleterious effect of mating with males is known as "sperm competition," because it affects the success of other males mating, but the term does not necessarily mean that it is mediated directly by sperm. Instead, peptides in the seminal fluid (accessory gland proteins) that are transferred along with sperm are responsible for the negative effects on females.[38] Even more interestingly, the behavior and "attractiveness" of females to males is affected by the seminal fluid peptides. In fact, one component of seminal fluid known as "sex peptide" is particularly important, because it induces several of the post-mating female behaviors that are exclusively beneficial to the male, including increasing egg-laying rates and reducing their receptiveness to future mates.[39]

Seminal fluid and sex peptide's effects are remarkable because none of the effects are *necessary* for reproduction—they are not mediated by the sperm or genetic material itself—but instead are accessories, functions that were added on simply for the purpose of benefitting the male's reproductive success, largely at the expense of the female. Male flies use seminal fluid proteins to manipulate their partner's fate and behavior, decreasing the chances that she will mate with a different male later. This war between the sexes is seen across the animal kingdom in different manifestations, but always with the male's goal of making sure that his own sperm gets used preferentially; the females are reproducing while trying to avoid immediately succumbing to these deleterious effects.

Worms also display sexual antagonism, but it's all a little less subtle. One day in 2012, Cheng Shi, a graduate student in my lab who was studying a mutant that extends reproductive span, *repx-1* (for **rep**roduction **ex**tension), came into my office to share a shocking observation: in the course of his experiments, he realized that mating was causing the mothers to rapidly shrink and die. He was mating hermaphrodites with young males to track how long the mothers could reproduce ("mated reproductive span," see chapter 11) when he saw that the mated worms looked absolutely terrible after a couple of days, whether he was looking at wild-type or *repx-1* worms. While the unmated worms were happily growing slightly larger and would live for another few weeks, the mated worms looked like sad, deflated balloons after a few days. By a week, they had shrunk and wrinkled, and then they died, which was not surprising given how awful they looked. My first reaction was one of disbelief; after all, people had been working with *C. elegans* for more than 40 years, crossing them to get double mutants for their genetic studies, and I had never heard of this shrinking business before—and it feels like the kind of thing that people would have talked about. But Cheng knew me well, and had brought with him not only photos but detailed measurements and statistics on the worms. He showed that there was no doubt—within a week of mating, the hermaphrodites had shrunk by 30%–40%, a significant difference from their unmated counterparts. Even the superhero *daf-2* worms couldn't fight against this mating-induced demise! Cheng and I agreed that he should drop the *repx-1* project and figure out what was going on with this mating-induced shrinking.

The first thing Cheng did was test whether sperm are required for the shrinking effect, by using a mutant that specifically stops sperm formation (*fer-6*). Remarkably, even though Cheng could see that the hermaphrodites had mated with the spermless males, they remained as large as the unmated controls.[40] He also found that sperm transfer causes an increase in proliferation of germline stem cells, and by eliminating the germline of the hermaphrodites (using germline-less *glp-1* worms), he showed these germline stem cells are necessary for the shrinking effect. In fact, the effect is dynamic—blocking proliferating germline stem cells with a DNA-intercalating drug at any time during reproduction can stop the shrinking in its tracks, like a light switch. So the males transfer something via sperm that activates the germline, and that in turn causes the worms to shrink.

Back in 1996 David Gems had reported that mated worms lived shorter, but he didn't report anything about shrinking.[41] Gems did test whether sperm were involved, but because the *fer-6* sperm mutant didn't change the result, he concluded (as we did) that sperm are not involved. Instead, he theorized that males caused physical damage to the hermaphrodites, as he later proposed with the lifespan of males in a group. He mentioned but didn't test whether seminal fluid was required for the lifespan decrease. To test the role of seminal fluid, Cheng used a mutant called *gon-2*, which has no sperm or seminal fluid but can still physically mate; he found that these worms had a totally normal lifespan—which means that the seminal fluid is to blame for lifespan shortening. (This also means that Gems's "beating up" theory was wrong, since the worms could still physically interact.) The deleterious effects of mating, shrinking and shortened lifespan, were not only caused by factors transferred from the male to the hermaphrodite during mating but also act through genetically separate pathways: sperm transfer activates the germline of the female, which in turn causes the worms to lose their protective glycogen, fat, and water, causing them to shrink, while seminal fluid transfers a factor that regulates the mother's lifespan through the insulin signaling pathway.[42] Mating with males essentially condemns the mother to death, because all of her protective mechanisms, including DAF-16/FOXO activity in the nucleus, are turned off. Even the mighty *daf-2* mutant can't avoid this fate.

While germline activation and subsequent shrinking may be an unavoidable cost of reproduction, seminal-fluid-induced demise appears to be an act of pure evil on the part of the male, a strategy to kill the mother as soon as she births that male's progeny. This is a sledgehammer version of sperm competition: instead of changing the female's receptivity toward males to reduce the chances of her mating again, the elegant approach that flies use, the worm mother cannot mate with another male because she is already dead! (To be fair to the worms, hermaphrodites already run away from males as much as possible, since they don't need them until they run out of self-sperm, and we found that mating does cause true female worms to switch from attraction to avoidance of males.)[43] We also found that these pathways are shared with other *Caenorhabditis* species, as mating also causes shrinking and death, but, importantly, crossing two species together, which does not result in cross-progeny, does not induce shrinking or death,[44] suggesting that there is a species-specific component to each pathway. This again implies that these mating-induced processes are not something that is necessary for reproduction, but more likely evolved to be matched to the males' specific partner. My favorite write-up of my lab's work appeared in *Jezebel*, which described our results in an article titled "Sex Is 'Kiss of Death' for Female Worms because PATRIARCHY."[45]

Coincidentally, right around the time that Cheng was discovering that mating causes shrinking and death, Anne Brunet's lab had made a similar discovery, that having males around constantly caused hermaphrodites to die early.[46] In addition to the effects of mating, they also discovered that when hermaphrodites

are exposed to a lot of male pheromone—the chemical milieu that males secrete, mostly consisting of molecules called "ascarosides"—the hermaphrodites
died faster. If many males are placed onto plates and then removed, then hermaphrodites are added to these "male-pheromone-conditioned plates," the
shortened lifespan of the hermaphrodites is obvious.[47] Our papers appeared
back-to-back in Science, accompanied by a third paper by Scott Pletcher's group
showing that male flies who were presented with female pheromone but were
prevented from mating die earlier (of disappointment, apparently).[48]

All the experiments that Cheng did involved mating spermless females for
24 hours, but later Lauren Booth in Anne Brunet's lab found that a short
mating, just 2 hours, revealed more subtle effects.[49] Lauren found that young
wild-type animals were resistant to mating-induced demise, but older ones
succumbed. This could seem like just an effect of getting older, but in fact it
turned out to be a clever strategy used by hermaphrodites to slow down the
effects of male sperm: the presence of self-sperm in the mother protects her
against male-seminal-fluid-induced demise. In a collaboration between our
two labs, we found that the self-sperm regulate levels of a protective insulin-
like peptide antagonist that keeps DAF-16/FOXO in the nucleus; those levels fall as the hermaphrodite's own sperm are used up.[50] Upon mating, male
sperm induce the expression of a "bad" insulin agonist that activates DAF-2
and moves the protective DAF-16 out of the nucleus, making the mothers
susceptible. This elegant insulin agonist/antagonist mechanism prevents the
hermaphrodite from dying before she has produced all of her own self-
progeny, then switches off when there are no self-sperm left for her to use
(and this lack of sperm also induces pheromones that attract males to use up
the remaining oocytes).[51] Mating with males induces the insulin agonist,
which further hastens the mother's demise. In fact, the male-induced-demise
pathway coordinates most of the players you have already heard about in the
regulation of lifespan: insulin signaling, DAF-16/FOXO, PQM-1, mTOR,
and the TFEB homolog HLH-30, a transcription factor that is important in
the regulation of lipid metabolism.[52] That so many of these molecular players
had been found previously suggests that in fact these so-called "longevity"
pathways evolved to regulate mating-induced death and to optimize the
reproductive period—not to help worms live forever. (Someday the field will
realize this.) The molecular changes parallel the changes in behavior and attractiveness that are also mediated by self-sperm,[53] providing another layer of
regulation of interactions between males and females.

The battle between the sexes rages on.

What about the Guys?

Almost every time I gave a talk about Cheng's results, someone in the audience would ask me, "What about the males?" and I'd give some flippant reply about how sexist I am to explain why we didn't know anything about males, but eventually Cheng came through again. You'd think that, since mating is all that male worms are interested in, it would do them no harm, but, in fact, mating is almost as dangerous for them as it is for females—they also experience shrinking and a significant reduction in lifespan.[54] You might remember I mentioned that the lifespans of solitary males are longer than those of grouped males in germline-less mutants—this is how we discovered that males just have to "think" they are mating successfully in order for their lifespans to be shortened, but they do need a germline to do it.* But unlike the divided sperm/seminal-fluid pathways that regulate shrinking and lifespan, respectively, in hermaphrodites, they both shrink and decrease lifespan through the same pathway, one that requires a germline and the PQM-1 transcription factor. While this sounds like the same story I told you about females and hermaphrodites, obviously the males are not getting something from the females, so it struck us as a little weird. It gets a little complicated, but basically what we found is that males die primarily from two things: germline activation upon mating and male pheromone. But only the first one, germline activation, is conserved across all *Caenorhabditis* species and sexes.

First, let's tackle male pheromone (MP). MP presented a conundrum. Anne Brunet's lab had shown convincingly (and we confirmed) that when large numbers of males (30–50) are put onto plates and the males are removed, then the hermaphrodites lived shorter.[55] While it is clear that exposure to high doses of MP was bad for the hermaphrodites, I couldn't quite understand the logic of it—as I've mentioned before, it's pretty rare for there to be a lot of male *C. elegans* around, except briefly after mating, since half of those progeny will be males, but most of the time the population is almost completely hermaphroditic. It was hard for me to envision a case in the wild where this might be useful to the worms—when would there be really high numbers of males in a population? And more to the point, what would be the benefit of indiscriminately killing your mating partner even before you've mated, especially in the

---

* More surprisingly, when "masculinized" hermaphrodites were grouped, they lived shorter than the solitary controls, suggesting that neuronal masculinization of the hermaphrodites is sufficient to induce the production of male-like pheromone in these hermaphrodites, and that neurons are key for male-pheromone-mediated death.

gonochoristic species (those with equal numbers of males and females) where the whole population should be swimming in MP? Like homicide detectives, we wondered, *Who is male pheromone really meant to kill?*

With this question in mind, we set out to test the effect of MP on male lifespan.[56] The first tool in our arsenal was the mutant *daf-22*, which is defective for making most ascaroside-based pheromones, including MP. Our first finding was that, unlike wild-type worms, groups of these pheromone-less *daf-22* males aren't short-lived even though they do clump together, suggesting once again that MP, not physical damage, is responsible for the short lifespan of grouped males. Then we tested dosage of MP and found that lower levels of MP that don't faze the hermaphrodites at all still kill the males. In fact, the MP secreted onto a plate by *just one male* was enough to kill a poor pheromone-less male that we added later. This all suggested to us that the real target of MP is not the hermaphrodite but *other males*—which makes sense if they truly are competing with one another. At high enough concentrations hermaphrodites will still die, but more likely as collateral damage, not as the main target of MP. Our further experiments revealed that the germline is necessary for MP to do its dirty work: males that lack a germline (or are treated with the cell proliferation blocker FUdR) are not affected by MP, even at high doses. But one of the craziest things we found was that hermaphrodites who had just their neurons "masculinized" not only succumbed to MP like males do, but also secreted MP that kills other masculinized hermaphrodites and alters their attractiveness to males.[57] In sum, males are exquisitely sensitive to male pheromone, which seems to kill them through a pathway that involves both neurons and the germline.

Next, we wanted to ask whether mating is as bad for males as it is for hermaphrodites. Sadly, it's not good news for the males—just adding a single hermaphrodite to a plate to make a nice worm couple will cut a male's lifespan in half. The good news for the males is that harems don't make it worse—but the longer the males are paired with any females, no matter how many, the shorter they live. But still, that seemed weird—unlike hermaphrodites, who also live shorter after mating because of the seminal-fluid-induced activation of insulin signaling, the males shouldn't be receiving anything from the hermaphrodites, and the hermaphrodites don't secrete a male-killing pheromone, either. It turns out that germline activation from mating causes both shrinking and lifespan shortening in the males,* and, most bizarrely, proteins called vitellogenins that are necessary for carrying the lipids into oocytes that they will need for

---

* This process is dependent on PQM-1—that is, *pqm-1* mutant males don't die early when mated.

development and that are usually made only in females aberrantly get turned on in males when they get the "I've mated" signal.[58] This can't be good for the males, since these vitellogenin proteins aggregate into gloppy messes that kill worms when they start to accumulate outside of oocytes.[59] Both male pheromone and mating can kill male worms, but through different pathways.

Revenge of the Hermaphrodite

The last piece of the male mating puzzle came from looking at the effects of mating and MP on different species. *C. elegans* is *androdioecious*, meaning that it has hermaphrodites and a few males, while *gonochoristic* species have equal numbers of males and true females, no hermaphrodites. By testing several of these species for both mating-induced death and MP-induced death, a clear pattern emerged: all of the worms exhibited mating-induced death, whether male, female, or hermaphrodite and no matter the species, but only the males of androdioecious (hermaphroditic) species were killed by male pheromone.[60] The gonochoristic species also make MP, but they seem to use it for a more logical purpose: to guide them in choosing a partner to mate with. We found that *C. remanei* males are very attracted to supernatant from female worms and are repelled by male supernatant, while *C. elegans* males can't seem to figure out who's who nearly as well[61]—and why should they, when the odds are good that they will always run into a hermaphrodite? By contrast, the species that are 50:50 male to female need more information to pick who they should mate with.

Why would MP be lethal only to males of the species that are primarily hermaphroditic? Remember I mentioned that hermaphroditism evolved from true *Caenorhabditis* females at least three different times, so the fact that MP kills only males in these species doesn't seem like it could be an accident, but rather possibly useful in some way. Hermaphrodites have their own sperm and oocytes, so they don't need males in order to reproduce. In fact, they try to attract males only when they run out of their own sperm,[62] and those late progeny propagate only half of their genome, because it is diluted by the male's genome. Mating with males also cuts the hermaphrodite's lifespan short, so it's really just a last-ditch measure to use up her oocytes when she gets old. Put plainly, males are dangerous to hermaphrodites, who need them like a fish needs a bicycle. So what would be better than programming males to *get rid of themselves* when they are no longer useful to the hermaphrodite? Even better if what you use is super toxic specifically to males and acts in a dose-dependent manner—the more males you have, the more pheromone they'll make, and the more likely they are to die. Agatha Christie couldn't have dreamed up a better plan.

But killing males after they've already reproduced seems somewhat pointless, also. With this in mind, we tested another theory: perhaps MP affects the odds of making *more* males. In all of the experiments I've told you about so far, adult animals were treated with MP, but in the wild, animals would be exposed to MP their whole lives, from the time they hatched onward. We found that MP treatment during development not only shortened male lifespan even more but also affected male fertility, decreasing the number of progeny from mating and thus the number of males around.[63] If there were a brief time when hermaphrodites needed males around—say, in times of stress when outcrossing can increase the odds of survival—it would be hard to get rid of the males when they are no longer useful, because half of their progeny are males, and they mate, and so on. But the more males there are, the more MP will be around, so the less well they are able to reproduce. Within a few generations, the hermaphrodites could get rid of those pesky males, back down to the 0.1% that we usually find. We named this process of hermaphrodites trying to return to an all-hermaphrodite population *androstasis* (for homeostasis of the androdioecious state). The fact that the males produce their own poison and do not have their own Y chromosome to do anything about it is genius on the part of hermaphrodites, if you ask me—the revenge of the hermaphrodites.

Mate Choice, Pheromones, and Behavior

Above, when I described the effect of mating with males on hermaphrodite health, I told you that she dies soon after mating—which is true. However, I should note that even before the mother dies, there are some changes in behavior that you can see only if you work with "true female" worms in other *Caenorhabditis* species that are 50:50 male to female (gonochoristic). Perhaps unsurprisingly, if you watch the behavior of a worm couple, hermaphrodites pretty much ignore and even run away from males whether mated or not—it's actually called the "sprinting assay," and it's fun to watch them completely blow off the males. But true female (self-spermless) worms actually wait around for males and don't run away when males come near them. Once they have been mated, however, they switch their behavior completely and act like the hermaphrodites, running away when the males come near them. That is very much like the reduction in receptiveness that was reported originally in flies. But there is a cool twist: Maureen Barr's lab had previously reported that older hermaphrodites become more "attractive" to males (what she laughingly calls "the cougar effect") and this is controlled by the level of self-sperm that the hermaphrodites contain.[64] As soon as they use up their self-sperm, they secrete signals that are attractive to males, likely allowing them to use up their remaining oocytes. Cheng found that mating causes a sperm-dependent decrease in attractiveness to males; that is, mated females or hermaphrodites are downright repellant to males, like there's a signal that they are already taken.[65] Therefore, sperm content inside the female, whether self-sperm or from males, ultimately decreases attractiveness and attraction to males,[66] suggesting that it modulates the levels of secreted or surface pheromones that influence the behavior of the opposite sex. Responses to sperm levels allow the animal to make the best choice—run away or stay—depending on her own mating status.

———

I've spent a lot of time here talking about worms, and you may be wondering how this might relate to other organisms, or even your own situation[67]—will that pheromone perfume you bought online actually attract the one you want? And will that affect their lifespan?

Like the gonochoristic (male and female) *Caenorhabditis* that use male pheromone to identify who's who, other animals, particularly insects, use pheromones to attract and distinguish partners as well. For example, bombykol is a powerful pheromone produced by female silk moth (*Bombyx mori*) to attract males. Cockroaches, such as the African species *Nauphoeta cinerea*, have dominance hierarchies and use different pheromone components to distinguish the dominant from subordinate males. A paper with the *best* title, "Females Avoid Manipulative Males and Live Longer" (which sounds suspiciously like the advice from supercentenarian Jessie Gallan), lays this all out. Weirdly, female cockroaches preferred the subordinate males, or at least their pheromone. In fact, exposure to the "non-preferred" male, or its pheromone, actually decreased lifespan. These mothers with longer lifespans had progeny that developed more slowly; exposure to the pheromone from subordinate males slowed development, protecting the mothers from later re-mating, while exposure to the non-preferred, dominant male pheromones sped up development. In fact, many of the females who died did so in childbirth ("parturition," which we would have censored as matricide in worms, but they were counted as deaths here), which increased with rapid progeny development.[68] Therefore, female cockroaches appear to prefer mates that will allow them to avoid what the authors term "costly male manipulation." These observations in cockroaches underscore the fact that, throughout the animal kingdom, progeny production is dangerous for females, from worms to

insects to women, and in most species the male has a vested interest that affects the female.

One of the positive roles of male pheromone might be to give the female valuable information about his health status, particularly aspects that are less obvious
than outwardly visible characteristics, to influence her choice of mate. In addition
to passing on her own genes, selecting mates with "good" genes might improve
the chances of their offspring surviving. The ability to fight infection might be one
such invisible quality marker that could influence mate choice.[69] Male pheromones of the mealworm beetle, *Tenebrio molitor*, not only affect female behavior
and promote mating but also report on the immunocompetence of individual
males, allowing a female to choose the male that will best protect her progeny
from future infections. Likewise, several vertebrate species' females use olfactory
cues to choose mates with major histocompatibility complex (MHC) genes that
differ from their own, perhaps to allow greater pathogen surveillance. This theory
has been supported in a study of a small, genetically isolated human population,
the Hutterites; the MHC haplotypes of Hutterite couples were less likely to be
shared than by chance, suggesting that there may be some mechanism by which
humans detect this information in one another.[70]

In mammals, secreted pheromones are detected by tissue in the nasal cavity
called the vomeronasal organ (VNO) that relays signals to the hypothalamus
and other areas of the brain, releasing gonadotropin-releasing hormone and
subsequently affecting reproduction and sexual behaviors. Whether human
pheromones function similarly has been vigorously debated, but there is evidence of sex-specific differences in activation of regions of the brain in response
to male and female pheromones.[71] Application of female pheromone to the
upper lip had previously been shown to affect timing of ovulation and menstruation, but the concept of menstrual "synching" was subsequently countered
by a study of menstruation patterns in the Dogon, a population in Mali, West
Africa; no evidence of such timing coordination could be found.[72] As far as
male pheromone goes, the application of androstadienone, a male steroid, to
the upper lips of 40 female volunteers appeared to alter some psychometric
characteristics and to activate neurons in the VNO, suggesting at least the possibility that male pheromones could be detected.[73] Interestingly, the authors
suggest that androstadienone application reduced negative states like nervous-
ness and tension, which I think was a surprising result. Later studies of the same
steroid suggested that androstadienone has effects on males, including increasing cooperativity between men.[74] Perhaps the most interesting question about
human pheromones is whether they influence mate choice, as the perfume

industry would like you to believe. Here the contributions of pheromone and the VNO have been debated,[75] and whether there is an influence on mate choice has been studied. Further, whether human pheromones can affect lifespan is still unknown.[76] Perhaps once we know more about whether pheromones actually influence behavior, and the molecular pathways that are used to do so, we can then determine whether they affect lifespan in humans.

What about SEX?

Now for the scary question: Does sex decrease human lifespan? I'll be honest, I don't know the answer here—a Google search would probably be unwise, and the experimental setup is not ideal. But you might recall the long-lived Korean eunuchs, where lifespan appears to be increased when the germline is removed. What would be the opposite of that situation—can we ramp *up* germline activation in humans and see an effect on lifespan, perhaps through increased mating?

This is a bit of an awkward question, but Cheng, who is also an amateur Chinese historian on the side, started his lab presentation one day with a painting of the Forbidden City. He proceeded to tell us a bit of history about the lifespans of Chinese emperors, who had the best living conditions and medical care consistently through the years, so at least offered some sort of control conditions. The only people who lived in the Forbidden City were the emperor, his family, eunuchs, and many women. Naturally, the emperor could do whatever (and whomever) he wanted. Since records of the emperors' lives were kept, it is possible to know not only how long each one lived but some information about their lifestyles and proclivities, including whether they would be considered particularly promiscuous—for example, it was notable when some emperors demanded that *additional* women be brought in from the outside. ("The Emperor [Wuzong] was promiscuous, decadent, only interested in pleasure and had a lust for women. . . . The Emperor set off on a decadent pleasure trip trying out all the women-folk he could find, inside the Palace (Forbidden City) and later outside of it.")[77] With this information, an emperor could be classified as normal (the majority) or "promiscuous."

Cheng collected the information for 2000 years of Chinese emperors (210 BCE–1908 CE),[78] and he removed ("censored") any cases where the emperors died unnatural deaths, such as being killed in war or murdered, ruled less than a year, or were younger than 18 years old. He also controlled for other factors, like extreme alcohol use, which was documented in both groups. With the

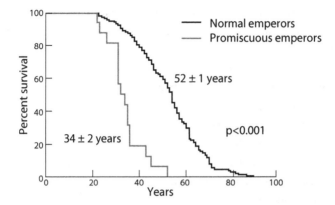

FIGURE 6. Cheng Shi found that normal emperors—that is, those who were not described as being particularly promiscuous—lived an average of 52 ± 1 years, while those described as being particularly promiscuous lived an average of only 34 ± 2 years. The difference in lifespan is significant ($p < 0.001$). Both groups were noted to drink alcohol and had conditions consistent with sexually transmitted diseases (Shi, Runnels, and Murphy 2016).

remaining emperors, he identified those who were noted for being exceptionally promiscuous, and compared their lifespans to those of the non-promiscuous (or, at least, not recorded as such in history) emperors.[79] The promiscuous emperors lived a whopping 35% shorter than the rest (figure 6)!

When we paired fathers and sons to better control for genetics (brothers were often deliberately killed, so it was not possible to compare directly) and standards of care at the time, the same difference emerged between promiscuous and non-promiscuous.[80] Of course, it's hard to extrapolate—did these men die of sexually transmitted diseases (which the non-promiscuous emperors were also likely exposed to)? Or other risky behaviors? Or really because of a lot of sex? But it seems that *something* interesting was going on. It's impossible to know now, and it is not immediately obvious to me how we can really test this in humans, so knowing how mating affects molecular pathways in model systems may be very helpful in understanding such effects. At the very least, it made for an amusing lab meeting.

———

Which of these principles of mating-induced and sex-specific effects of longevity are shared across all organisms, and which are specific to only certain animals or organisms? Pheromones are used to distinguish individuals in the same group, so it's unlikely that pheromones from one species can affect another, since receptors that detect them are required. Signals like seminal fluid sex peptides (*Drosophila*) and seminal fluid components (*C. elegans*) are species specific as well, but the general effects and functions of pheromones and other signals seem to be the same in many animals. We just don't know yet whether our reliance on other cues, particularly vision, has diminished human reliance on pheromone detection in choosing partners or greatly influencing behaviors. Germline activation and usage upon mating seems unavoidable, and the effects of the increase in germline cell usage might be the best conserved across species, but we will have to see what molecular mechanisms emerge from studies of animals beyond insects and animals that have a large number of progeny. It's possible that, despite the tantalizing studies of Korean eunuchs and promiscuous Chinese emperors, species with relatively low numbers of progeny, like humans, may be less affected by anything other than drastic changes in germline function. Other mechanisms are likely to allow the members of each sex of a species to be optimized for their own genetic success, regardless of the well-being of their mates.

The Jessie Gallans of the world may have found their own secret to longevity, but it's one that is unlikely to be the answer for most of us—no matter what happens in the ongoing battle of the sexes.

13

I See Dead Flies

NEURONS AND SENSORY REGULATION OF LONGEVITY

I see dead people.

—COLE SEAR, *THE SIXTH SENSE* (M. NIGHT SHYAMALAN)

IT SEEMS QUITE INTUITIVE THAT DIET, exercise, metabolism, and genetic factors figure into the complicated equation of lifespan—after all, those all modulate how well our cells function and maintain themselves. But would you be surprised to discover that what you smell, or taste, or even see might also affect how long you live? Or that losing one of your senses might actually extend your lifespan? The idea that the world around us might influence our longevity through our senses seems odd, unsettling even, but there is growing evidence that animals are not only sensitive to their surroundings but even have built-in systems to use this information to tune their longevity accordingly.

Losing Your Mind to Save Your Life: Killing Neurons Can Regulate Lifespan

In 1999, Cynthia Kenyon's lab announced a surprising discovery: getting rid of *C. elegans*'s neurons made the worm live longer.[1] That is, when Kenyon's graduate student Javier Apfeld compared the lifespans of wild-type worms and mutants with defects in the function of a subset of neurons that are responsible for smelling food, the mutants were longer-lived. (It's important to note that this was one of the first examples of "nonautonomous" signaling regulating lifespan—that is, cells that signal to other cells to control their metabolism and systemic longevity—a well-accepted notion now, but at the time it was groundbreaking.)

The neurons that Apfeld and Kenyon studied are called "ciliated neurons" because they have microtubule-based structures called cilia at their ends, which is fairly common in cells throughout the animal kingdom. Worms with defects in their sensory cilia, or in the neurons that convey the signal that food is being smelled, or in the signaling pathways that convey this information all lived longer. In fact, outright killing ("ablation") of the ciliated neurons using a laser, as Kenyon's group had done with the germline precursor cells (chapter 11) had the same lifespan-extending effect as the mutants did. The researchers were able to eliminate the trivial explanation, that lacking these neurons caused the worms to eat less and therefore they were just calorically restricted, since the cilia mutants had normal feeding behaviors. Their data suggested that worms normally sense some environmental cue that essentially shortens their lifespan, and loss of that sensation makes them live longer. Why in the world would the *lack* of a sensory signal make worms live long? And is this just a weird worm thing? (Short answer: no.) To understand this surprising phenomenon better, it helps to think about what was known about sensory function already.

Sensing One's Environment Is Critical for Survival Decisions

Animals use the information around them to make decisions all the time. Is there enough food and water here? Is it too crowded? Too hot? Too dangerous? The solution for most animals is to migrate until they can find food, water, whatever they need to improve their odds of survival. Another strategy is to shift metabolism into low gear and switch into a salvage mode until times are better, as in dietary restriction conditions, which slows reproduction and metabolism. While I've just described what most organisms can do, some animals are *really* good at surviving hard times. Mammalian hibernation and insect "overwintering" are versions of such adaptations, slowing and shifting metabolism to survive cold conditions. Other animals go into "diapause" states when times are tough—developmental arrests that allow the animals to resume normal function once the conditions, like temperature or food availability, are better. *C. elegans* is a champion in this area; it has multiple starvation-induced arrest states during development. Under extreme conditions it will go into an alternative developmental dauer (from the German for "lasting") diapause state before it develops its germline. It can survive in this pre-reproductive state for months, then resume development once conditions are more favorable. The worms make this developmental decision in the larval stage right after they hatch from eggs. By surveying their conditions to determine whether there is enough food, whether the temperature is acceptable, and whether it's too crowded, they decide either to develop normally or to commit to going down the dauer developmental path. If they choose the latter, they will adjust their metabolism to favor use of stored sugars, remodel their outer coverings to become tougher, close their "mouths" to stop feeding, and even change their neuronal dendrites.[2] All of these adjustments help them survive harsh environments. They even have a special escape behavior called "nictation" that causes them to stand up stiffly on their tails, allowing them to hitch a ride to a new location if an animal or insect happens to come by. (In one of my favorite papers ever, Junho Lee's lab (Seoul National University) and colleagues used a special light-activated protein to activate the neurons that cause the worms to nictate—when they shined light on the optogenetically engineered worms, all of them stood on their tails at the same time, like zombies. And then they showed that those nictating worms could get carried to a previously empty plate if they added flies to a different chamber, showing for the first time that this idea of nictation allowing animal-based "migration" of worms might be correct.)[3]

The decision to go into dauer or not is a critical one for the worms: if they make the wrong decision in either way, they lose. Going into dauer will protect them from all kinds of poor conditions. So why not do this all the time? Because it's costly. In addition to the energy they spend to remodel their tissues entering and exiting dauer, if they choose to enter dauer unnecessarily, then they lose out in the evolutionary sense of competition, compared with the animals that chose to plow on with development and start reproducing, putting their own genomes out into the world. Thus, the dauer decision is one that is not taken lightly.

Neurons are at the heart of this dauer decision. To choose correctly, worms must take in cues from their environment, sensing food odors, pheromones (to measure crowding), and temperature. As a postdoc in the lab of Bob Horvitz (who won a Nobel for his early work on *C. elegans*), Cori Bargmann carried out a groundbreaking study showing that just a few of the worm's 302 neurons are responsible for this dauer decision. She used a laser to kill neurons that sense odors, temperature, chemicals, and multiple cues, then measured the propensity of those animals to go into dauer. Her elegant experiments (amazingly, executed before fluorescent labeling of neurons had been developed, so she had to identify the neurons only by light microscopy, an impressive feat) revealed that just four specific neurons—called ADF, ASI, ASG, and ASJ—determined whether the worm would go into dauer or not.[4] On the surface, this seems like a story about worms, but what it really tells us is that there are simple networks of neurons that sense the environment and then use that information to determine the most important developmental and metabolic decisions. Worms just happen to do it with a couple "pro-dauer" and a couple of "anti-dauer" neurons that take in the sensory cues; the neurons calculate the output to shift the likelihood of dauer formation through the regulation of molecular signaling pathways. Exactly *how* this calculation is made is still not fully understood, but the output includes both the insulin signaling pathway (like *daf-2* and *daf-16* function) and the transforming growth factor beta (TGF-beta) pathway, though a gene called *daf-7*. DAF-7 is a TGF-beta "ligand"—that is, a protein that is secreted from one cell and binds to a receptor on another cell. DAF-7 is made in the ASI neuron, and signals to other cells about the dauer decision status.

I've just told you a ton about dauer regulation, but what does that have to do with lifespan? I usually hesitate to even mention dauer, because immediately someone might reasonably think that this is just a weird worm phenomenon that requires dauer formation, and thus the whole pathway is irrelevant to animals other than *C. elegans*. But we already know from Dillin and Kenyon's knockdown of insulin signaling only in adults (see chapter 6) that worms don't need to go into dauer (a developmental phase) to live long. Other animals don't go into diapause, so the connection to longevity regulation may at first seem tenuous—and as I mentioned before, the fact that *C. elegans* can form dauers, while helping us understand worm longevity regulation via mutants that both live long and increase dauer formation, may have slowed the understanding that longevity regulation can happen outside of worms. However, both of the major pathways that regulate dauer formation, insulin signaling and TGF-beta, are well conserved in higher animals and also regulate lifespan, and in both cases the genes in these pathways can affect lifespan independently of the animals entering dauer. Understanding the neurons involved in this process is helpful.

The fact that *loss* of neurons could extend lifespan at first may seem counterintuitive to most of us—don't you need all of your neurons to be functional in order to live long? While the fact that animals who have lost the function of some of their neurons live longer seems surprising, the reason that Apfeld and Kenyon were able to guess correctly that neurons might be responsible for these lifespan decisions is because of the dauer studies done by Bargmann, which already hinted at the role that neurons might play in regulating lifespan—that is, worms use these neurons to decide whether conditions are good enough to proceed on to reproduction, or are bad and therefore they should wait it out in dauer until times are better. Perhaps the simplest way to think about it is that the worms use these neurons to regulate the insulin signaling pathway in this *go/no-go* decision. If they perceive that food is present, then they communicate with the germline and somatic gonad that times are good enough to reproduce, so they should proceed full blast ahead—but if the conditions are not good (for example, if there's no food around, or there are too many other worms—lots of pheromone), then it might be good to slow things down and wait to reproduce when conditions are better. These signaling pathways seem to be used not just in the dauer decision pathway, but also to decide whether to proceed forward through reproduction, even in adults.

Because Apfeld and Kenyon could do the same neuron-killing treatments on other mutants, and combine the sensory mutants with other longevity mutants, they were able to show that much of the sensory neuron lifespan effect depends on our old friends, insulin signaling and DAF-16/FOXO, the transcription factor that mediates insulin signaling regulation of lifespan. Perhaps most surprisingly, loss of somatic gonad signaling, which normally shortens lifespan (remember that it "undoes" the lifespan-extending effect of germline ablation) did not shorten the lifespan of sensory neuron mutants, meaning that the two pathways are redundant with one another. Following in the footsteps of Bargmann's neuron ablation studies of the dauer decision, Joy Alcedo, a postdoc in Kenyon's lab, went on to carry out painstaking laser ablations of individual neurons to determine which ones are necessary for normal lifespan, and which ones might work in opposition.[5] She found that several of the best-studied worm neurons that sense food-like odors like butanone and isoamyl alcohol (the AWC, AWB, and AWA neurons) had effects on lifespan even when just that one specific neuron was killed. She also determined that two neurons, the ASI and ASG, normally act to shorten lifespan, so killing them increases longevity, while two other neurons, the ASJ and ASK, act in the opposite way, and together they all act through the insulin signaling pathway.

The Regulation of Lifespan by Neurons is Cell Non-autonomous

The Kenyon lab's studies firmly established that sensory systems can regulate lifespan. It is important to note that this was one of the first examples of "non-autonomous" signaling from somatic cells that regulates lifespan—that is, cells that signal to other cells to control their metabolism. This was a really big deal because it suggests that some tissues, neurons in particular, could make longevity decisions for the whole animal, rather than each tissue just doing its own thing and an organism dying when whichever tissue failed first. Such a system-wide nonautonomous effect was of course obvious from the dauer decision work, but the revelation that these same decisions impact longevity had more implications for other animals. Later work by Natasha Libina in Kenyon's lab revealed that the tissue-specific, cell-nonautonomous functions of DAF-16 control longevity;[6] the insulin signaling pathway plays a key role through the expression of insulin-like agonist and antagonist peptide signals in remote tissues. The downstream target of this nonautonomous signaling is primarily DAF-16. For example, overexpression of the heat shock factor HSF-1 in neurons extends lifespan via its activation of DAF-16 in peripheral tissues, particularly the intestine.[7]

Other groups went on to find that the ASI neuron is also the site of dietary regulation of lifespan, through a key transcription factor called SKN-1.[8] Further investigations of neuronal signaling pathways have blossomed in the field, leading to an ever-growing understanding of the ways in which neuronal signaling is conveyed to other cells to affect longevity. A dizzying web of insulin-like peptides, biogenic monoamine signals (e.g., serotonin, dopamine), and AMPK/TOR signaling from neurons to intestinal cells regulate stress response gene transcription, metabolism, proteostasis, mitochondrial stress responses, and autophagy in downstream tissues, which ultimately determine how long the animals are able to survive the stress of living[9]—and these pathways all appear to be conserved from worms to flies to mice.

Even microRNAs play a role in conveying signals from neurons to the rest of the animal to regulate longevity. Thorsten Hoppe's group (CECAD) carried out a clever screen of miRNA mutants to identify ER-stress-induced mechanisms of cell-nonautonomous regulation of longevity: they used a worm that would report on breakdown of the ubiquitin-mediated proteasomal turnover pathway induced by ER stress—mutants that stop the proteasomal process produce worms with bright green guts.[10] The screen revealed that the miRNA *mir-71* is required for proper proteostasis, because it regulates the expression of *tir-1*, which encodes a receptor in the AWC olfactory neurons that sense food, and in turn, TIR-1 regulates gut proteostasis and longevity. This same miRNA, *mir-71*, is also implicated in communication between the neurons and the somatic gonad, as its neuronal expression in neurons is required for germline-mediated DAF-16 nuclear localization in the intestine and subsequent longevity.[11] *C. elegans* can even distinguish different types of food through its neurons; a receptor called neuromedin U (NMUR-1) is responsible for a lifespan-shortening effect when worms are fed *E. coli* rather than other bacteria.[12] This suggests that the NMUR-1 receptor plays a role in sensing and distinguishing food sources, affecting longevity.

Most recently, even astrocyte-like glial cells—which were long considered to be "support cells" for neurons but are now being recognized for their own functions—have been found to play a role in longevity regulation.[13] Expression of a key regulator of the ER stress unfolded protein response (UPR^{ER}) in just four "cephalic" glial cells increases lifespan, through regulation of neuropeptides that convey ER stress status to the rest of the animal. This nonautonomous mechanism is very much like those shown to function in neurons. We are still discovering all the ways in which sensory cells signal to peripheral tissues to modify longevity in response to nutrient cues, but it is clear that the sense of smell is critical for longevity regulation.

Smelling and Tasting Food Also Shortens Lifespan of Flies—and Mice

Once the principles of neuronal regulation of longevity were shown in worms, it didn't take long for them to be shown in *Drosophila* as well. Although flies don't have as large a family of insulin-like peptides as worms have—only 8, compared with about 40 in *C. elegans*[14]—many of the same rules seem to apply—for example, with signaling from neurons via their *Drosophila* insulin-like peptides.[15] Flies lacking a sense of smell or taste have increased lifespan, paralleling the results in *C. elegans*.[16] Scott Pletcher's lab showed that just the smell of food-derived odors, such as the smell of yeast, could "undo" dietary-restriction-induced longevity—that is, even when the flies didn't eat, so technically were still on DR, they didn't reap the benefits.[17] (Can you imagine going to all the effort to calorically restrict yourself, and then realize that the scent of your spouse's bagels has undone all your hard work?) And they knew which receptor this undoing of DR by smell required, because if a specific odor receptor (Or83b) was depleted, then the effect went away.

A different but important question is whether forcing animals to consume a particular diet has the same effect as allowing them to choose their own diets. (This may have implications for humans to actually buy into adopting DR.) Jenny Ro and Pletcher developed new methods to study food choice in fruit flies, adding a new layer beyond just simple intake of nutrients and longevity output that involves food choice and its regulation by biogenic amine (serotonin) signaling.[18] Flies can even have their lifespans shortened by disappointment (see males and lack of mating),[19] but the principle still holds: the sense of smell affects lifespan even in animals that can't go into dauer, suggesting that the notion that sensory function regulates lifespan is not just a worm phenomenon.

OK, this is all fine and good in worms and flies, but could the scent of food affect the lifespan of a mammal? In mice, the functional equivalent of the worm and fly olfactory neurons are OSNs, or olfactory sensory neurons. When OSNs were disrupted in adult mice, the mice became lean and resistant to diet-induced obesity, even when fed a high-calorie or high-fat diet.[20] In fact, loss of smell due to OSN disruption improved insulin resistance, fat mass, and lipolysis. The adipose tissue of the healthy, hyposmic mice "browned"—that is, mitochondrial number and respiration increased, burning calories. Sounds great, right? Maybe not. In a twist worthy of O. Henry, you could eat whatever you want without gaining weight . . . you just can't taste anything. On the flip side, loss of the IGF-1 receptor in OSNs reversed all these effects, as one might expect in insulin-resistant animals.[21] It seems that the sense of smell does in fact regulate metabolism in peripheral tissues, but whether there was an effect on lifespan wasn't mentioned here.

Heat and Pain Sensation Affect Longevity

The smell of food isn't the only sensation that plays a role in longevity regulation—thermosensation can also regulate lifespan. This is perhaps unsurprising, since the dauer decision also depends on temperature, but Seung-Jae Lee and Cynthia Kenyon showed that the AFD neuron, which senses temperature, regulates lifespan via the interneuron AIY and another dauer pathway, the daf-9/daf-12 nuclear hormone signaling pathway.[22] The CREB homolog CRH-1 appears to also play a role in the regulation of lifespan at warm temperatures by the thermosensory neuron AFD, through the now-familiar insulin-like peptide INS-7 and a neuropeptide called FLP-6.[23] The fact that everything lives longer at lower temperatures might at first seem to be due to slower metabolism, but Shawn Xu's lab (University of Michigan) showed that worms can sense noxious cold via the TRPA-1 receptor, and this also affects how long the worm will live.[24] These data show that regulation is a deliberate process, not just the result of passive changes in metabolic rates. Amazingly, the human version of this channel, TRPA1, can functionally replace the worm version, suggesting that the two must work very similarly.

Of course, you know where this is going next—mice. But mammals are not cold-blooded, so perhaps there is less of a range of longevity regulation by external factors. Moreover, many senses are associated with dangers—avoiding death by predation will certainly increase survival, for example—but may also cause a chronic stress response, which may shorten lifespan. But connections between senses, particularly olfactory cues, and metabolism might make sense. AMPK regulation of mTOR is one of the conserved mechanisms of longevity regulation, and another factor, CRTC-1, acts downstream of AMPK in worms, along with the calcium-sensing calcineurin.[25] Not only does this AMPK/CRTC-1 pathway act in worm neurons to regulate lifespan, but this same pathway acts in mouse neurons, downstream of a pain receptor.[26] The TRPV1 ion channel senses noxious stimuli, including pain sensation—for example, from capsaicin, the chemical that is found in hot peppers. (This may sound familiar if you read about the 2021 Nobel Prize in Medicine, which was jointly awarded to David Julius [UCSF] and Ardem Patapoutian [Scripps Research] for their work on pain and pressure sensors.)[27] The TRPV1 pain receptor is present in neurons of the central nervous system that innervate the pancreas, which in turn regulates the secretion of neuropeptides, calcitonin gene-related peptide (CGRP), and insulin. For that reason, mice with TRPV1 knocked out, or mice with olfactory neurons ablated, are less susceptible to diet-induced obesity caused by a high-fat diet and are leaner owing to higher energy expenditure.[28] Celine Riera and Andrew Dillin showed that these TRPV1-knockout mice are also long-lived because of metabolic changes. The mice also show lower rates of inflammation, improved glucose tolerance, and better energy expenditure. So eat those hot peppers to live longer!

Just Breathe: The Sensation of Gases Affects Lifespan

Oxygen is, of course, critical for the survival of aerobic animals, so it is perhaps intuitive that sensing of oxygen and carbon dioxide might regulate longevity. (*C. elegans* actually prefers a lower oxygen level than ambient air—5%–12%

instead of 21%[29]—perhaps to better identify sites of active bacterial metabolism.) Under low-oxygen conditions, cells go into a state that helps them protect themselves, using the hypoxia induced factor, or HIF-1. In normal conditions (normoxia) HIF-1 is "off" because it is degraded through an oxygen-dependent proteasomal degradation process, but in low oxygen HIF-1 becomes stabilized.[30] Loss of the E3 ubiquitin ligase that promotes degradation (VHL-1) causes constitutive stabilization of HIF-1 and increases lifespan.[31] This longevity could have been caused by a cell-autonomous effect—all in the intestine, for example—but HIF-1 activity in *C. elegans* neurons is able to activate the xenobiotic detoxification flavin-containing monooxygenase FMO-2 in the intestine, through a serotonin signaling pathway and the activity of the HLH-30 transcription factor, resulting in improved proteostasis and longer lifespan.[32] With such a system, if an animal's neurons sense low oxygen, a protective longevity mechanism gets activated cell nonautonomously.

But hypoxia sensing in mothers can also affect the lifespan of their offspring, as my lab found somewhat accidentally: we discovered that there is a trade-off in hypoxia survival from generation to generation that is regulated by the zinc-finger transcription factor PQM-1. This curious factor is the "anti-FOXO," as it usually acts in opposition to DAF-16/FOXO downstream of the DAF-2 insulin/IGF-1 receptor.[33] PQM-1 and DAF-16 trade places in the nucleus, shifting the cell's output to "development" (promoted by PQM-1) or "survival" (promoted by DAF-16), like a cellular teeter-totter. Naturally we wanted to understand that better, by figuring out what proteins might interact with PQM-1. In the course of growing a ton of worms on very large plates for biochemistry experiments, Thomas Heimbucher, a postdoc in my lab at the time, accidentally allowed condensation to seal the plates around the edges, creating a little hypoxic chamber for the worms by mistake. He first found that worms overexpressing PQM-1 had a strange, zebra-striped (or, in Princeton terms, tiger-striped) pattern, the result of some intestinal cells storing more fat. But Thomas also found that *pqm-1* mutants survived this hypoxic condition better than wild-type animals did.[34] Normally we think of treatments that help animals survive better as good, but Thomas also noticed that their progeny didn't look so great. Through a series of genetic and genomic experiments, he found that *pqm-1* mutant mothers survive longer under hypoxia by hoarding fat instead of properly provisioning their developing eggs. Their progeny can't survive hypoxia because their selfish mothers stole all of their protective fat. Worms don't usually get any credit for being good moms, but these results tell us that PQM-1 normally functions to help mothers sacrifice a bit for the good of their children.

The ability to sense carbon dioxide has also been linked to lifespan regulation. When flies lack a receptor for CO_2 called the gustatory receptor 63a, or Gr63a, they live longer, store more fat, and reproduce longer.[35] Similarly, if the neurons in which this receptor is normally expressed are killed, they also live longer. Why might that be? One clue is that when presented with the smell of live yeast, their major food source (and remember, this yeast smell was enough to undo caloric-restriction-induced longevity), flies lacking the Gr63a receptor still lived long. This tells us that one component of the "yeast smell"—at least to flies—is the scent of carbon dioxide. It further suggests that flies might use this as a way to distinguish live yeast from dead yeast. They prefer the former, and CO_2 is a good marker of live yeast, since dead yeast stop respiring. While normally we would expect such a pathway to overlap with DR pathways, the fact that Gr63a mutant flies could live even longer if dietarily restricted suggests that these are separate, possibly additive mechanisms to extend lifespan (or that each pathway is not fully activated, so can look additive).

In general, the pathways that regulate *C. elegans* longevity fit into the category of neurons that sense food satiety, hunger, oxygen, temperature, stress, and crowding, and do so through the genetic pathways that either help these neurons to develop or function. Their activity conveys the status of the neurons to downstream cells that more directly affect metabolism and lifespan.[36] These studies established some basic principles: an important survival feature (food, temperature, etc.) is sensed by neurons, those neurons or neural circuits then regulate the output of molecules that transmit the information (e.g., neurotransmitters such as serotonin, or neuropeptides, like insulin-like peptides and the neuromedin U peptides involved in food sensation), in turn they signal downstream to larger tissues that carry out both cell-autonomous (e.g., proteostasis and other cell-preserving biochemical functions) and cell-nonautonomous (signaling) functions that coordinate a systemic response. This response results in either high reproduction and short lifespan or delayed reproduction and long lifespan, depending on the signal from the sensory neurons. For example, insulins in the intestine are regulated by DAF-16 to signal to neighboring cells, allowing the nonautonomous, systemic regulation of insulin signaling.[37] Steroid hormones also perform such a role; the nuclear hormone receptor DAF-12 is activated by its ligand, which is made by the DAF-9 cytochrome P450. We are still discovering new sensory pathways that result in system-wide survival decisions.

I See Dead Flies

Smelling or tasting food or sensing oxygen and carbon dioxide can affect the lifespan of worms and flies and maybe mice. Temperature regulation may also affect your lifespan, particularly if you are a cold-blooded animal, and pain sensing, or at least activation of pain receptors, can also affect lifespan. But what about what you can *see*?

Amazingly, Scott Pletcher's lab (University of Michigan) found that when fruit flies see other dead flies, it makes them live shorter.[38] Basically, *Drosophila* can sense dead kindred, and that is so stressful to the flies, they die. As crazy as that might sound, what's even nuttier is that they have to *see* them for this effect to manifest. Most of us in the field would have expected the flies to have sensed death or infection through odor, but the researchers disproved this in several different ways: first, flies with impaired odor sensing still showed the same behavior, and dead flies behind plexiglass had the same lifespan-shortening effect, while flies that couldn't see—either genetically or because their eyes were painted over—didn't show the behavior. Even creepier, the flies cared about the dead flies only if they had died by what looked like "natural causes," such as by old age or infection; flies that were flash frozen *looked* alive and didn't cause other flies to die faster. On top of this, only flies of the same family caused the effect; apparently, they don't care if some other species of fly is dead. The serotonin pathway is required for these death-vision lifespan effects, essentially reversing the kind of serotonin-mediated lifespan extension we see in DR and hypoxia-mediated proteostasis regulation, linking these different sensory inputs to a shared pathway of longevity regulation. All in all, like the kid in *The Sixth Sense*, these flies could see dead flies, and it stressed them out so much that it caused their early demise.

What's the logic to this system? Well, if you're a fly and you see a bunch of dead flies, you might take it as a cue that this particular location is not a great place to be (perhaps there's a pathogen or noxious agent) and you should take the hint and scram. One can imagine that being forced to remain in the same place, gazing at your dead compatriots without being able to escape, is super stressful.

Do You Hear What I Hear?

Shawn Xu (University of Michigan) has made a number of surprising discoveries regarding the ability of worms to sense things we didn't realize before—but did you know they can hear? In 2021, his lab found that worms are sensitive to airborne sound through a mechanism in the skin that communicates vibrations to neurons through acetylcholine receptors,[39] kind of like how vibrations act on our eardrums. Whether sound can affect worms' longevity is not yet known, but considering how sensitive they are to everything else, it wouldn't surprise me.

―――――

We don't yet know how many of these sensory pathways might affect human lifespan, but the wealth of data from invertebrates and some work in mice implicating shared genetic pathways suggest that it is not beyond the realm of possibility. That might make a lot of sense; your environment has information that's important to your health and survival—either because it's relevant to your food intake and future metabolism and reproduction, or because it might warn you about dangerous conditions. We know that the smell and taste of food triggers hormone responses, including insulins, so the parallels with model systems is not so far-fetched. And it's already recognized that stressful conditions can have both immediate and long-term effects that may impinge on survival in both positive (hormetic—"what doesn't kill me makes me stronger") and negative (chronic stress) directions. Perhaps additional work in vertebrates, like killifish, mice, and primates, will uncover mechanisms that will suggest whether there are links between sensory systems and regulation of longevity in humans.

14

Don't You Forget about Me

WHAT WE ARE LEARNING ABOUT
COGNITIVE AGING AND HOW TO SLOW IT

If we can't make memories, we can't heal.

—LEONARD SHELBY, *MEMENTO* (CHRISTOPHER NOLAN)

What I lost is, in the grand scope of things, almost . . . negligible. It's true that there's grief: it wakes me in a cold sweat thinking, *Who was I? What did I care about? What did I find funny, sad, stupid, painful? Was I happy?* All of those memories I accumulated, gone. Which one, if there could have been only one, would I have kept?

—NICOLE KRAUSS, *MAN WALKS INTO A ROOM*

AS CLICHÉD AS IT MIGHT SOUND, the loss of memory is perhaps the greatest decline that we experience with age. The notion that we are made up of our life experiences, and our ability to remember them, is fundamental to our understanding of ourselves. Who are we if we can't remember our lives, our personal history? Being able to run and walk into my old age is important, but perhaps not as important as being able to think, create, reason, and remember. All of these abilities are crucial to my identity, not only as a scientist but as a human being. Watching someone we love—and who otherwise might seem totally healthy—slip away from us as they lose their ability to remember is undeniably one of the worst things about aging. This intertwinement of memory with our sense of identity is an ancient, if debated, concept; Locke explored the connections between identity and memory, theorizing that as memory faded, so did one's identity. As Borges's *Funes the Memorious* reminds us, we know that exceptional memory doesn't necessarily equal intelligence or success, but we'd still like to keep what memory we have functioning for as long as we can.

Ironically, as we successfully treat other previously deadly age-related disorders, like cardiovascular disease and cancers, the number of elderly who suffer from neurodegenerative disorders and dementias with age is rising simply because more people are living long enough to experience these declines. Therefore, slowing cognitive aging is one of the most imperative goals of aging research, in my opinion. Despite the obsession by some billionaires and the popular press with increasing lifespan, extending our ability to think and reason with age is far more important both for individuals and for society. As anyone who has spent time caring for a person with cognitive impairment can tell you, this is one of the most draining and expensive aspects of aging.

Because of the distinctions between non-pathological cognitive aging and age-related neurodegeneration, it can be frustrating to realize how little we know about *normal* slowing of our brains with age. Research into (and funding for) normal cognitive aging has, until very recently, fallen into the cracks between aging, neurobiology, and neurodegeneration research. (For full disclosure, this issue is so important to me that in 2019 I became the director of the Simons Foundation's Collaboration on Plasticity in the Aging Brain (SCPAB), which is funding pioneering research into cognitive aging.)

Part of our lack of knowledge about cognitive aging is due to the fact that changes in learning and memory with age are harder to study than simply looking for longer lifespan. Another issue is that many longevity researchers believe that treatments to extend lifespan will automatically improve cognitive function, so it's not necessary to distinguish the two. The assumption that it's unnecessary to specifically study cognitive decline when examining longevity extension is a dangerous (and, frankly, lazy) view. In fact, we have found that some lifespan-extending measures, like reduced insulin signaling, also help maintain cognitive function with age, while others extend lifespan with no measurable effect on learning and memory—and, worse, some have deleterious effects. Recently, we have also found a few ways to improve memory with no effects on longevity in model systems. Finally, as I'll discuss in detail, much of what we know about cognitive decline is strictly through studies of Alzheimer's disease, which has limited what we know about cognitive decline in the context of normal aging.

In this chapter I'll explore what we currently know about cognitive aging, provide a brief update on the state of the Alzheimer's field, and present some promising directions that might not only help with normal cognitive decline but even slow other age-related neurodegenerative disorders, including Alzheimer's disease.

What about Normal Cognitive Aging?

At this point, because of work done by many labs on humans and model system animals, we know quite a lot about what it takes for a neuron to grow and develop and to function. Some human neurons might last a lifetime without being replaced, and therefore understanding how to keep them working is critical. (Replacement of neurons in aged humans is hoped for, but its possibility is hotly debated.) So why don't we understand how to keep neurons functioning with age? The early, stereotypical and relatively fast stages of a cell *becoming* a neuron have been well studied, but age-related falling apart isn't understood in quite as much detail yet. Rather than an ordered reversal of developmental processes, aging is a separate process all unto itself, guided by stochastic inputs, the slings and arrows of everyday misfortune. When *exactly* a neuron's function starts to decline is not known, and it can be different for different neuron types—and may first occur after decades in humans. Moreover, what exactly falls apart first, molecularly speaking, might be different in different neuron types. Overall, studying the process of neuronal aging is a different problem from studying how cells develop into neurons; a process that is inherently temporally variable, starting at random times over a huge time span, is extremely difficult to study. This means that we need to collect a lot more data to fully understand it, compared with the often-stereotyped, tightly regulated, and rapid processes that occur in early development. Therefore, it's important to distinguish what's necessary for a neuron to develop from how it breaks down with age—and that process during normal aging is likely to be distinct from neurodegenerative processes that are induced by specific pathologically aggregated proteins.

Neuronal Aging at the Cellular Level

What happens when a neuron fails? Not surprisingly, the answer is "lots of stuff." At the subcellular level, when organelles like lysosomes stop working (that is, their pH no longer changes, so they can't properly process and degrade incoming damaged proteins), then autophagy stops and the usual cleanup processes that cells rely on—the trash collectors of the cell—stop working. Broken proteins and other cellular trash starts piling up, aggregating into useless and even toxic clumps, including the seeds of several neurodegenerative plaques. If transcription factors stop getting their usual activation signals, then they may no longer enter the nucleus at the right time, or may bind to DNA without properly activating transcription, or mistakenly transcribe non-neuronal, developmental genes. The cell stops making the correct new mRNA transcripts and the corresponding proteins, including those that might replace damaged proteins and others that might help in the process of repairing damage. On the other hand, the nucleolus—that factory in the nucleus where ribosomes are made—swells in old cells, perhaps because they are desperately trying to make new ribosomal components to address the ongoing functional protein shortage. Any chromatin or double-stranded DNA leaking from damaged nuclei will be detected by the cGAS-STING alarm pathway in the cytoplasm, signaling cellular distress and inducing senescence. Damaged mitochondria will not only fail to make sufficient ATP (which neurons need a *lot* of) but also may generate damaging oxygen radicals as they become leaky and send distress signals that may usher in further destruction.

But those are all events that could happen in any old cell—what fails specifically in neurons with age? Of course, one thing that is special about neurons is the synapse, the specialized compartment that communicates with other neurons.[1] To work, synapses need not only the proteins that are resident there but also a constant supply of new proteins, receptors, mRNAs, neuropeptides, and neurotransmitters delivered to the synaptic compartment. Loss of any of the critical components at the synapse will cause synaptic dysfunction,[2] which of course means that the neuron can no longer sense and convey information. These proteins and mRNAs are delivered via a highly regulated system of delivery of cargo down axons to synapses, as well as trafficking and turnover at the synapse itself. Motor proteins known as "kinesins," such as KIF1A/UNC-104, carry vesicles filled with cargo destined for synapses, walking along microtubule tracks in axons. In old worm neurons, KIF1A/UNC-104 stops working, so those vesicles and cargo don't make it to their destination—but in long-lived *daf-2* worms, which have extended learning and memory with age, KIF1A/UNC-104 keeps working longer and helps maintain function of the synaptic traffic system.[3] The downregulation of *daf-2* signaling allows these worms not only to live longer but also to keep their neurons functioning better

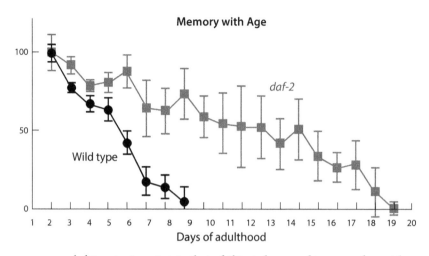

FIGURE 7. *daf-2* mutants maintain their ability to learn and to remember with age better than wild type. This ability exceeds their proportional extension of lifespan (author's lab, unpublished results).

with age. In fact, *daf-2* mutants extend their memory ability proportionally longer than they extend lifespan (figure 7).

With age, neurons stop acting like neurons. Just like a person with amnesia, neurons can lose their transcriptional "identity" and essentially forget who they are with age—and cells that don't remember who they are don't work quite right. Gene expression studies (including my own lab's work in old worm neurons)[4] show that "housekeeping" and other generic cell-function genes may continue to be expressed, but that neuron-specific genes get turned off with age, while developmental genes that have been silenced during adulthood get aberrantly turned back on. This phenomenon is interesting, because it suggests that specific transcription factors that regulate neuronal gene expression might be particularly susceptible to aging. Gene expression changes in aging neurons might be well conserved between animals, since we see similar functions decline with age from worms to humans.[5] We are hoping that by documenting these changes in gene expression in brains at multiple ages in many different animals (*C. elegans*, killifish, mice, nonhuman primates, and postmortem brain tissue in humans) we might identify some of the very early and highly conserved gene expression changes that may lead ultimately to neuronal failure, and that these might be good targets for clinical interventions.[6]

While neurons are generally post-mitotic, one possibility is that they may undergo replacement at some rate through the activity of neural stem cells

(NSCs).[7] Increased NSC activity might help slow age-related cognitive decline, particularly in the hippocampus, the major site of memory storage.[8] This field is a bit contentious; while adult neurogenesis is measured frequently in rodents (mostly using immunostaining techniques), several studies using single-cell RNA sequencing of postmortem human brain tissue have suggested that humans lack neurogenesis in the hippocampus in adulthood[9]—but the debate will go on, likely until live imaging of neurogenesis can be done. The hope of course is that there is some NSC activity in humans that might be harnessed or boosted to help slow cognitive aging or to replace neurons lost in neurodegenerative disorders. Techniques that could induce even a very tiny number of NSCs to become active and renew the aging brain might be helpful.

Another possibility is that cells could be reprogrammed to regenerate to replace aging neurons, particularly in the case of age-related decline or in neurodegenerative diseases, using "induced pluripotent stem cells" (iPSCs). These iPSCs can be derived from patients' own somatic cells, reprogrammed by adding "Yamanaka factors"—transcription factors that allow a cell to become pluripotent (able to make all other cells)—and then differentiated to form new neural and glial cells (see chapter 10 for more info).[10] Recently, retinal cells transfected with three of the four Yamanaka factors were able to rescue the functions that decline in retinal cells with age, including sight.[11] Even more excitingly, neurons of the mouse dentate gyrus were also amenable to reprogramming, suggesting that there may be a possibility of reprogramming neuronal cells of the brain (not just peripheral neurons like retinal cells) to replace dying neurons.[12] Perhaps one day we will renew our brains by inducing regeneration of new neurons. How these new cells will maintain our long-term memories is still unclear, but since at least a few neurons might be replaced in adult brains,* understanding how they do this in a natural setting may help us develop new methods to increase regeneration with age.

The Vasculature and Age-Related Neuronal Failure

What we really need to know now is, what is the *very first thing* that starts the neuronal aging process, the snowball that launches the avalanche? Just as coloring gray hair does not reverse aging, pathological end points that are the most obvious characteristics of aged or postmortem brains are easy to identify and study, but their elimination is unlikely to be an effective strategy in slowing

---

* This is actually still heavily debated.

aging. What we need to find are the causes of neuronal aging, so we can treat those rather than downstream effects.

The aging microvasculature is presently an underappreciated source of cognitive decline.[13] Recent mammalian studies suggest that problems with the blood-brain barrier and vasculature may precede other more obvious changes in the aging brain. The protective blood-brain barrier can fail with age, leading both to leakage from inside and increased susceptibility to cells that shouldn't be there coming in from the outside, such as mis-activated inflammatory cells. Because the vasculature is responsible for feeding oxygen to the brain, its failure is both a cause and a consequence of brain aging and can trigger a whole cascade of downstream neuron failure. Many of these problems in the vasculature happen at the micro level, but ultimately result in macro-level problems, such as arterial stiffening; narrowing of arteries, which then causes plaque buildup; and weakening of arterial walls, which in turn leads to inflammation and ischemia (lack of blood flow and oxygen delivery). We often think of diabetes, high blood pressure, and high cholesterol (and smoking) as problems for the rest of the aging body, but in fact, these vasculature diseases and clotting problems can damage blood vessels and increase the chance of stroke, which can in turn cause vascular dementia. A lack of oxygen coupled with an invasion of immune cells, bacteria, and other deleterious factors can seriously damage neurons. As the immune system increasingly becomes dysfunctional with age, this susceptibility might be particularly problematic, and the inability to properly clear cellular waste can also be damaging. Therefore, strategies to prevent blood-brain-barrier breakdown and vasculature problems may be more useful than is generally acknowledged. This concept makes intuitive sense, since we know that the brain needs a lot of oxygen to function, but it may have been unappreciated because it's easy to study only neurons when trying to understand neuronal aging, while the vasculature is made up of non-neuronal cells. Slowing problems with the vasculature might be key to preventing cognitive decline in many people. Vasculature- and microclotting-related cognitive problems may become even more of an issue as long Covid—which seems to have long-lasting effects on the vasculature in general—affects more and more people.[14]

Exercise is an excellent way to keep our vasculature functioning with age, but another, even easier way to protect our brain is to sleep. Sleep loss is associated with increased chronic disorders and increased cell senescence.[15] We accumulate metabolic garbage during our waking hours, and sleep is now being recognized for its importance in "cleaning" the brain via the activity of glial cells and the lymphatic system (or "glymphatic" system).[16] Sleep may also help repair synapses and generally "reset" the brain. Given these important functions, it's becoming clearer that loss of sleep duration and the disruption of regular sleep/wake cycles might not just be increased with aging but actually cause the brain to accumulate harmful metabolites that lead to further loss of function.

Can Worms Learn? Measuring Cognitive Decline in Model Systems

As a species, we like to imagine that we are special and uniquely intelligent, so you might be surprised to hear that, not only are we *not* the only ones who can learn and remember information, but humans are also not the only animals who suffer from cognitive aging. Of course, depending on their predator environment, in the wild most animals will not survive long enough for cognitive aging to arise, though; once an animal can no longer care for itself, it will be unable to survive—and an animal that is past reproductive age has no evolutionary pressure to avoid cognitive aging. In fact, genes that promote growth and early survival may become deleterious with age (remember the antagonistic pleiotropy theory from our earlier chapters), which may explain why we see the increase in appearance of these dementias in older people. But in the lab, we observe that aging animals do show many of the same declines that we are interested in slowing in humans, including losses in learning and memory ability with age.

How do we measure learning and memory in model organisms? Of course, we are a lot smarter than mice and flies and worms, so we're not asking them to solve differential equations or even put together an Ikea bookshelf. But they are all animals who have been successful on an evolutionary scale. This success implies that they have learned how to survey their environments, make decisions about food, pathogens, predators, and mates, and remember that information as long as they need to use it. You remember Pavlov's dogs, who salivated at the sound of a bell that they thought would mean steak was coming, right? We have used the same strategy to design learning and memory tasks for our favorite model organisms. For example, rodents can learn various tasks, as long as they receive rewards (water or food pellets), or sometimes negative stimuli (e.g., foot shocks). In much the same way that we can't forget the awful taste of that seafood that gave us food poisoning, foot shocks and other negative stimuli are often more strongly remembered than positive cues, which makes sense if you're trying to avoid a lethal situation—and in the lab, these negative learning tasks can even be models for post-traumatic stress disorder (PTSD). Mice can learn to navigate mazes, find a platform that is submerged under milky-colored water based on visual cues, or anticipate a smell or sound if paired with a mild electric shock to their feet. Testing how quickly they can learn these tasks and how long they remember them gives us metrics for their cognitive abilities, and we can then measure how these abilities change with age. Such behavioral tasks are the cornerstone of cognitive function analyses in model systems.

Like mice, invertebrates can sense odors and other cues and pair this information with positive (food) or negative (starvation) cues, which allows us to test them for these Pavlovian associations. Flies exhibit many learned behaviors and can be used to study various forms of short-term and long-term associative memory; for example, you can put odor in tubes and let them choose which arm of a T-maze to go down. Genetic screens in flies have been used to identify many of the important components of learning and memory.[17] Work in the sea snail *Aplysia* was critical in establishing the role of the cyclic AMP response element (CRE) binding protein CREB transcription factor in long-term memory.[18] People always seem surprised when I tell them that *C. elegans* can learn and remember, but of course these animals wouldn't be evolutionarily successful if they couldn't take in information and process it; what's remarkable, as it turns out, is how evolutionarily well conserved the molecular pathways for learning and memory are between worms, flies, *Aplysia*, and mammals.[19]

Worms don't care about much in life besides eating, so in my lab we train them to expect to get food (bacteria) when they smell an odor that they usually don't care much about (e.g., the odorant butanone at a neutral concentration).*[20] We grow worms to adulthood on normal bacterial food, then starve them for an hour, then refeed them bacteria but put drops of butanone on the lid of the plate, so they eat while smelling butanone. After we have presented this butanone odor with their food to the worms for an hour, or for several training sessions, we can then test their ability to form the food-butanone association by performing a "chemotaxis assay": we place all the worms at a starting point, and then after an hour we count the number of worms at the butanone spot versus the number at the control spot, minus any worms that never left the origin, to avoid confusing impaired locomotion from memory problems. If they still travel toward the butanone, we know that they still remember the

---

* NB: these are all hermaphrodites, who care about finding food so they can feed their developing progeny; all that males care about is mating, so we are developing new memory assays that depend on pheromone sensing.

association between food and butanone. Worms are smarter than you'd think; we have to use a paralytic chemical to trap them at the spots, because otherwise when they get to the butanone and find out there's no food, they figure it out and leave, simultaneously retraining themselves. By simply leaving the worms on a plate of food without butanone for a given period of time after training, we can study how long they remember that original odor-food pairing.[21]

These relatively simple assays in worms, flies, *Aplysia*, and other invertebrates have allowed us to test the contributions of individual genes and proteins to the complicated processes of learning, short-term memory, and long-term memory, which turn out to be molecularly and cellularly distinct from one another.[22] For example, we know that all animals need to transcribe new mRNAs and synthesize new proteins to build long-term memories, and it takes several hours and multiple training sessions to achieve this task. That's because long-term memory requires the activity of a specific transcription factor, CREB, to make new mRNAs, and then these mRNAs must be translated into proteins to form long-term memories. This slow creation of long-term memories only after repeated training makes sense—why learn and remember something (like pairing of an odor or location with food) for a long time if it is only fleetingly true? Better not to waste the energy on transient information. Short-term memory takes only a single training session, like associative learning, and requires the mRNAs that are already poised for translation into proteins, mostly at the synapse.[23] Learning is more instantaneous, using just the proteins that are already made and functioning—for example, right at the synapse. One of the crazier things we found is that halting protein translation prevents forgetting, suggesting that short-term memories are actively overwritten, perhaps to allow animals to replace out-of-date information with new information. This suggests that we could maintain some memories longer if we could understand how to slow some protein translation. And relevant for the question at hand, all of these processes decline with age, but at different rates and ages—generally the more complicated the process, the faster it is lost with age.[24] Using these assays, we can explore what it might take to make old animals retain their abilities longer with age, independent of studies of neurodegenerative diseases.

Despite the relative simplicity of these animals, they are immensely useful in studying learning and memory, because the analogies with learning and memory in "higher" animals are so strong. One of the most surprising things we learned while studying associative learning and memory in *C. elegans* is just how well conserved these processes are at the molecular level, all the way up through mammals. I guess we shouldn't be surprised every time we rediscover that evolution happened, but there is something humbling about finding out that little worms use the same transcription factors, neurotransmitters, neuro-peptides, and synaptic proteins to remember things they've learned (like that "butanone means food's coming!") as we do. It's even more surprising to scale that information to time and realize that if they can remember this information for a day, which is about a twentieth of their lives, it's years on our lifespan scale. And they do it by turning on the same transcription factor, CREB, in just one pair of neurons, while we use CREB in our hippocampus to remember information as well. I think that we all thought that since worms have so few neurons, they'd somehow use different machinery to do the same jobs—but they don't. Again, they are not necessarily doing the kinds of calculations we do, but they have been evolutionarily successful because they detect, learn, integrate, and remember cues—both good and bad—that help them make the decisions that allow them to survive, so maybe we should have expected the similarities all along.

While humbling on the species level as humans, this is great news for us as researchers. It means we can use those learning and memory tests not to just identify which genes are involved in these analogous processes but also to discover what fails with age, and then do rapid tests with mutants and with small molecules to try to prevent this decline. Finding out exactly which genes and drug candidates might slow neuronal aging, or rescue memory with age, will be accelerated if we can first do it in worms, flies, and mice, before starting the ethically fraught and very slow process of testing in humans.

What we know at the functional level is growing with every experiment. So far, we know that all of these animals lose their ability to learn and remember with age, and that these functions fail at different rates. For example, we found that worms can remember the food-butanone association and remember it overnight when they are young, but by the end of their reproductive period, they are no longer able to do this. Why? Levels of the critical transcription factor CREB fall with age, and if we boost those levels either by making more CREB or by turning on the activators of CREB, we can rescue old worms' long-term memory ability for a few more days.[25] But people obviously don't totally lose their memory right after reproduction ends, and our short-term memory seems to be the first one to go on the fritz (at least, that's the one we notice right away when it doesn't work right.) But we do eventually lose all kinds of learning and long-term memory abilities with age. Since mammals also use CREB for memory, and CREB levels also decline with age, we are currently testing whether we can use a similar approach to rescue memory in old mice, by making more of an activator of CREB called Gnaq—the mammalian equivalent of the gene we tested in worms[26]—in the mouse hippocampus, and then testing whether we see improvements in memory. Since short-term memory and learning require different proteins, are regulated differently, use different neurons to do so, and fail at different ages to long-term memory, we must discover what is limiting with age for each of these events, and then ask whether boosting those declining genes and proteins in aging people could similarly rescue learning and short-term memory. The power of model systems is obvious in trying to find solutions to our loss of memory with age.

Systemic Regulators of Cognitive Aging

Another, perhaps more surprising approach to boosting CREB activity has come from studies of blood-borne factors. You remember those crazy parabiosis experiments where researchers stitched together mice to share a circulatory system, and the old mice benefitted (and the young mice suffered)? Tony Wyss-Coray's (Stanford) and Saul Villeda's labs (UCSF) have carried out more sophisticated versions of this experiment. Instead of stitching the mice together, old mice are given blood plasma from young mice. Villeda discovered that the animals provided with young plasma have better learning and memory than their age-matched counterparts.[27] Interestingly, several different factors seem to contribute to these improvements.[28] Some of these blood factors help maintain CREB levels; others improve neural stem cell regeneration, helping replace old neurons with new ones; at least one of these "youth factors" may be a factor shared with XX animals (as females live longer and retain function longer in general); and others seem to help with blood flow and metabolism. The list is probably not yet exhausted, though, and some of the factors may work combinatorially to improve brain function.

Like hormones, blood factors work through "nonautonomous" effects— that is, signaling from one cell type that affects the function of other cells. Another source of nonautonomous effects on brain aging and neuronal decline is the immune system, through maladaptive inflammation during aging, or "inflammaging." While we've discussed this in terms of other aging phenotypes, emerging evidence suggests that inflammation is hugely disruptive to brain function.[29] Activation of microglia and astroglia causes nonautonomous neuroinflammation, a runaway process that damages the brain. In this spirit, studies to understand how the function not only of neurons but also of the cells that were thought of as "support" cells—glia—might affect cognitive decline are ongoing, as are studies of the positive effects of sleep in "cleaning" the brain through the glia-lymphatic—or "glymphatic"—system.[30]

Villeda's group made the exciting discovery that exercised mice produce proteins in their blood plasma that are beneficial to aged, sedentary mice.[31] They found that a particular liver enzyme called Gpld1 increases in the blood plasma of both exercising mice and active humans, and correlates with cognitive function. When they increased levels of Gpld1 in old mice, the mice showed better regenerative and cognitive functions. Wyss-Coray's lab later found that plasma from "runner" mice can improve memory and reduce neuroinflammation in mice via a complement cascade inhibitor called clusterin; patients with cognitive impairment treated with structured exercise had higher levels of clusterin.[32] Exercise plasma, or more likely a cocktail of purified factors from exercise plasma, might be key to improving both normal and pathological cognitive decline. Someday you might benefit from "blood boy" proteins to boost your memory function in old age, even if you are a couch potato yourself.

Cognitive Aging and Alzheimer's Disease

Work on normal cognitive aging has lagged behind the study of aging and longevity in general, despite great interest and generous funding for Alzheimer's disease research. In fact, the devil's advocate might argue that our huge investment specifically in AD research—for which there are still only symptom-treating medications, but no cures—has in fact *slowed* work on normal cognitive aging, because the majority of NIH research funding in cognitive aging is allocated specifically to AD research, and historically to only one aspect of AD. However, normal cognitive aging is what almost everyone will suffer from if we live long enough, but until recently it has not been considered a research priority. If we are lucky, some of the findings for age-related neurodegeneration might be generalized to non-AD cognitive decline and mild cognitive impairment—but the vast majority of AD research focuses on one particular idea, that amyloid protein plaques are the cause of AD (the "amyloid hypothesis"), so while it is not impossible, it is unlikely that this will translate into treatments for non-amyloid-dependent cognitive decline.

Although the two terms are often used interchangeably, cognitive decline and AD are molecularly and cellularly distinct processes. For example, people who are not suffering from neurodegenerative diseases and who never show any accumulation of beta-amyloid plaques can still suffer from cognitive decline (and vice versa). Nevertheless, most of what we know about the aging brain comes from the intense research focus on AD, and to be fair, much of our fear of aging arises because of the risk of developing AD. Here I'd like to touch briefly upon what we know about AD at this point, explain why AD might be distinct from other forms of dementia and processes of cognitive decline found with age, describe both promising and disappointing therapeutics, and suggest how we might change research directions in the future.

Alzheimer's Disease: Important, but Not the Only Thing that Goes Wrong with Age

In addition to AD, there are other types of serious neurodegenerative disorders, and several have at least some age dependence. Most neurodegenerative diseases are thought to arise from some sort of protein aggregation. For example, Lewy body dementia, ALS (which you may know as Lou Gehrig's disease, and strikes mostly younger and middle-aged men), Huntington's disease (which has a strong genetic component, affecting whole families, even at young ages), and Parkinson's disease all have different suspected sources of disease-causing protein aggregates. Parkinson's has both familial and spontaneous forms, and while the familial form has been linked to genes like α-synuclein, parkin, and PINK1, the causes of the spontaneous forms are not yet understood, and may be various.[33] There are also differences in the age of onset, brain regions they affect, and the disabilities these neurodegenerative diseases cause. Parkinson's and AD are both associated with advanced age, but the former is associated primarily with motor defects, with cognitive defects arising much later, while AD is marked by dementia, often in the absence of other age-related declines. These differences are thought to be caused by the different susceptibility of specific brain regions to each disease. Thus, these diseases are serious, important to study, and molecularly distinct, but they are also separate from the kind of normal age-related cognitive decline that will affect many of us as we grow older. Because an entire book can easily be written about each of these diseases, I will not be covering them in detail here.

Although there is much debate about this point, AD is generally associated with aggregates of amyloid-beta, called plaques, and/or neurofibrillary tangles (NFTs) made up of hyper-phosphorylated tau protein (a microtubule stabilizing protein encoded by *MAPt*) that appear inside cells—but whether either of these is truly causative rather than just a biomarker of AD has not yet been resolved. Certainly, these two pathologies, amyloid-beta plaques and tau NFTs, are abundant in the brains of postmortem AD patients, and for a long time this postmortem diagnosis was the only way to distinguish AD from other forms of dementia. Expression of the truncated form of A-beta peptide is toxic to cells, and NFTs are also toxic. Their expression in model systems also causes a range of defects, from cell death to behavioral problems, suggesting that they may be causative in AD.

But which one, if either, is the cause of AD? The amyloid hypothesis has been the dominant theory in the AD field for many decades, and much of the focus in AD research has been on identifying methods to eliminate beta-amyloid plaques. While there has been a lot of support for the amyloid hypothesis, there's plenty of reason to be skeptical, as well. In many experiments, beta-amyloid plaques do not cause problems in memory, suggesting that amyloid protein is not the cause of AD, it's just a downstream biomarker of the disease. In fact, work in several animals suggests that the oligomeric form of A-beta is more toxic than its final plaque form,[34] so altering that balance by eliminating the relatively benign plaques and forcing the formation of more toxic oligomers might even be detrimental, if those plaques are "sinks" for more dangerous oligomers—definitely a concern if the therapeutic approach in question is designed to eliminate plaques and shift the balance toward A-beta oligomers.

What exactly amyloid precursor protein (APP) is supposed to do in neurons is (shockingly) not yet well understood, although it is thought to play a role in in neural migration and may have functions at the synapse, including plasticity, axonal growth, and cell survival.[35] In fact, some of these aggregation-prone proteins are likely to be important for normal neuronal function, so their loss—not their aggregation—might in fact be the true root of the problem. For example, APP seems to be required for normal function of endolysosomes in fly neurons, and elimination of APP caused neuronal cell death.[36] Fly APP might mediate interactions between neurons and glial cells as well. The jury is still out on what APP should be doing in cells, and how APP and its fragments are involved in AD, although one proposed therapeutic strategy is to knock down APP, since it seems not to be essential for survival (seems like a bad plan to me, though, if APP actually does something important).[37]

I'm going to make some enemies here, but it seems to me that this debate between entrenched parties—"BAPtists and Tauists"—who support only research in their own area has slowed progress in the field. That is, if you have spent decades promoting the amyloid hypothesis of AD, then you may be naturally invested in showing that A-beta is the root cause of AD—so data suggesting that getting rid of these plaques does nothing for AD patients is probably not welcome. In fact, paying attention to both A-beta and tau, and their temporal and spatial expression and potential direct interactions, might provide an important missing link.[38]

Recently, the NIH has recognized that new AD research ideas are needed beyond the most popular, 30-year-long, well-funded focus on the amyloid hypothesis, and so they called for research on new directions in AD. Some of these new hypotheses include a focus on inflammation and viral or microbial infection as the root cause of AD, and may still involve A-beta or tau, but not necessarily as the causative agent. Connections between AD and other metabolic diseases, particularly diabetes, are also interesting, as they share many characteristics—including the potential damage to the vasculature—and type 2 diabetes might be a significant risk factor for AD.[39] Perhaps these new research directions will finally unravel the AD mystery.

Inflammation's Effect on AD and Cognitive Aging

Just as we discussed for cognitive aging, there are connections between inflammation and AD, and further with infection. Studies of neuroinflammation and activation of microglia—as indicated by the expression of markers such as IL-1beta, IL-6, CCL2, CXCL1, and CXCL10—suggest that microglia are "primed" to produce IL-1beta through their proximity to amyloid plaques.[40] Additionally, acute infections in AD patients alter the brain neuroinflammatory profile. The connection between infection and AD should be studied more to determine whether there could be a causative role for infection and subsequent inflammation that increases one's risk of AD.

Another Direction for AD Research: Tau

While much of the work in the AD field and subsequent drug trials (see below) have focused on A-beta and amyloid plaques, tau is the other aggregating protein that might be a major driver of AD and other neurodegenerative diseases known as "tauopathies."[41] Tau is a microtubule-binding protein or microtubule associated protein (MAP). Microtubules are abundant in most cells but particularly in neurons, where they are used as both structural components and railways in the axons for synaptic vesicle delivery to the synapses. How tau might be pathogenic—through the formation of aggregated tau in the form of NFTs or paired helical filaments—is an active area of research. It is easy to imagine that tau dissociation from microtubules might have deleterious effects, even without its accumulation elsewhere, owing to possible destabilization of microtubules. This is essentially a loss-of-function phenotype that could be imagined leads to neuronal dysfunction. However, NFT pathology has been linked to neuronal atrophy and cognitive decline, suggesting that pathological tau accumulation or aggregation of tau is toxic.[42] For example, the accumulation of tau can lead to synaptic loss. Recent studies suggest that tau may be able to spread between cells in a prion-like manner, or via extracellular vesicles. Therefore, unprotected extracellular tau might be a target for immunotherapy.[43]

Dissociated tau has been found to be mis-localized to various parts of the neuron, including synaptic vesicles.[44] Tau has also been shown to be associated with the mis-activation of glia observed in neuroinflammation. Tauopathies are associated with specific pathogenic mutations in tau (e.g., the P301S mutation).[45] An additional phenotype in tauopathy models is the loss of synapses,[46] which of course would lead to inactivation of neurons, but the relationship between tau, synapse loss, and neuroinflammation is not completely understood. A recent mouse study suggests that removal of the synaptic protein synaptogyrin-3, which tau binds to, might prevent the loss of synapses observed in tauopathies.[47] If such an approach translates to humans, synaptogyrin-3 could be a good pharmaceutical target to test.

The Intersection of Aging, Tau, and Transposable Elements

Transposable elements (TEs) have recently emerged as an important component of aging, rising in activity with age and then causing further age-related damage. These potentially dangerous, parasitic DNA elements (see chapter 10 on DNA homeostasis) are normally "silenced" by the cell to keep them from getting expressed and inserted elsewhere in the genome.[48] Keeping them silenced is important because activated TEs can insert into DNA and interrupt genes that are important for normal function, which can then lead to cellular dysfunction. It's been observed in various model organisms and human clinical samples that TEs become active in aging cells, as the silencing mechanisms fail. This general mechanism is referred to as the "transposon theory of aging." Neurons are not protected from this effect, and TE activity has been associated with at least four neurodegenerative disorders, including AD, particularly in hippocampal and cortical neurons, the major sites of AD pathologies.[49] Age-dependent TE transcripts have been found in mouse brains, and a large study of postmortem brains revealed that activation of TEs corresponds with NFT burden, suggesting some sort of correlation between tau and TEs.[50] Of course, correlation does not mean causation, but several studies suggest that age and tau induction lead to activation of TEs, particularly endogenous retrovirus (ERV) retrotransposons.[51] Overexpression of tau in aging fly brains led to loss of heterochromatin and activation of expression of ERVs, as seen in postmortem clinical AD brains, suggesting that the link may occur at the level of the loss of silencing mechanisms. Although the molecular mechanisms linking tau overexpression and AD-related pathogenic forms of tau with chromatin are not yet clear, the correlation between tau, loss of heterochromatin, and the activation of TEs appears consistent across model systems and human AD brains.

The exciting aspect of this new hypothesis, TE-induced AD, is that it suggests a somewhat novel treatment for AD: antiretroviral therapy. You have probably heard of antiretrovirals as a treatment for the virus that causes AIDs, HIV, and more recently, the use of Paxlovid, a combination drug of two antiretroviral medications, to treat Covid-19. If suppressing TE activation could slow the onset of cognitive decline in humans, then antiretroviral therapy could be a powerful new alternative for treating AD—and the idea that HIV drugs might even be useful for AD is even being tested.[52] In fact, there is some recent clinical trial data for the treatment of ALS, another aggregation-based neurodegenerative disorder, with antiretrovirals.[53] Perhaps a similar treatment with similar antiretrovirals would have a positive effect on AD patients.

APOE alleles, AD, and Cognitive Decline

In addition to A-beta and tau, a few other genes have been associated with AD in genome-wide association studies (GWAS) and other genetic tests. These include presenilin (an enzyme that processes APP) and apolipoprotein E (APOE). APOE is one of the genes that keeps popping up in GWAS of exceptional human longevity (chapter 3), because certain alleles seem protective, while other alleles are associated with short life. The ε3 allele is most common and is therefore considered the "reference allele" for GWAS. The ε2 allele—the mostly good one—has been associated with slower cognitive decline with age and a reduced risk of AD. The ε4 allele is bad news, particularly when both of one's inherited parental APOE alleles are ε4, but even when the other allele is ε2. The ε4 allele is associated with increased risk of early- and late-onset AD, increased tau pathology, greater A-beta deposition, increased vascular cognitive impairment and synaptic pathologies, and a host of other dementias (Lewy body dementia, Parkinson's disease dementia, and TDP43-related dementia).

Why would one protein be so dangerous? APOE is a secreted and truncated glycoprotein that is expressed in lots of neuron-associated cells, like microglia, astrocytes, and vascular cells, as well as stressed neurons.[54] APOE binds to lipids to form lipoprotein particles, which can be found in cerebrospinal fluid (CSF). You might remember that I also mentioned this theme of cholesterol and lipid metabolism in chapter 3, on human longevity, because several of the most important longevity genes—including APOE, but also CETP and others—regulate cholesterol metabolism. *How* they affect longevity is still being unraveled, however. Different alleles of APOE appear to have different effects on this process. It is thought that A-beta seedlings bind to these lipids, perhaps increasing the formation of plaques, which might in turn affect the vascular wall. Cerebral amyloid angiopathy is a downstream consequence of these plaque accumulations on vasculature.

Of course, it is possible that there are multiple causes of AD, like the seed-and-soil hypothesis for cancer; that is, a particular genetic allele might predispose you to AD, and the second factor (cholesterol levels or poor cardiovascular health) might tip the balance. For example, while early-onset AD has been strongly linked to APOE, less is known about the genetic causes of late-onset AD and other types of dementia. There may be other, non-APOE genes that emerge from GWAS, human brain gene expression, and other related studies. Being able to rapidly test these genes is a bottleneck, since it's impossible to do knockdown experiments in humans, and aging mouse studies can be slow (even with CRISPR); moreover, knocking down a gene too early in life might screw up a mouse brain's development, which is interesting but might not effectively model the progression of AD in aging humans. Yousin Suh, a human geneticist at Columbia, and I are hopeful that by combining human GWAS and gene expression studies with rapid testing of learning and memory in *C. elegans*, we can uncover new genes that play a role in AD. Identifying more early biomarkers for AD that are accessible in blood would also speed both diagnosis and therapy.[55] Moving the AD field away from amyloid-beta tunnel vision might allow us to finally discover a way to slow this and other forms of age-related dementia.

Current Treatments for AD

Right now there are no drugs on the market that can cure AD.[56] The best treatments currently available essentially manage symptoms and try to keep neuron signaling active longer but will not eliminate or reverse AD (but might be a good idea for non-AD cognitive decline as well). For example, slowing the breakdown of acetylcholine, a critical neuronal signaling molecule, can extend AD patients' cognitive function. Acetylcholine is normally broken down by cholinesterase, and cholinesterase inhibitors slow this degradation and can stimulate the release of more acetylcholine through the upregulation of nicotinic receptors. This is the mechanism for several different medicines prescribed for AD, including Aricept (donepezil), rivastigmine, and galantamine. Another approach is to block effects of excess glutamate using an N-methyl D-aspartate agonist, like memantine. Combinations of these drugs are also often prescribed. Acetylcholinesterase inhibitors may also help with vascular dementia, which is the most common form of dementia in the elderly. While there are both high costs and various side effects to these drugs, and none can cure AD, slowing the onset of impairment is critical to improving quality of life. This can be measured by brain imaging and cognitive testing; for example, a 2005 study of Japanese AD patients with and without donepezil treatment concluded that the acetylcholinesterase inhibitor does slow the rate of hippocampal atrophy.[57] But, more practically, even if these drugs cannot reverse AD, early diagnosis and treatment can delay the requirement for full-time care by two to five years.

It is increasingly being recognized through "pharmacogenetics" that cytochrome P-450s, enzymes that can metabolize drugs and thus alter their efficacy and side effects, might vary with each patient and increase the heterogeneity in responses to treatment.[58] Perhaps the increasing use of genetic profiling in combination with drugs will maximize the efficacy of AD treatments.

Clinical Trials for New AD Drugs: Targeting A-Beta

Since the amyloid-beta hypothesis has been the leading theory about the cause of AD for several decades now, one would have expected that by now we'd have a drug that gets rid of those plaques and cures AD, right? Wrong. In fact, several promising drugs aimed at eliminating amyloid plaques have crashed and burned in clinical trials, either showing no efficacy or, in at least one tragic

While I was writing this chapter in the summer of 2021, the FDA made the surprising announcement that it had approved aducanumab, aka Aduhelm—against the wishes of 10 of the 11 members of the advisory committee.[a] Within days, several members had resigned from the committee, with one advisor, Aaron Kesselheim, calling it "probably the worst drug approval decision in recent US history."[b] Why all the vitriol against a drug that we so desperately need to slow AD? Despite the fact that aducanumab does seem to reduce amyloid plaque, clinical trials of the drug did not show any evidence of improvements in cognitive function—that is, the patients still showed memory loss. In fact, some brain scans showed abnormalities, increasing safety concerns. Advisors were rightly concerned that the lack of improvement in cognitive function means that there is little point to administering the drug, even if amyloid plaques are reduced. Another concern is the cost of this monthly infusion of Aduhelm. While normally one would argue that the price of a drug is likely to be cheaper than the cost of care, if the patient still experiences memory loss and dementia, then there is no point—but a family caring for an AD patient will feel morally compelled to provide this treatment, of course. Adding this ethical dilemma to the burden of caring for AD patients seems particularly cruel. This decision has diminished confidence in the FDA, which ostensibly should approve only drugs that are both safe and effective—and aducanumab appears to be a failure on both fronts: shockingly, up to 40% of clinical patients developed dangerous brain side effects (bleeding, swelling). And the FDA is not asking for clinical results for nine years, so if there are no positive effects, patients will have been taking this expensive and possibly even dangerous drug for many years for no reason. My hope is that another, more obviously beneficial drug will be found soon, making this FDA misstep irrelevant for AD patients. In December 2021, the European Union's equivalent of the FDA refused to authorize Aduhelm, reasoning that it did not show clinical improvements and was not sufficiently safe.

case, making the patients' cognition even worse.[59] There's no doubt that finding effective therapies is extremely difficult, no matter what the disease is, but AD has had a particularly bad drug trial track record. These drug trials include tests of inhibitors of the beta- and gamma-secretases that chop up amyloid, such as the gamma-secretase inhibitor semagacestat, and both direct and passive immunotherapies against A-beta (such as solanezumab). While it's certainly possible that the drugs were flawed, doses were wrong, or even more likely that the enrolled patients had progressed too far along with the disease for the drug to make a difference, it's still not clear that the ultimate target, amyloid plaque, is the right choice. Maybe aducanumab, the latest antibody drug targeting plaques (see box), will finally prove the BAPtists right, but its efficacy was also challenged even before its FDA (Food and Drug Administration) approval, raising alarms.[60]

Perhaps the development of drugs targeting both tau NFTs and A-beta plaques, rather than treating the two proteins as competing teams, will be the key. For example, AADvac1 targets pathological tau, and is undergoing clinical trials now.[61] In any case, a new direction in AD research is sorely needed—at least it seems that way to this outsider.

Timing Is Everything

One of the big hurdles in clinical trials of candidate AD drugs that has not yet been overcome is early treatment. It is possible that some therapeutics might have actually succeeded in slowing the progression of AD, if only the drugs could have been administered at an earlier stage of the disease. For example, if a drug would be effective at a very early stage of the molecular process, such as prior to the nucleation of pathogenic aggregates, but can't do anything at a later stage when the aggregates have already formed, then that drug will look totally useless in a clinical trial that is administered only to late-stage AD patients . . . but between the risks of such treatments and, more importantly, the inability to diagnose AD early, we won't ever know.

The key to getting patients into drug trials before their disease has progressed too far will be figuring out ways to identify patients who will go on to definitely develop AD well before they demonstrate symptoms. At the very least, trials that include patients who are much earlier in the disease progression should be attempted, since the tested drugs may have no chance of working in patients with more advanced disease if they are effective only at an early stage of AD. This will require better identification of early-stage AD dementia and biomarkers that will distinguish AD from other types of dementia, and FDA approval to carry out such testing. New, noninvasive measurements of cognitive aging—including blood tests, cognitive tests, brain scans, and measurements of eye movements, particularly for those for whom language barriers can impede testing—combined with genetic and preclinical markers of

A surprisingly simple test that might accelerate AD diagnoses has become popular since its inception in 2013: the peanut butter test.[a] Because parts of the olfactory cortex are thought to be where the disease might first manifest, and often there is more damage on the left than the ride side, AD patients may demonstrate an asymmetry in their ability to detect odors through one nostril versus the other.

AD might finally allow clinicians to identify patients who are most at risk, and enroll them in clinical trials for new AD drugs. More sensitive tests for mild cognitive impairment and for the aberrant presence of tau and A-beta proteins in CSF will help with this. Genetic tests to identify at-risk individuals with the high-risk APOEε4 allele, which seems to promote the onset of AD (see above), and testing of cognitive tasks, brain scans, and so on, that precede frank dementia could be combined to identify high-risk individuals to enroll them in drug trials at an earlier stage of the disease. Other CSF and blood factors in addition to tau might also be good predictors of AD. Using these factors together to predict the onset of AD would be a major step forward in enabling the execution of earlier clinical trials.

Race and Environmental Factors that Might Affect AD and Cognitive Aging

We generally think of AD as a disease caused by internal (genetic) factors, but external factors might also have an influence. Caleb (Tuck) Finch, a researcher at USC and one of the founders of the modern aging field, has been interested in the role that air pollution, sex, and APOE alleles might play in neurotoxicity, the development of AD, and subsequent cognitive decline. Perhaps the association between the vasculature—which brings oxygen to our brains—and APOE explains the observation that air pollution, which is already a major risk factor for mortality, is also associated with neurodegeneration, cognitive decline, and AD.[62] Whether these same factors might contribute to general cognitive decline is not known, but it would be surprising if they were specific to AD.

Considering environmental factors might also address some of the racial disparities that have been observed in the onset of AD and other dementias.[63]

Generations of city planning that ultimately destroyed Black neighborhoods have led to the disproportionate exposure of minority and low-income populations to environmental pollutants.[64] For example, razing Black neighborhoods to build highways in cities resulted not only in the loss of homes and of thriving businesses but, combined with exclusionary housing practices ("red lining"), resulted in areas with restricted mobility and little access to other areas. (Think, what is between the overpasses in a city?) These zones trap families in poor neighborhoods with high rates of air pollution from highways. Similarly, factories that produce dangerous air pollutants and waste runoff are more likely to be near low-income housing, and housing in racially segregated neighborhoods is on average older and more likely to have problems (mold, lead paint, etc.). Finally, disadvantaged neighborhoods are less likely to have access to grocery stores, healthy foods, parks, and safe exercise options, and therefore are associated with increased risks of metabolic diseases.[65] Childhood asthma, low birth weights, premature births, and other disorders have been linked to neighborhood pollutants.[66] Stressful life experiences also have a negative effect on lifespan and on cognitive aging, and African Americans and other historically disadvantaged populations in the United States experience stressful events over a lifetime at higher rates than whites do.[67] In fact, stress factors accounted for differences in the rates of cognitive decline better than other factors (e.g., APOE alleles, age, education, etc.) in some studies. AD and cognitive decline have been associated with disadvantaged neighborhoods, and states with high infant mortality rates among African Americans had much higher risk (80%) of dementia, specifically in the Black population.

According to the Alzheimer's Association, there are severe racial disparities both in susceptibility to AD and its diagnosis.[68] African Americans are twice as likely and Hispanic patients 1.5 times as likely to develop AD as white patients, respectively. (Asian Americans are at slightly lower risk than whites.) The Washington Heights–Inwood Columbia Aging Project found that while 30% of the white population over age 85 had been diagnosed with AD, more than 60% of the Hispanic and African American population had AD. Additionally, patients in these high-risk groups are more likely to be first diagnosed at a later stage, with more severe dementia—making it harder to slow the progression even with the admittedly limited currently available drugs mentioned above. Some of the differences in susceptibility to AD may be related to increased risks for high blood pressure and diabetes, which may be related to cognitive decline. Thus, racial disparities in diseases, including AD, might have at least as much to do with where one lives (and thus the pollutants one is exposed to), the odds of being properly diagnosed, and stressful life experiences as with genetic factors.

———

How Might We Slow Cognitive Aging?

Given what we know so far about normal cognitive aging and AD, how might we go about using what we've learned to slow cognitive decline? Since that is really the point of what we are doing, asking how we are progressing on this front is one of the more important questions we can address in this field, beyond simply extending lifespan. Studies of model systems and even basic cell mechanisms have given us some hints about how to slow cognitive decline.

Lessons from Longevity Mutants and Dietary Restriction

As is always true when studying aging, we can learn a lot from longevity mutants in model systems. Not only do some of the long-lived mutants look better with age and extend their healthspan, but several (not all) also maintain their cognitive function longer. For example, using the Pavlovian assays I described earlier in the chapter, my lab found that the long-lived *daf-2* insulin signaling mutant performs better in all ways: not only do these worms have better short-term and long-term memory when they are young, but they also maintain these abilities longer with age.[69] Some people in the field had speculated (and had even written, but without supporting data) that *daf-2* mutants should perform less well than wild-type worms (I guess again with the idea that you can't have it all!), but our data showed us that insulin signaling mutants manage to even extend their memory ability disproportionately longer than one might expect from their lifespan extensions. Even though these animals live long and one might expect some sort of "trade-off" in function, they still performed better, even when scaled to their long lifespan.

We now have a fairly good idea how young *daf-2* animals maintain their cognitive function: by regulating the expression of a neuron-specific set of DAF-16 targets.[70] This allows them to keep their neurons functioning longer. We also know that there is higher expression of CREB, the limiting factor for long-term memory ability, in *daf-2* animals.[71] Perhaps more excitingly, we also found that by turning "on" a gene that activates the CREB transcription factor (EGL-30), we can completely rescue the lost memory ability of old wild-type animals.[72] Because the ortholog of EGL-30 is also present in mammalian brains, we are currently testing whether a similar boost in an old mouse brain can rescue its memory ability. (*Spoiler: it does!*) If this were to work, it would mean we could explore drugs that might have the same effect. For a *C. elegans* researcher like me, the possibility that we could go from designing a memory assay for worms—an animal that people doubted even had long-term memory before we did our experiments—to finding that it uses the same molecules we use, to finding that its memory declines with age as ours does, to finding multiple genetic "fixes," and then testing it in mammals, and hopefully finding drugs through high-throughput small molecule screening in worms is a dream come true. It's the whole reason I got into this business in the first place. I'm quite hopeful that this approach will yield therapies that might never have come to light if we were limited to looking at old postmortem human brains.

In mammals, circulating levels of IGF-1 decline with age, and with them, various defects in memory (e.g., deficits in spatial and working memory) arise.[73] Of course, this sounds the opposite of what we find with *C. elegans* and the better function in *daf-2* (insulin/IGF-1 signaling) mutants, but the mammalian situation is much more complex. Lack of IGF-1 in aging mammals is correlated with worse cerebrovascular function and changes in glial cells that may lead to the impairment of delivery of oxygen and other factors to neurons. This drop in IGF-1 is also associated with loss of the blood-brain barrier, decreased blood flow, increased inflammation, and other problems. While some problems might be due to decreased circulating IGF-1 levels, some IGF-1 is made by the central nervous system itself and could have specific and local effects, such as on synaptic function. This is more parallel to the situation in *C. elegans*, where we observe that neurons of *daf-2* mutants have a completely different set of transcriptional targets than other somatic cells, and those neuronal targets—not the longevity genes acting in other cells—are critical for extended learning and memory with age.

Not everything that could make you live longer will help your memory; in our learning and memory assays, other longevity mutants fared quite poorly. Mitochondrial gene knockdown in early life extends *C. elegans* lifespan,[74] but we found that these mutants have essentially no ability to learn or remember.[75] We also tested the *eat-2* dietary restriction mutant, which lives about 30% longer than wild-type worms; these worms took longer to learn and had only about 60% of the level of memory in young animals. On the other hand, around day

4—when wild-type worms have lost their memory—*eat-2* worms were still at the same level as on day 1.[76] This correlates with their CREB levels, which are lower to start with but are maintained at that same level longer. I don't want to overinterpret the worm data because our learning assay requires eating bacteria in order to form the association and might not work so well in an animal that can't chew its food—but the fact that these mutants can maintain memory later does suggest that DR late in life might be helpful . . . I just wouldn't start it too early! Some data on DR in other animals seem to be quite positive; for example, calorically restricted rhesus monkeys exhibit more gray matter and have greater insulin sensitivity, their hippocampi seem to be more protected from aging, and their brains seem more active in PET scans that measure glucose usage.[77]

Pharmacological Treatments for Slowing Cognitive Decline

Finding treatments to slow cognitive decline is a priority, and one way to look for them is to tap into pathways that are already known to be involved in aging and longevity. The "integrated stress response" (ISR) is a pathway that is separate from the unfolded protein response and other stress responses. As the name implies, this pathway integrates signals from various cellular stresses; in response, translation initiation rates are reduced, and the expression of a small number of specific mRNAs is increased, resulting in changes in protein translation and, eventually, in cell death. The ISR seems to be a sort of toxic, last-ditch reaction to multiple insults. Counterintuitively, *inhibition* of this integrated stress response can actually help organisms survive and repair after injury (this still strikes me as odd). Susanna Rosi and Peter Walter (UCSF) postulated that perhaps the ISR is a causative mechanism underlying various forms of neurodegeneration,[78] and therefore blocking the ISR might help improve cognitive aging. The connection here seems to be through protein synthesis, which you might remember is required for memory formation but also can become defective with age. Because the ISR shuts down all but a few proteins from being translated, the ISR blocks long-term-memory formation. The authors then went on to find a small molecule that inhibits the ISR (ISRIB) and found that treatment of old mice with ISRIB seems to reduce age-associated changes in hippocampal neurons, reduce inflammation, and slightly increase memory.[79] ISRIB seems to have a small but significant impact on age-related cognitive decline—an exciting step in the right direction, opening the door for the screening of other small molecules that might have similar or greater effects on cognitive maintenance.

Manipulating the Immune System and Metabolism to Slow Cognitive Decline

Figuring out how the immune system and metabolism affect cognitive aging will give us more tools in our arsenal to slow or even reverse cognitive aging. For example, one recent study from Katrin Andreasson's lab (Stanford) focused on the role that dysfunctional metabolism can play in driving cognitive decline by altering the activity of immune cells and affecting mitochondrial function. The lipid messenger PGE_2, which is a regulator of inflammation through the COX2 pathway, increases with age, and with it mitochondria become dysfunctional. These dysfunctional mitochondria shift their metabolism to store energy as glycogen instead of producing ATP from glucose, and with this shift, spatial memory declined with age. The deleterious effects might be due to this shift in metabolism that robs ATP from the neurons that need it, and instead leads to defective inflammation.[80] The logical hypothesis was that reduction of PGE_2 signaling or inhibiting the ability of the cells to shift to glycogen storage might help slow cognitive aging. To reverse this deleterious metabolic shift, the authors used a variety of approaches to reduce PGE_2 or to block its receptor, called EP2, in monocyte-derived macrophages, or to cut off the pathway to glycogen storage—all aimed at reversing this shift in metabolism that seems to be triggered by increased PGE_2. Inhibition of either this myeloid COX2-PGE_2-EP2 pathway or the shift to glycogen storage rescues mitochondrial function and glucose metabolism, rejuvenates neurons, and rescues memory. This promising approach of restoring metabolic function in aging cells might not only apply to immune cells, but to other aging cells as well.

Metabolism, Aging, and Treatments for Cognitive Decline

Although many aspects of neuronal aging are specific to neurons, cognitive aging is not completely independent from the metabolic and genetic pathways that control longevity. Individuals with diabetes and prediabetes develop cognitive deficits (e.g., verbal memory, attention, executive function), and type 2 diabetes is a risk factor for Alzheimer's disease as well—possibly owing to problems with the vasculature, as I mentioned before. Therefore, it is possible that drugs aimed at treating diabetes might be helpful in the treatment not only of adiposity, hyperglycemia, and insulin resistance but also of cognitive impairments. Diabetes patients (BMI > 30 with impaired glucose tolerance, both groups on metformin) were treated either with lifestyle change advice (control) or with the glucagon-like peptide-1 receptor analog liraglutide for three to six months until they lost a set amount of weight, then they were tested with a battery of cognitive and biochemical tests.[81] The liraglutide-treated patients exhibited a small increase in short-term memory and memory domain scores for the same degree of weight loss.

Metformin is the AMPK activator that has long been used in diabetes treatment; a meta-analysis of metformin treatment found that cognitive impairment and dementia incidence were reduced for diabetes patients treated with metformin. However, this study did not find support for a cognitive benefit for individuals without diabetes, although we might learn more about the cognitive aspects of metformin treatment once the large-scale TAME trial is completed.[82]

Conclusions

Maintaining our memories and our ability to function with age is fundamental to our quality of life. Beyond frank neurodegeneration and Alzheimer's disease incidence, cognitive decline that accompanies "normal aging" is important for us to study so that we can maintain a high quality of life with age. I'm optimistic that the new clinical possibilities that are emerging as a result of multiple, complementary research directions to better understand how aging causes cognitive decline will identify useful treatments soon. These treatments for the maintenance of cognitive function with age will likely make a far greater impact on our daily lives than any efforts aimed at simply living longer—but the two goals might also be complementary.

15

Lamarck's Revenge?

TRANSGENERATIONAL INHERITANCE, THE "MOLECULAR CLOCK," AND THE EPIGENETIC REGULATION OF LONGEVITY

> How happy is the blameless vestal's lot!
> The world forgetting, by the world forgot.
> Eternal sunshine of the spotless mind!
> Each pray'r accepted, and each wish resign'd.
>
> —ALEXANDER POPE, *ELOISA TO ABELARD*

IS IT POSSIBLE THAT the experiences of your grandparents can affect your health, behaviors, and longevity now? We don't often consider how our actions might affect our future progeny, or how our ancestors' experiences might affect us now, but perhaps we should. While genetics underlies the regulation of aging, epigenetics—modification of the genome that causes heritable changes without altering DNA itself, which Denise Barlow once called "all the weird and wonderful things that can't be explained by genetics"[1]—may also play a role in the regulation of aging. Whether conditions that modify one's grandparents' chromosomes could affect one's lifespan is still not known, but model-system research suggests that it might—or that key pathways exist to erase these transgenerational memories. After all, the ability to respond and adapt to one's environment is a key function of most "longevity" pathways. The loss of various kinds of plasticity happens to cells with age, and the major function of epigenetic mechanisms is to modify cellular plasticity. Finding the epigenetic marks associated with lifespan and aging might identify mechanisms to erase short-lifespan epigenetic marks and improve health.

One of the most famous examples cited as the source of a possible transgenerational effect—that is, a parental experience that affects progeny, or even grandprogeny—came from a sobering "natural experiment," a tragic historical event that tested conditions we could never ethically carry out in humans. From late 1944 until May 1945, in retaliation for a strike by Dutch railway workers (a failed effort to prevent the Nazis from taking a bridgehead across the Rhine), the Germans prevented food and fuel from reaching Amsterdam and the western region of the Netherlands. An exceptionally harsh winter further complicated food transport. As a result, over 4.5 million people were subjected to malnutrition, with some receiving as little as 400–800 calories per day. The "Dutch Hunger Winter" ended when the Allies liberated the western Netherlands and air drops of food began.

Not only did the famine kill an estimated 18–22,000 people but there were also long-lasting effects on those who survived, as well as their children—particularly those born to mothers who were pregnant during the famine.[2] Children who were in utero during this time were reported to have experienced health problems at greater rates than their siblings, eliminating a genetic component to the effects, and instead pointing to a "critical period" in development. These health problems include susceptibility to diabetes, obesity, cardiovascular disease, and schizophrenia. Some of the metabolism-related disorders make sense: it might be smart to shift to an energy-hoarding metabolism if you have experienced starvation at critical times of development in utero, so obesity and diabetes are not surprising in some ways, even if they are a result of "maladaption" to the abundant food supply after the famine. Psychological effects, such as the reported increased rates of schizophrenia in children of mothers starved during the second trimester of pregnancy, are less well understood. Perhaps even more intriguing is the fact that there is now evidence that even the *grandchildren* of women starved during pregnancy were impacted. For example, these grandchildren were smaller than average, and there seems to be some increased incidence of psychological issues. (Keep in mind that girl babies in utero during the Dutch Hunger Winter also already had all of their future eggs, so if starvation could affect germline cells, the grandprogeny generation would have also been subjected.)

A 1972 retrospective study of 19-year-old men enlisting in the Dutch military suggested that the in utero effects of the Hunger Winter did not extend to cognitive function, as these individuals showed no statistical differences in mental ability or general intelligence. However, some of the long-term effects of in utero starvation might have manifested only later, as rates of age-related diseases were increased in this cohort.[3] For example, type 2 diabetes and coronary heart disease risk were found to be increased in a study of this same cohort in their mid- to late 50s. Furthermore, despite showing no significant differences in cognitive function at a young age, this same group suffered from accelerated cognitive aging compared with their peers.[4] A study of the 56–59-year-old cohort found that while general intelligence (again) was not affected, the group performed less well on a selective attention task, indicating decreased cognitive function.[5] Finally, this cohort had an increased mortality rate compared with those born immediately before or after the famine. Together these data suggest that in utero experiences might have affected rates of aging, possibly through reduced resilience that is only revealed at advanced ages. Whether similar effects on aging and cognitive decline might manifest in the grandchildren of the starved mothers of the Dutch Hunger Winter remains to be seen.

The Great Chinese Famine of 1959–61 was perhaps the largest famine in history: 30 million people died and fertility was greatly affected. Increased rates of tuberculosis, blood-borne infections, and sexually transmitted diseases have been observed into the grandprogeny's generation.[6] Although the Great Chinese Famine had less sharp boundaries than the Dutch Hunger Winter, and is therefore less amenable to interpretation, these results suggest that prenatal famine exposure and low birth weight may contribute to a multigenerational effect on immune function.

Historical Trauma:
Transgenerational Inheritance of PTSD?

The controversial idea that traumatic experiences may not only have effects on the generation that experienced the trauma but might also affect their children (intergenerational) or even later generations beyond the first (transgenerational) is known as "historical trauma," and was first postulated by Maria Yellow Horse Brave Heart in the 1980s to describe the long-lasting effects of trauma experienced by the Lakota community. The Lakota, like most Native Americans, experienced many forms of trauma, including colonization, relocation, genocide, assimilation, family separation, suppression and erasure of culture and language, and harrowing experiences in American Indian boarding schools. Yellow Horse Brave Heart suggested that in addition to ongoing stress, descendants may also suffer from the trauma of previous generations.[7] Similarly, descendants of enslaved African Americans, Cambodian refugees, and Holocaust survivors have also been reported to suffer from historical trauma, which can manifest itself in higher rates of substance abuse, depression, suicide, and post-traumatic stress disorder (PTSD) than their non-descendent counterparts. Native Americans have worse health than any other group in the United States, and African Americans also suffer poorer health than whites with similar socioeconomic status. Children of Holocaust survivors and Cambodian refugees have been diagnosed with PTSD, anxiety, and depression at high rates, suggesting at least an intergenerational inheritance of trauma.[8] Historical trauma not only affects the mental health of the young but has also been reported to affect aging rates and health of the elderly, including increased rates of dementia.[9]

Of course, the causes of some of these health risks are difficult to disentangle from ongoing oppression, poverty, violence, and other systemic effects, including structural and systemic racism. For example, low-income neighborhoods are often subjected to poorer air and water quality because they are forced to be near highways or factories that negatively affect health. But there is now some recognition that there may be heritable factors that contribute to PTSD. If we can learn more about the molecular mechanisms underlying inter- and transgenerational trauma, we might be better positioned to treat them and stop the cycle of inherited and childhood trauma. For example, work by Dr. Nadine Burke Harris, a pediatrician and currently the surgeon general of California, has suggested that chronic childhood stress (adverse childhood experiences) can be a major contributor to toxic stress and neurodevelopmental problems, but can also be treated through interventions.[10]

Inter-versus Transgenerational Inheritance

Some phenotypes could be explained by metabolic changes, as well as upbringing and parental care, and are actually *intergenerational* rather than transgenerational—that is, are seen only in the first generation of progeny but not in later generations. While the words sound similar, the biological mechanisms at work can be entirely different. One can imagine that the conditions in utero might have long-lasting effects in that individual, but it still might not affect the following generation. For example, depression and psychological stress during pregnancy are associated with an increase in the risk of childhood asthma, even controlling for factors like age, smoking during pregnancy, BMI, and history of asthma, suggesting that the in utero state may affect the child's later health—although the mechanism is not understood, and it might not last beyond one generation.[11] A stress response in the uterus during gestation includes increased levels of the "stress hormone" cortisol and changes in other biological components (such as reductions in transport of nutrients, oxygen, growth factors, and impaired mitochondrial function) within the uterus, which in turn could directly affect development of the fetus. These changes in utero may have an influence on epigenetic mechanisms that affect both the developing fetus and later generations as well. Interestingly, levels of oxytocin rise in mouse mothers that have had pups, which alters their behavior;[12] how behavior of mothers might also affect their children or future generations is another aspect of transgenerational inheritance that we might be able to probe with molecular approaches.

Lifespan and mortality are closely linked with metabolism and development of disease with age. An adverse environment in the uterus has been linked to poor health outcomes later in life, as pointed out by the Dutch Hunger Winter studies and supported by mouse studies. These adverse effects are intergenerational. But some of these changes may also affect epigenetic mechanisms. For example, NAD^+ is effectively a nutrient sensor, and affects histone acetylation; S-adenosine methionine is another intermediary metabolite and is required for methylation, an epigenetic mark. In both cases, there is a link between nutrients, intermediates, and epigenetic mechanisms. While we have focused on the in utero conditions, there is growing evidence that sperm can pass on epigenetic information as well.[13] Differences that manifest in the third generation after a traumatic event—removed two generations from the experience—are less likely to be connected directly to the conditions at the time, and suggest that epigenetic mechanisms must be involved in the inheritance—that is, an inherited effect that is due to something that is not strictly genetic in nature.

In a remarkable (if controversial) study, Dias and Ressler (2014) showed that not only did mice subjected to odor fear conditioning become sensitive to that odor, but their progeny (F1) and grandprogeny (F2) also exhibited increased sensitivity—an example of transgenerational inheritance of a learned association.[14] These results suggest that learned aversions could be passed down through the generations. The authors suggested that hypomethylation of the gene for the receptor for the odor could be passed on through the male germline, a truly Lamarckian finding. Later work in worms by Rachel

Posner and Oded Rechavi (2019) suggested that small RNAs in neurons might control chemotaxis behavior transgenerationally, as well.[15]

Aging and Late-Onset PTSD

Whether or not we can see effects in just one generation or many, adverse environmental conditions and trauma should be considered in our analyses of age-related declines. Disturbingly, some symptoms of trauma (whether direct or intergenerational) that might have been managed in early and mid-adulthood can reemerge in aged individuals, a phenomenon known as "late-life onset stress symptomatology."[16] The emergence of PTSD in the elderly has been noted particularly in veterans of wars; because such a large fraction of the elderly male population is made up of war veterans, war-related distress has been called a "hidden variable" in aging research. For example, in 1996, 76% of all men aged 70–74 were military veterans; of course, this will shift given the large number of soldiers in World War II, the Korean War, and the Vietnam War, but the resurgence of a population that experienced traumatic conditions increased again post 9/11. If we ignore the late-life effects of trauma or the emergence of PTSD with age, we will be missing part of the picture. Some of the regulatory mechanisms that keep PTSD-type responses silenced early in life might be relaxed and reemerge late in life, suggesting an epigenetic change.

How Epigenetics Lost Its Luster: Lamarck, Lysenko, and the Dark History of Epigenetics

Before we delve further into the underlying biology, we have to acknowledge the elephant in the seminar room. Until relatively recently, epigenetics had a bad reputation because of its historical and political connotations. On the one side, we have Mendelian genetics, Darwin, and natural selection, the clear winners in scientific history; on the other side, we have a discredited theory of adaptation, a Soviet Communist obsession with eradicating the prevailing ideas about genetic inheritance, and the resulting crop failure and famine.

The French naturalist Jean-Baptiste Lamarck (1744–1829) is associated with the notion that characteristics we observe in animals can be traced to physical changes in organisms during their lifetime that have been transmitted

Caveats to Reports of Human Transgenerational Phenomena

Before we progress further into the mechanisms underlying trans-generational inheritance, it's important to point out that there remains skepticism about the concept of transgenerational epigenetic inheritance of trauma in humans, and maybe even work done in mice aimed at modeling transgenerational trauma. Both Ewan Birney, director of the EMBL-European Bioinformatics Institute, and Kevin Mitchell, a neurobiologist and author of *Innate*, have pointed out (in great depth) that many of the studies on the Dutch Hunger Winter and historical trauma that I have mentioned above suffer from small sample sizes and statistical problems, lessening the confidence we should have in these studies, despite the compelling stories they present.[a] This is of course difficult to fix in historical studies, where one is obviously limited for the number of study "participants." For example, a famous historical study of residents of Overkalix, Sweden, and their children and grandchildren investigated the relationship between mortality rates and food availability. The study reported complicated sex-specific relationships between BMI, smoking rates, and food supply that the authors conclude supports the idea of transgenerational inheritance and parental imprinting, particularly in specific life periods. However, Mitchell notes that this study lacks a prior hypothesis, has only 317 people in the starting group, and no multiple hypothesis testing is done—so it's hard to know what to believe.

Mitchell went on to point out that it's not just human epidemiological or historical studies of transgenerational inheritance that suffer from these statistical flaws; several mammalian studies that purport to model transgenerational stress also make similar errors. As an example, he shows that a paper from Isabelle Mansuy and colleagues titled "Epigenetic Transmission of the Impact of Early Stress across Generations" in fact appears to suffer from "significance fishing," by focusing only on the result that showed a barely significant difference, and is not corrected for multiple hypothesis testing. Mitchell points out similar flaws in another paper from the same group ("Implication of Sperm RNAs in Transgenerational Inheritance of the Effects of Early Trauma in Mice"), which in theory should model some of the historical trauma effects described above. It's a shame that the work doesn't seem solid, since it's important to test these concepts in genetically

Continued

tractable model systems that one could use to discover the underlying molecular mechanisms. Although it is understandable that human studies are limited by real-world issues of small numbers and lack of control of environmental conditions, and so on, the whole point of carrying out studies in mice should be to overcome these types of limitations and truly test hypotheses with sufficient statistical power. As much of a model-systems proponent I am, it's hard to justify weak studies, so we will need to rely on better work to test our models of transgenerational epigenetic inheritance. To that end, I'll focus more on the work done in worms and flies, and mouse studies that seem a bit more solid. Certainly, work in plants and diverse metazoans, including *Drosophila*, butterflies, nematodes, *Daphnia*, and lizards suggests that this Lamarckian notion is not entirely far-fetched, but mammalian conclusions broadly need better and statistically stronger support than may have been shown in some human or mouse studies at this point.

to their offspring—"inheritance of acquired characteristics." The adaptation of the giraffe's neck to eating leaves in high branches is the oft-quoted illustration of this notion. To be fair, some of Lamarck's ideas have been misrepresented over time. He wasn't the first to propose this model of inheritance, but he is perhaps most associated with the idea that an experience or need in an animal can then be carried on to its progeny. Of course, this principle requires that effects on that animal, even if they are somatic traits seen only in adults, be transmitted through the germline; given what we now know about genetics, that seems like a stretch. Countering this idea, August Weismann (1834–1914) argued that there is a barrier—now known as the "Weisman barrier"—that prevents changes in the soma from being transmitted to the germline, resulting in the continuity of the germline and relatively constant phenotypes over generations. (This is the same Weismann who first postulated that aging is caused by the evolutionary pressure to get rid of old animals in the population—"altruistic aging"—and later he suggested that aging is programmed.)

The notion that learned information—in this case, a conditioned reflex—could be inherited was tested by Ivan Pavlov—yes, that Pavlov of the dogs, steak, and saliva. He tested the concept in mice, and later others tried to show inheritance of learned information in different ways, with little success. At first, it seemed there was some supporting evidence that a conditioned response could be inherited (William McDougall, 1930) but these conclusions were later refuted (Francis Crew, 1932) and Pavlov later also retracted his findings on the matter. Other experiments tested the heritability of experiences, learned information, or conditioned reflexes in a range of animals (e.g., guinea pigs, goldfish, rabbits, rats, moths, and mice) in the early part of the twentieth century, with mostly inconclusive and irreproducible results. Thus, the notion that a response to conditions (such as heat or humidity) or a learned experience could be inherited seemed unlikely. More recently, Dias and Ressler's finding that male mice could pass on an olfactory-conditioned fear to their offspring (2014) was controversial, as it suggested once again that experiences could be inherited, and as with the Mansuy papers, again Kevin Mitchell has expressed skepticism due to weak statistics. In general, the "Larmarckian" view that experiences can affect inherited features was largely discredited through genetic experiments. The principles of Mendelian genetics were supported by more experimentalists, particularly *Drosophila* researchers like Thomas Hunt Morgan, who established that inheritance depends on chromosomes.

It seems a bit bizarre in this day and age to consider the political implications of each of these viewpoints, since we now know the underlying molecular components and quite a bit about how they work (below). But these questions were being asked in the late 1800s and early 1900s, a time when even the genetic material wasn't known or understood, so these biological questions were still somewhat philosophical in nature. To put these ideas in context: at the time, the proteins coating the chromosomes were thought to be *the* inherited factor, because proteins are complex, and DNA was thought to be too simple and boring, so it was assumed (obviously!) that only proteins could encode such complicated information as would be needed to pass along hereditary information. It was only in 1928 that Frederick Griffith showed that a component that could survive heat killing (that is, *not* protein) was the "transforming principle"—suggesting but not proving that DNA was the inherited factor. And it was not until 1944 that Oswald Avery, Maclyn McCarty, and Colin MacLeod showed through a purification approach that the component was very likely to be DNA. And it wasn't until 1952 that Alfred Hershey and Martha Chase performed their experiments with radioactively labeled proteins and DNA that proved that the inherited material is DNA.

So perhaps it is not so strange, given the context of these discoveries in the early decades of the twentieth century, that there was considerable interest in, and then discounting of, Lamarckian mechanisms of inheritance—except in the USSR, where neo-Lamarckism took hold because of Stalin's opposition to genetics, which to him seemed counter to the communist viewpoint. To the adherents of Stalin's form of communism, Lamarckism was preferred because it suggested that one could control one's own destiny (and I guess one's descendants' destiny, as well?), while the principles of genetics were a threat, as the implication was that one was stuck with what one inherited; genetics was considered "bourgeois." To eradicate the conflict with communist ideals, geneticists and evolutionary biologists were violently removed from Soviet science through dismissal, imprisonment, and in some cases execution. (Before we start to feel too smug about American science and politics and the obvious wonders of genetics, keep in mind that several US geneticists supported the malevolent principles of eugenics in this era; at least the *Drosophila* geneticist Thomas Hunt Morgan was one of the critics of the eugenics movement. And the "Scopes Monkey Trial" took place in Tennessee in 1925, so Darwin wasn't exactly winning here, either.) As a result of Stalin's distrust of genetics, all biological research in the USSR was replaced with Lamarckism. The pinnacle of this ideological thrust was the disastrous, famine-causing crop failure brought about by the Lamarckian views of Trofim Lysenko, Stalin's head of biology. Lysenko insisted that plants and their offspring should be able to respond flexibly to extremes of temperature and humidity (known as "phasic development") and thus should be able to survive extreme cold through a process called vernalization—where seeds were sprouted in extreme cold, and their offspring would inherit cold resistance—but instead of testing this concept rigorously, he plowed ahead and implemented these principles on a massive scale. Tragically, it turns out that hatching seeds in the frigid cold of winter or treating them with acid or sandpaper doesn't make them stronger, it just kills them. Although he did make some legitimate contributions to plant research, Lysenko's Lamarckian legacy to Soviet science— "Lysenkoism"—was an agricultural "revolution" that ended up ruining wheat crops and causing widespread famine by the 1930s.[17] These crackpot agricultural ideas even spread to China in the late 1940s, resulting in the Great Chinese Famine (1959–61) that I already mentioned.

It's understandable, then, that the idea of nongenetic mechanisms of inheritance lost favor as we gained a better understanding of how DNA functions and how genetic information is passed on. It's worth pointing out, though, that at least one scientist was on the mark regarding the future of epigenetics, even before the structure of DNA was known: Barbara McClintock, using maize as a model organism, not only discovered transposable elements (see below) and experimentally proved Thomas Hunt Morgan's theory of genetic recombination via crossing-over (1931), but also was the first scientist to propose that genes can be expressed and silenced—an epigenetic mechanism—during mitosis. In more recent years it's become apparent that as we look closer, there are indeed examples of heritable changes that don't involve modifying the DNA code itself, and at least some scientists think that epigenetic mechanisms may account for some of the "missing heritability"—effects that are not easily explained by Mendelian genetics—particularly in complex phenotypes. While genetics largely underlies the regulation of aging, epigenetics—modification of the genome that causes heritable changes without altering DNA itself—also plays an important role. Whether the conditions or experiences that modified one's grandparents' chromosomes could affect one's lifespan is still not known, but model-system research suggests that they might. Finding the epigenetic marks associated with lifespan might identify mechanisms to erase short-lifespan epigenetic marks and improve health.

Epigenetic Marks: Molecular Signatures and Regulators of Epigenetic Function

I've now described a lot of possible cases of epigenetics and how they might affect aging in humans, but what does this really mean at the cellular level? What might be the molecular causes of these epigenetic events? And how can we think about passing on information from one generation to another, and how these molecular events might be affected by aging—or how they themselves affect aging?

To understand and test how events that cause historical trauma might cause long-lasting effects and get passed on to later generations, we need to identify the relevant underlying molecular mechanisms and, if possible, use model systems to test these functions. Early work in plants reported observations consistent with epigenetic mechanisms of inheritance (for example, throwing of "rogue" peas) as long ago as Bateson and Pellew's work in 1915, and some of the ideas put forth by E. E. Just, the noted African American embryologist, also hinted at epigenetic mechanisms of inheritance at work in the 1930s.[18]

Several types of epigenetic regulatory mechanisms have been discovered in the past few decades. The currently known mechanisms of epigenetic regulation include transposable elements (TEs), DNA methylation, modifications of histones (the proteins that wrap up and compress DNA), and the expression of various noncoding RNAs. In addition to probing how these are involved in transgenerational inheritance of experiences, we can also ask whether they exhibit changes with age, and how these epigenetic mechanisms regulate longevity.

Transposable Elements: Genomic Parasites that Become Active with Age

Barbara McClintock (Cold Spring Harbor Laboratory) first discovered and defined the role of TEs—that is, pieces of DNA that could "hop" to change their position in the genome—while studying chromosomes in maize in the 1940s. She had studied genetic recombination and cytogenetics starting in the late 1920s,[19] and in 1950 she reported that there were chromosomal elements that could "transpose" to different sites in chromosomes and then regulate physical characteristics; these could be passed from one generation to the next, a form of epigenetic inheritance. Unfortunately, McClintock's work on TEs met considerable resistance at first from the research community, and so she stopped publishing on the topic.[20] (Remarkably, she did this work at a time just before the structure of DNA had been determined.) But her brilliance was later recognized, and in 1983 McClintock received the Nobel Prize in Physiology or Medicine for her discovery of mobile genetic elements. In the 1960s, other researchers found similar transposition events and transposable elements in organisms ranging from bacteria to yeast, *Drosophila*, even human cells. Maize was a particularly serendipitous—or wise—choice of model system for McClintock's studies, as almost 90% of the maize genome is made up of TEs. In fact, almost half of our own genome is composed of TEs, often mistakenly referred to in the popular press as "junk DNA."

TEs are generally considered "genomic parasites" that need to be silenced to prevent deleterious insertion events, which is likely one of the reasons that epigenetic mechanisms of silencing initially evolved, such as RNA interference (double-stranded RNAs that lead to degradation of mRNAs) and piRNAs (very small RNAs that silence the expression of TEs). Sometimes this silencing itself can be deleterious as well, if the silencing "spreads" instead of remaining restricted to the TE, turning off the expression of nearby genes that need to be active. However, we may also thank TEs for our current state, as they likely played a role in evolutionary adaptation, including speciation events. In fact, TEs, in particular endogenous retroviruses (ERVs), are likely responsible for the evolution of the placenta through the modulation of transcriptional networks in the tissues that make up the maternal-fetal interface.[21] So TEs are not all bad, but the unregulated activation of TEs in individuals can cause problems. The impact of TE activation on human health became more apparent when it was realized that some TEs, such as the "long interspersed element-1," or LINE-1 (L1), are active and contribute to mutations in cancer, as well as to some diseases, like hemophilia, muscular dystrophy, and immunodeficiencies.

With age, TE-silencing mechanisms become defective and TEs start to be mis-expressed. Epigenetic and epigenomic "drift" with age can affect how cells function by altering their normal regulation of gene expression, but the transposition of these DNA pieces can also generate double-stranded DNA breaks, which is of course very bad for cells. Senescent cells are particularly vulnerable; recent work by John Sedivy, Jill Kreiling (Brown University), and others suggests that TEs may become activated with age, particularly in cells that have become senescent.[22] Once the retrotransposon is activated, the type-I interferon immune response is triggered by the L1's complementary DNA (cDNA). This age-related, inappropriate immune reaction, known as "inflammaging," is part of why senescent cells are so dangerous to their healthy neighboring cells. Luckily, the immune response can be stopped by inhibiting the enzyme that makes the cDNA, the L1 reverse transcriptase; old mice that were treated with this inhibitor slowed inflammaging in many tissues. Caloric restriction also appears to slow the inappropriate activation of retrotransposable elements with age, adding to the slew of positive effects of caloric restriction on aged animals.

Inappropriate activation of retrotransposons may also promote and accelerate neurodegeneration and contribute to intracellular toxicity. In fact, work from Josh Dubnau (State University of New York at Stony Brook) suggests that retrotransposon activation in flies may cause the spread of toxicity.[23] The Gypsy ERV may cause DNA damage, leading to the toxic spread from glia to neurons. In *Drosophila* glial cells that express the ALS-associated protein TDP43, retrotransposon activation can lead to not only death of those cells but also the spread of toxicity to nearby neurons, leading to their demise. Whether this spread of toxicity happens through a signaling mechanism, or through the actual spread of protein is currently unknown, but it offers another glimpse into possible epigenetic mechanisms of age-related neurodegeneration.

Prions: Heritable Spread of a Protein Conformation

While the spread of a toxic protein from cell to cell may sound like science fiction, the "prion" hypothesis of disease is exactly this idea, that a protein can take on a toxic form that induces other proteins it meets to do the same, resulting in aggregation. Prions may be thought of as another form of epigenetic regulation, since the genome encodes one protein that may take on either a normally folded form or a toxic, infectious form that induces this change in its neighbors. Diseases such as scrapie, Creutzfeldt-Jakob disease, fatal familial insomnia, kuru, and "mad cow" disease (bovine spongiform encephalopathy) are all examples of prion diseases. They are particularly pernicious when transmitted from one animal to another, as in mad cow disease. It has been speculated (but not yet proven) that several age-related neurodegenerative diseases may be caused by prion-like spreading and aggregation.[24] The proteins associated with Alzheimer's disease, A-beta and tau, form amyloid and neurofibrillary tangles, respectively, and might be subject to this kind of aggregation.

While it would be easy to think of prions as all bad, since our awareness of them arose through disease, the fact that prions exist in so many types of cells and organisms suggests that there might be a cellular function for them. For example, some yeast prions may exist to allow responses to environmental cues that may require phenotypic diversity, and seem to have been maintained by purifying selection[25]—that is, if prion proteins had no useful function, by now they should not exist. As far as epigenetics goes, prions are poised to transmit information in a nongenetic manner from cell to cell and transgenerationally, and recent work suggests that a yeast DNA-binding protein, Snt1, that acts in a histone deacetylase complex can form a prion-like conformation, infect other yeast cells, and act transgenerationally to activate yeast genes through the activated chromatin state.[26] Even more excitingly, there is emerging evidence from flies and *Aplysia* (also known as the California sea hare, or *A. californica*—a snail that is used in memory studies) that prions may have evolved to be important for long-term memory formation and long-term facilitation, as the prion form of some synaptic proteins is required.[27] The existence of a regulated form of prion might also suggest the existence of mechanisms to "undo" the toxic forms.

DNA Methylation and the "Aging Clock"

One of the major epigenetic marks on chromosomes is the methylation of nucleotides of DNA. Methyl groups can be enzymatically added to the pyrimidine ring of the nucleotide cytosine (the C in the A, T, G, and C that make up DNA), resulting in 5-methylcytosine; methyl groups can also be added to adenine, observed mostly but not exclusively in prokaryotes. Cytosine methylation is often at "CpG" dinucleotide sequences, which has served as a useful tool for examining differences in methylation during development, in cancer, and with age, as these methylation changes can greatly affect which genes are expressed and thus the developmental trajectory of a cell.[28] Marking a strand of DNA with methyl groups allows the cell to distinguish self from nonself, and to control gene expression. TEs are silenced via DNA methylation, as is the extra X chromosome in females; methylation patterns are also critical for imprinting during early development. Methylation of DNA in the promoter region of a gene can prevent transcription of that gene, and methylation in a gene body may alter splicing of RNA, so even though the DNA code itself is the same, it can change the output. Importantly, this can be undone through "demethylation," so the process is reversible, which also suggests that the processes are regulated.

The utility of epigenetic programming is particularly striking in eusocial insects that use castes to divide labor (queen, workers, foragers, nurses, etc.), such as honeybees and ants. In these species, the genome is the same for all the individuals, but differences in DNA methylation may be the key to specifying an individual's role in the colony. For example, in one of my favorite papers, all the nurse bees were removed from a hive "using a strategy of hive trickery" while the foragers were out; when the foragers came back to the hive, since there were no nurse bees around, the foragers switched to become nurses.[29] The researchers then characterized the differentially methylated regions in these bees to compare nurses, foragers, and foragers-turned-nurses. This analysis revealed that the gene expression differences caused by DNA methylation changes were mainly due to alternative splicing, producing different versions ("isoforms") of these genes.

The biological role of DNA methylation (DNAm) in aging is not well understood at the functional level, but DNAm has become a useful biomarker of aging. In a groundbreaking, single-author paper, the UCLA mathematical biologist Steven Horvath found that methylation of just 353 sites in the whole genome roughly correlates with—and can essentially predict—an individual's age.[30] A large number of DNAm datasets were publicly available because of the ubiquity of DNAm analysis to study cancer; Horvath used the unaffected healthy controls in his study. Using a machine-learning approach to analyze data from different tissues and ages, he found a set of methylation sites that best report on age. Of the more than 21,000 sites that are present and probed on the Illumina CpG arrays, only 353 of them being necessary to predict age is remarkably few. Some tissues do better or worse than others for this DNAm predictor (and sperm didn't work well at all), and the predictor works slightly better in adolescents than adults, but there is still enough resolution to predict age relatively well. As a proof of concept, Horvath found that centenarians accumulate these DNA methylations more slowly, and Werner's progeria patients accumulate them more quickly, suggesting that these sites of DNAm might be evidence of an "epigenetic clock" of aging. Interestingly, reprogramming cells to become induced pluripotent stem cells (iPSCs) essentially resets this clock—again, emphasizing Weismann's point about the pristineness of stem cells. Simultaneously, another team built another DNAm predictor, the Hannum clock, based on whole blood.[31] Since these 2013 papers, there have been additional DNAm clocks built to address aging in mouse tissues, since the human DNAm clocks worked only in human and primate tissues. (What this last point means biologically is still not clear.) Additionally, newer clocks that better reflect biological age and aging phenotypes now exist, and have been used to study not only biology but also how socioeconomic factors affect epigenetic aging.[32]

Morgan Levine and Horvath used the DNAm clock to address an interesting question: does rapid aging cause menopause, or does menopause cause aging?[33] To do so, they looked at age acceleration—that is, where a woman's blood or saliva DNAm (her "biological age") looks as though she's "older" than her chronological age—and when this might happen. To avoid the chicken-and-egg-type problem of cause and effect, they also examined factors that might speed or slow menopause effects. That is, their initial data correlating menopause with age acceleration could not distinguish cause and effect, since a woman who is aging more rapidly might go through menopause earlier. Therefore, they also analyzed women who were treated with menopausal hormone therapy (MHT), which might slow effects, and women who had undergone surgical ovary removal (oophorectomy), which causes early-onset menopause because ovarian hormones are removed. They found that MHT slowed DNAm aging, and that surgery-induced menopause accelerated the DNAm aging. Thus, Levine and Horvath were able to use DNAm to probe a previously unanswerable question about the cause and effect relationship between menopause and aging. Although it is not exactly clear *how* the DNAm sites that best correlate with aging are related to aging biology, it is a useful tool to predict aging, and has great potential in characterizing tissue aging. There is no doubt it will be used to study potential anti-aging therapies in the near future, and some of these "molecular clocks" are already being marketed to consumers who are curious about their own epigenetic age.

If I Could Turn Back Time

A recent breakthrough study went beyond DNAm clocks to instead use single-cell RNA sequencing in the subventricular zone of the mouse brain—where neurogenesis happens—to ask whether transcriptional information could be used as a biological aging clock. Anne Brunet's and Tony Wyss-Coray's labs (Stanford) used a huge number of single-cell transcriptomes from 28 mice, across ages from young to old, and from these data were able to build both chronological and biological clocks.[34] Using these clocks, they could see particular transcriptional signatures for specific cell types; this aspect is far more interpretable than DNAm clocks, where the biological effects related to specific DNAm marks are often unknown. They then used these clocks to assess a few kinds of rejuvenating therapies, heterochronic parabiosis and exercise, to assess the extent to which they could "reverse" the transcriptional clocks. Although this is technically not an epigenetic study, I mention it here because I believe this single-cell transcriptional clock, because it is biologically more informative than DNAm clocks, will be important in our understanding of aging and rejuvenation treatments.

Histone Code Changes with Age and the
Epigenetic Regulation of Longevity

Our chromosomes are not just naked strings of DNA flailing around in our cells—and in fact, if it were floating freely it wouldn't all fit; it's estimated that we all have enough DNA to reach to the sun and back more than 300 times! Instead, DNA is able to be so compact because it is wrapped around proteins called histones, and those histones are further organized into octamers that make up a "nucleosome," lending a "beads on a string" appearance in electron micrographs, which was first recognized by husband and wife team Don and

Ada Olins in the 1970s.[35] Packing and unpacking of DNA onto nucleosomes is critical for gene expression, and the "code" for this regulation of gene expression is thought to be through modification of histones, such as methylation and acetylation of specific amino acids, particularly lysines. Recently, even more new modifications have been discovered, such as serotonylation of glutamine in histone H3 of serotonergic neurons, which might further refine expression regulation.[36] The exact code of the marks is not completely understood, but, in general, certain marks are associated with repression of expression, while others are activating. The most prominently studied marks are acetylation of lysine 27 of histone H3 (H3K27ac), which is associated with active enhancers, and trimethylation of lysine 4 of histone H3 (H3K4me3), which is associated with promoters.

Histone modifications can change with age, altering gene expression; in fact, changes in these histone modifications have been recognized as one of the major, conserved hallmarks of aging, and have been documented in yeast, worms, flies, and mammals. As these marks change with age, they lose coherence, leading to increasing variation and "epigenetic drift," which in turn affects gene expression.[37] Those changes in gene expression due to changes in histone modifications can affect cellular and systemic functions. For example, at the cellular level, loss of proper expression of metabolic genes with age can lead to loss of metabolic homeostasis.[38] At the systemic level, these changes with age can induce inflammation, another cause of "inflammaging." Like DNAm marks, the histone code can distinguish ages and can be used to predict transcriptional changes with age in multiple species, including humans, rats, and killifish.[39]

Histone modifications can also regulate lifespan: loss of the ASH-2 trithorax H3K4 trimethylation complex (ASH-2, WDR-5, SET-2) extends lifespan, acting in the germline of *C. elegans* to regulate somatic lifespan.[40] Similarly, reduction of UTX-1, the histone demethylase specific for lysine 27 of histone H3 (H3K27me3) increases *C. elegans* lifespan through activation of DAF-16/ FOXO,[41] although in different contexts overexpression of UTX-1 and other histone methylation regulators can also increase lifespan.[42]

Perhaps most intriguing is the fact that changes in histone modifications can affect lifespan for several generations, known as transgenerational epigenetic inheritance, or TEI—a finding that is both exciting and frightening for its implications.[43] Anne Brunet's lab (Stanford) found that not only are *C. elegans* mutants of the COMPASS complex of H3K4me3 histone modifiers long-lived themselves but when they are crossed with wild-type worms so that the progeny are *genetically* wild type, they are also long-lived, a true epigenetic inheritance of longevity. This longevity lasts for four generations before returning to the normal wild-type lifespan in the fifth (F5). Brunet's lab found a similar epigenetic effect of COMPASS complex on fat regulation that lasted a few generations, then disappeared; that is, loss of COMPASS seems to extend lifespan through its regulation of monounsaturated fatty acids, and this could be mimicked by dietary addition of such fatty acids.[44] It's important to note that all of these results were caused by the *loss* of histone modification regulators, either by mutation or RNA interference knockdown—so in the wild, these marks are likely reset every generation. In fact, mutants of *wdr-5* require many generations to become long-lived, and reduction of the H3K9me2 demethylase JHDM-1 can mimic this effect, suggesting a heterochromatin spreading effect at work in these mutant conditions.[45] Resetting the epigenome is important for maintaining the pristine nature of the germline, so it might not be surprising that organisms go to great lengths to maintain these marks—even if their loss might increase the longevity of the animal in some cases.

Small and Noncoding RNAs in TEI: RNA Interference, piRNAs, and Noncoding RNAs

In addition to messenger RNA—the kind of RNA that gets transcribed from DNA and then translated into amino acids to make proteins—cells also make several different flavors of "noncoding" RNAs. These RNAs can be long (lncRNAs) or short (microRNAs and small RNAs), and they each carry out different tasks in the cell. MicroRNAs (miRNAs) are generally repressive, either blocking translation of mRNAs or preventing transcription; they were originally discovered (in worms, of course, by the Ambrose and Ruvkun groups) through their well-defined roles in the regulation of development.[46] miRNAs have been implicated in the regulation of longevity, for example, by interacting with components of the insulin and FOXO pathways.[47] miRNAs are sometimes mis-regulated in disease and cancer, and are known to change with age in many organisms.[48] Long noncoding RNAs are abundant but less well understood, and also play a role in regulation of transcription and post-transcriptional mRNA processing. They may interact with the same chromatin remodeling complexes that play epigenetic roles, and telomere regulation, as well.

If we were to try to make a direct comparison between the Dutch Hunger Winter and a similar situation in *C. elegans*, we could look at the progeny of starved animals.[49] The worms are masters at surviving low-nutrient conditions, and starvation at various times in their development can cause arrest— not just the famous dauer arrest, but also L1 arrest and even "adult reproductive diapause."[50] In fact, starvation can have transgenerational effects: grandprogeny of starved worms were more resistant to starvation themselves, and heat resistance lasted yet another generation, echoing the longevity results caused by COMPASS mutations but in the natural context of starvation.[51] Small RNAs might be key to some of these transgenerational effects: an analysis of endogenously produced small RNAs from young adult worms after either being starved or fed in early larval stages revealed that even though their great-grandchildren (F3 worms) had been fed for three generations, a sizeable fraction of the small RNAs that had been up- or downregulated in their starved ancestors were still affected in the F3 generation.[52] Moreover, those small RNAs regulate nutrition- and aging-related genes, and these patterns of gene expression were also maintained through the F3 generation. Finally, the great-grandprogeny of starved animals were long-lived. Like the COMPASS regulation of longevity over multiple generations, the starvation-induced longevity that acts through small RNAs is conveyed for at least three generations. Whether humans also change their endogenous RNAs in response to stressful conditions—like stress—and whether that in turn affects longevity and aging rates is still unknown, but these results in *C. elegans* suggest mechanisms that could be explored.

What Would Be Worth "Remembering" in the Lamarckian Way?

A few years after starting my lab, I gave a presentation at a meeting of Pew Scholars about the work my lab had been doing to develop learning and memory assays in worms, and using these tests to study cognitive decline. After the talk, Craig Mello (who, with Andrew Fire, had won the Nobel for his work on RNA interference) asked me a question I hadn't considered before: Did I think there would be something the worms would remember not just in adulthood, but into the next generation? I said that I didn't think that the worms would want to remember an odor for that long, since even with lots of training they forgot this food-odor association after about a day, and the food source would surely be gone by the next generation.* "But . . . maybe a pathogen?" I suggested, and shrugged. We talked about it for a minute longer, then I went to join the rest of the attendees, promptly forgetting about the conversation.

What I *should* have done is run back to the lab and immediately start working on this problem. (Note to self: next time a Nobel Prize winner asks you about an experiment, maybe get on it sooner.) Instead, it would be another six years before my graduate student Rebecca Moore, who was interested in how aging might affect pathogenesis, did a crazy experiment that tested this idea. She had been asking whether exposure to the human (and worm) bacterial pathogen *Pseudomonas aeruginosa* (PA14) affected development and dauer formation in progeny; she had already found that the progeny of mothers placed on PA14 for 24 hours would later go into an alternative diapause state at an increased frequency compared with progeny of control mothers, and that this effect lasted more than one generation. While chatting with Rebecca one day, we decided it might be interesting to ask whether the worms that she was testing for dauer and developmental changes—the progeny of mothers that had been on *Pseudomonas* overnight a few days earlier—already knew to avoid the pathogen when they encountered it. That is, would the children of mothers who had been made sick by the pathogen also override their natural attraction to *Pseudomonas*? We already knew from work by Yun Zhang and Cori Bargmann that worms love the smell of *Pseudomonas* (because it's a good food source at low temperatures), but when they are kept on it for 4 hours, which makes them sick, they learn to avoid it.[53] Bargmann's lab had tested this idea with the 4-hour-trained mothers, and there had been no effect in the next generation.[54] But when Rebecca tested the progeny of her 24-hour-trained mothers, surprisingly, they did avoid the *Pseudomonas*! That is, even though those specific animals had never seen *Pseudomonas* before and had never been sick, they already knew to avoid it because their mothers had learned to avoid it. Even more shocking was the fact that *their* progeny, and two more generations beyond that, all avoided *Pseudomonas*, even though the only worms to actually encounter the pathogen were their great-great-grandmothers. It was only in the fifth generation after the original pathogen exposure that the

* A 2010 single-author study reported the transgenerational inheritance of an "olfactory imprint" of an odor association for over 40 generations, but the study contained no mechanistic information, and as far as I know, has never been replicated. See Remy, J.-J., 2010, "Stable Inheritance of an Acquired Behavior in *Caenorhabditis elegans*," *Current Biology* 20:R877–78, https://doi.org/10.1016/j.cub.2010.08.013.

worms went back to their natural attraction to *Pseudomonas*. Thus, it seems that the worms had learned to avoid pathogenic *Pseudomonas*, and then passed this information on to their progeny for four generations, a true transgenerational process.[55]

Despite the many excellent reasons why this shouldn't happen that August Weismann had pointed out—since the adult worms were learning to avoid *Pseudomonas* and somehow passing this information on through their germlines, it was clear that it did. So how does it work? And does this process defy the rules that Weismann set out? Rebecca Moore and Rachel Kaletsky went on to delineate much of this complicated pathway using logic and a whole lot of experiments. By training candidate mutants with *Pseudomonas* and then testing to see if the mutant worms were able to learn to avoid *Pseudomonas* or not, and if they could pass this information on to their progeny, Moore and Kaletsky systematically tested the role of genes that function in neurons, in the germline, in RNA interference, and in the COMPASS complex. Amazingly, all of these systems are required. Additionally, they discovered that a huge number of piRNAs are changed up on PA14 training, so they tested the piRNA pathway—and found that this is required, as well.[56]

Echoing the classic experiments of Avery, McCarty, and MacLeod, Kaletsky and Moore next partitioned the *Pseudomonas* bacteria into their components to identify what could be the "transforming element"—that is, what the worms were eating that made them "know" they were ingesting a pathogen—and whether this affected the information they passed on to the next generation. Our first hypothesis was that the worms sense a bacterial metabolite, but the supernatant (where a secreted metabolite should be) turned out to not have any effect. Instead, the small RNAs that *Pseudomonas* makes only under virulent conditions did the trick: squirting the small RNAs isolated from pathogenic PA14 onto nonpathogenic *E. coli* (the worm's normal, nonpathogenic lab food) induced the avoidance learning in both mothers and their progeny—and out until the fourth generation of progeny, again—without ever making the worms sick.[57] With a bit more genomics and genetics, we figured out that just a single *Pseudomonas* small RNA called "P11" (which is uncharacterized but seems to be necessary for PA14's virulence and possibly nitrogen metabolism) mimicked the entire avoidance learning and transgenerational inheritance program until F5. The last piece of the puzzle was revealed by examining the sequence of P11; it has a 17-nucleotide perfect match to the sequence of a gene called *maco-1*, which is expressed in neurons; and sure enough, *maco-1* is downregulated upon *Pseudomonas* treatment, and a mutant of *maco-1* basically mimicked treating the worms with P11.[58] So the input to the program is the ingestion of P11 small RNA by the worms, and the output is modulation of the activity of a neuron that decides whether the worms are attracted to or repelled by *Pseudomonas*.

In between those steps that take place in the gut and neurons, respectively, lies the germline—the key to this learned avoidance not breaking Weismann's rules. We think that the germline is where the small RNAs are amplified, and where there is an interaction with piRNAs and P granules, germline granules that are centers for small RNA processing. This explains at least something (but not everything) about how the message is passed on to four generations of progeny: if a small RNA is amplified in the germline, and this pool of amplified small RNA products is above a certain threshold, then the neurons get the message to downregulate *maco-1*, and we see avoidance. (We later figured out an even crazier mechanism for the transport of this information from the germline to neurons, and even between individuals, that uses a retrotransposon called Cer1 that makes virus-like capsids.)[59] Whether this mechanism of transgenerational inheritance of a learned behavior fits better with Weismann or Lamarck is debatable: although we are training adults to acquire a learned behavior, and this becomes transgenerationally inherited through the germline, it seems that the direction (gut → germline → neurons rather than neurons → germline) does not violate Weismann's germline barrier, at least not from the neuron side.

So why did I tell you about this somewhat baroque mechanism of pathogen avoidance learning in worms, and what does it have to do with (possible) inheritance of information in humans, or with aging? First, it would make sense for animals to "learn" about a pathogen that they should avoid, so passing on this signal to progeny for several generations so that they know to leave a seemingly great-smelling food source would help the animals survive in the long run, even if they have to avoid a yummy food source for a while. So to answer Craig Mello's question, remembering a pathogen is probably important enough to pass on to the next generation, even if you have to temporarily give up a little food. (In fact, recent work in pregnant mice infected with *Yersinia pseudotuberculosis* suggested that their offspring inherited IL-6-mediated immunity, an intergenerational immune response.)[60] In the same way that starvation conditions might cause an animal to inherit a "thrifty metabolism" phenotype, inheriting information about pathogens and passing on information to your progeny to help them avoid or fight off the pathogen later is probably a good idea. Additionally, the fact that *C. elegans* can interpret *Pseudomonas* small

RNA coding is a demonstration that there can be informative "trans-kingdom" signaling—that is, signaling between different types of organisms that is not just inhibition of immune systems in plants by invading bacteria or viruses. Since worms can "read" the small RNAs made by bacteria, and we have bacteria in our gut, on our skin, and other places in our bodies (see chapter 16, the Microbiome and Aging), and we have the same molecular machinery that carries out this process, it's worth asking whether human cells, like worms, can similarly read the small RNA transcriptome of bacteria to alert our immune systems. And, if so, can humans encode small RNA information in their germline that is passed on? Right now it's unknown, but I think it's worth looking into.

———

Although it's always tempting to write off observations in invertebrates, especially worms, as being unique and therefore not applicable to humans, in fact, since much of the molecular machinery involved is shared with mammals, it is possible that some aspects of TEI might happen in humans as well. This is both a scary and a useful prospect. It is very likely that many of these mechanisms are in place to "reset" each generation with a pristine germline, and therefore most of these events and much of this information will not be carried on to their progeny. But what if, as in worms, there is a threshold above which something might be carried on, a mark that doesn't get erased or a small RNA that persists? Perhaps trauma inhibits our ability to properly carry out these erasures, leading to the trickling down of learned experiences to the next generation. Identifying such a mechanism might be the key not only to how transgenerational inheritance is carried on in the worm but also to how it functions—and how it might be reset—in higher organisms, like us. From metabolism to neuropsychiatric disorders, understanding how our ancestors' experiences might influence our health is worth understanding. Wiping the hereditary slate clean, whether from the stresses of starvation or trauma, might finally be possible.

16

Gut Feelings

THE MICROBIOME AND AGING

Well, I've been listening to my gut since I was 14 years old, and frankly
speaking, I've come to the conclusion that my guts have shit for brains.

—ROB IN *HIGH FIDELITY* (NICK HORNBY)

Who's Calling the Shots, Our Microbiome, or Us?

One of the ideas that has taken the scientific world by storm in the past few
years is the concept that our "microbiome"—the bacteria that colonize our
gut, skin, and other bodily crevices—might dictate more about our health
than we had previously realized. Bacteria are probably not "the multitudes"
Walt Whitman was referring to, but it's been observed that the almost 40 tril-
lion bacterial cells contained within us outnumber our own cells by several
factors[1]—*so are we more human or bacteria?* (I'd vote human, since our bacterial residents turn over quite frequently, but maybe that's just me.) The bacteria
that occupy our intestine are of particular interest, as the relative makeup of
species of the gut microbiome changes in individuals with different diseases,
with diet and obesity, and with age. The problem with interpreting much of
this correlation-based data on microbiome differences—no matter how compelling they may seem—lies with disambiguating cause and effect. As Kurt
Vonnegut wrote, "we germ hotels cannot expect to understand absolutely
everything"*—but we can try.

The idea of using probiotics to promote good bacteria has taken hold in the
popular psyche and is almost inescapable, at least if you are ever subjected to
yogurt commercials. And the notion that our microbiome is not just a fellow
passenger but actually the *cause* of many disorders is gaining steam. While this

---

* Vonnegut, K., 1997, *Hocus Pocus* (New York: Berkley Books), 207.

may sound like the mantra of new-age health bloggers, in fact it was promoted over a century ago by the "father of aging research," the 1908 Nobel Prize–winner Elie Metchnikoff. He was widely regarded as a pioneer not just in cellular immunology for his discovery of phagocytes but also in "gerontology," a term he coined. You might remember that I mentioned Metchnikoff earlier because he also proposed one of the evolutionary theories of aging, that of programmed aging, and he was one of the first scientists to carefully study aging.

Metchnikoff published some of these ideas regarding probiotics and a healthy gut microbiome more than a century ago in his 1907 book *The Prolongation of Life: Optimistic Studies.** (What's not to love about that title?) He was convinced that what we eat and drink could affect our longevity, mostly through bacteria in the intestine. Metchnikoff suggested that there were microbes in yogurts and sour milk that could be beneficial. In fact, his hypotheses went even further, blaming changes in the gut microbiome and "intestinal putrefaction" (*blech!*) for the development of "senility." He wrote quite a bit about the digestive system and its contribution to aging, even proposing fecal transplants (yes, you read that right—and they are being clinically tested these days) to help treat disease. If you like reading about constipation, fecal matter, enemas, fecal transplants, and other scatological topics, Metchnikoff's tome is the book for you. But be warned, despite the sunny title, the book opens with tales of how various societies traditionally eliminated (his word: "*assassinated*") old people. At a time when we didn't yet know anything about so much of the biology that we now take for granted—antibiotics, DNA, replication, RNA, protein structure, cellular signaling, the list goes on and on—Metchnikoff was proposing treatments that are now being revisited. He practiced what he preached, and reportedly drank sour milk to benefit from lactic acid fermentation of *Bacillus*, which he proposed has beneficial effects.

Metchnikoff's ideas eventually lost favor with the rising popularity of "germ theory" and the development of antimicrobials, with society and medical opinion swinging in the opposite direction, toward ultra-sterile, non-biotic cures for disease. And for good reason—one of the biggest contributions to the increase in life expectancy in the twentieth century was the elimination of bacterial infections through sterile technique and the miracle of antibiotics, since those now-preventable infections used to be one of the major causes of

---

* Metchnikoff, E., 1908, *The Prolongation of Life: Optimistic Studies . . . The English Translation Edited by P. Chalmers Mitchell* (New York: G. P. Putnam's Sons).

death. But more recently, resistance to those same antibiotics has increased, owing to a perfect storm of medical overuse, widespread antibiotic treatment of livestock, and overreliance on just a few antibiotics. Using antibiotics to wipe out all one's bacteria is not really healthy, as it turns out; this became apparent when antibiotic use was linked to increased *Clostridium difficile* infections and other problems, suggesting that some bacteria might even prevent the over-proliferation of "bad" bacteria. In general, there has been a revival in interest in the possibility that our microbiomes might actually be helpful.

To be frank, the field may have overcorrected, with some people attributing *all* of our health and disease to the microbiome. For example, there was some thought that the gut microbiome might contribute to autism spectrum disorder (ASD), but the most recent data strongly suggest that the dietary preferences of ASD patients—in particular, their reduced tolerance of a varied diet—is the true cause of the observed reduced diversity of ASD patients' microbiome.[2] Similarly, the differences in the microbiome of Hutchinson-Gilford progeria patients suggested that environment had a bigger influence on the microbiome composition than did disease.[3] Thus, we need to better understand cause and effect in microbiome studies and view the contribution of the microbiome to various disorders, including aging, with appropriate skepticism. Somewhere between a cure-all/cause-all and no role whatsoever is probably the right compromise. In this chapter, we'll explore what lessons we can learn about the microbiome and aging, when the microbiome might be an aging characteristic and when it might play a causal role.

Given the possibility that our microbiome might be more important than we originally thought, it's not such a leap to ask, Does the microbiome change with age? Can the microbiome—or something it produces—influence the rate at which we age? And can we somehow use this information to rejuvenate ourselves, or at least our guts?

What Is the Microbiome?

First, before we can ask how it influences longevity, what do we know about the microbiome? Thanks to great improvements in sequencing methods, the gut microbiota of humans, mice, fish, flies, and even *C. elegans* have now been sampled and compared under different conditions. The explosion of genomic methods, from simply sequencing 16S ribosomal RNA (which can be used to catalog which bacteria are present, the "operational taxonomic units" or OTUs) to more complex and complete genomic sequencing of both host cells

Although I'm usually quite enthusiastic about drawing parallels between *C. elegans* and mammals, worms are complicated to use as a model for the microbiome. This is because bacteria are their primary food source in the wild (think bacteria on rotting fruit), rather than existing in their guts solely as commensal (coexisting, nonpathogenic) bacteria, as is typical of the microbiomes of most other aging model systems. Additionally, in the lab, *C. elegans* is typically raised on *E. coli* bacteria, an enterobacterium that is most likely completely foreign to it, rather than its natural mix of environmental bacteria (primarily *Pseudomonas* species). Nor is it surprising that there are differences in lifespan on different types of bacteria, since *E. coli* isn't the worms' natural food. Thus, the analyses of the effects of worms' exposure to bacteria are harder to interpret as straight-up microbiome experiments; instead, they are analyses of their food sources, responses to pathogens, competition, and their microbiome, so I'm hesitant to draw too many parallels between the effects of the worm and human microbiomes on longevity without the appropriate disclaimers.

However, it should be noted that: (1) bacteria do accumulate in the guts of worms with age, and those bacteria eventually kill the worms,[a] (2) worms have different lifespans on various (nonpathogenic) bacteria (such as OP50 *E. coli*, HT115 *E. coli*, and DA1877 *Comamonas*),[b] (3) some bacteria (e.g., *Bacillus subtilis*) might promote longevity,[c] (4) some bacterial metabolites, such as colanic acid, affect lifespan,[d] and (5) chemical compounds, such as metformin, can affect bacterial folate metabolism, which in turn affects worm lifespan (at least in worms with defects in their own folate metabolism)[e]—so there clearly is a role for the worm's gut microbiome in the regulation of longevity. But the *C. elegans* gut lacks intestinal stem cells, and worms don't have the canonical adaptive immune system that we think of when we hear "inflammaging." These differences may somewhat limit the parallels that can be drawn with the human microbiome's effects on longevity. The worm microbiome is fascinating for its own reasons, however, completely distinct from whether we can infer effects on human health from it. For example, the presence of one species of commensal *Pseudomonas* bacteria, *P. mendocina*, can slow the colonization and killing of worms by a pathogenic *Pseudomonas* species, *P. aeruginosa*,[f] and worms spend a lot of energy avoiding bacterial pathogens. Nevertheless, I feel more comfortable interpreting and extrapolating the effects of the microbiome on aging from vertebrate model systems, like mice and fish, to humans, rather than from worm experiments.

and bacterial genomes, has allowed better identification of the species present in all kinds of hosts under different conditions, despite the fact that many gut bacteria cannot be cultured in the lab.

Although bacteria occupy many niches within our bodies, for the most part we will focus on this collection of bacteria in our intestine, the gut microbiota. Most animals have a wide variety of bacterial species in their guts, and this microbiome composition is remodeled from birth through development to adulthood and with age. Human and mouse gut microbiota are composed primarily of Firmicutes and Bacteroidetes, while zebrafish and killifish gut microbiomes include those bacteria plus Proteobacteria and Actinobacteria. Worms are an exception, containing primarily various Proteobacteria with some Bacteroidetes, Firmicutes, and Actinobacteria; the makeup of their microbiomes correlates with growth versus arrest,[4] but then again, bacteria are their food rather than "commensal" (co-living in the gut; see box). Flies have similarities with worms in their microbiomes, although their primary food source is yeast and they do have true commensal bacteria, like mammals do. Flies also have intestinal stem cells, and therefore can be used to study and test microbiome-based mechanisms of aging, such as intestinal proliferation dysfunction and intestinal barrier failure using a blue dye to visualize the breakdown of the intestine (the ingeniously named "Smurf assay").[5] The fly gut microbiome shifts with age toward expansion of Gammaproteobacteria and a decrease in Firmicutes; this shift precedes gut barrier dysfunction and other age-related declines.[6] The gut microbiomes of honeybee workers and queens differ, so understanding differences that might contribute to the much longer lifespan of queens (as much as ten times longer!)—including whether any of these differences are due to royal jelly or to their ability to "overwinter"— could be informative.[7]

The Microbiome Can Be Beneficial

Most omnivores host a diverse set of commensal bacteria that both produce metabolites and metabolize nutrients themselves. The gut microbiome has functions that are useful to its host, such as amino acid and vitamin synthesis (e.g., vitamin B12, biotin), metabolic roles in fiber catabolism, and xenobiotic detoxification.[8] Commensals also often provide some protection from opportunistic pathogens, such as *C. difficile*. It's not surprising then that the gut microbiome has a complex relationship with the host immune system. Some animals with restricted diets rely heavily on their microbiomes for food digestion—think koalas eating eucalyptus leaves and pandas eating bamboo, because both animals require cellulose breakdown by their resident gut microbial enzymes for their survival.[9]

While host genetics might play some role in the selection of our microbiome, perhaps it's not surprising that our environments and lifestyles have a much bigger influence on what survives in our gut—only a few bacterial taxa are likely to be heritable.[10] Many health phenotypes also appear to correlate with particular sets of bacteria, but it is less clear from this work which are cause-and-effect relationships between the microbiome and health characteristics; although we'd like to think that shifting our microbiome might be beneficial, the microbiome might simply reflect the health of the individual. The microbiome itself changes in response to different conditions—diet (which itself can be affected by ethnic/social factors, environment, illness, and socioeconomic differences), levels of "microbiota-accessible carbohydrates" (fiber), disease, host genetic variations, surgical procedures, medications—even the artificial sweetener aspartame in the diet of mice can change the composition of its microbiome.[11] Whether those differences cause *functional* changes in aging or merely passively reflect changes with age is now a major question in the field. Because the composition of the microbiome changes in response to so many inputs, even if it does not directly cause aging, the microbiome's composition might at the very least be a good biomarker for health status. One can imagine that changes in the microbiome composition might ultimately have effects on its vital metabolic and immune functions. If the "unhealthy" microbiome has deleterious effects on the host, then one can imagine a vicious downward cycle of further damage and dysfunction. The good news is that these results suggest that we can change our microbiome through our environment, perhaps overcoming genetic predispositions.

Reduced Diversity and a Shift in Bacterial Populations Are Hallmarks of the Aging Microbiome

Studies of the aging microbiome have now been done in many animals, including humans, mice, marmosets, bats, and invertebrates. A reduction in microbial diversity with age is a hallmark of aging across species. For example, aged mice and aged marmosets display lower microbial diversity and host more Proteobacteria and fewer Firmicutes with age.[12] The microbiome of progeroid (rapid aging) mice and humans differs from controls and is linked to intestinal dysbiosis in mouse models and in Hutchinson-Gilford progeria patients,[13] suggesting a link between accelerated aging and the microbiota. Like other animals, humans also have increased Bacteroides and Firmicutes with age, as well as obligate and facultative anaerobes, but generally display a decrease in microbial diversity with age.[14] By contrast, an analysis of the anal microbiome in exceptionally long-lived bats (words I never expected to type) revealed almost no differences between young and old bats, perhaps correlating with their extended youthfulness and general lack of physiological changes with age.[15] Invertebrates also show microbiota differences with age and conditions: in contrast to worker bees, who accumulate Proteobacteria in their guts with age, the much longer-lived queen bee shows a depletion of Proteobacteria but increased *Lactobacillus* and *Bifidobacterium* with age, correlating with her resistance to oxidative stress, and perhaps her ingestion of royal jelly as a primary food source.[16]

But do exceptionally long-lived people have a different microbiota from the rest of us? A study of Sardinian centenarians showed that their gut microbiome was more diverse than that of other elderly people.[17] Their microbiomes were enriched for glycolysis metabolism, correlated with clinical parameters of independence, and had more bacteria associated with a youthful microbiome signature. While this study did not explain how one could obtain such a microbiome, or disentangle cause and effect, it is intriguing that there was a detectable difference in centenarians that could not be easily traced to differences in diet or region, since the comparisons were within Sardinia. Moreover, specific bacterial species might be associated with healthy aging, including *Akkermansia*, *Bifidobacterium*, and Christensenellaceae species, while *Ruminococcus bromii* increased in centenarians.[18]

A grimmer 20-month study of centenarians in three regions of Hainan Province revealed that changes in the microbiome could distinguish healthy centenarians from dying centenarians; by identifying enriched and depleted species, the group found an inflection point in microbial diversity at the seventh month before death.[19] Whether this change preceded declining health or reflected changes in the diet as a result of illness (for example, a limited diet due to loss of mobility or impaired ingestion) is not clear.

Possible Mechanisms Underlying
the Benefits of a Healthy Microbiome

The microbiome adds another wrinkle to our understanding of cellular signaling and longevity. In previous chapters, I've discussed what we know about cellular signaling pathways, such as insulin signaling, TOR, AMPK, and various pharmacological agents, such as metformin and rapamycin, and their roles in regulating aging. It's challenging to think about how the microbiome affects longevity, because many of its effects might in fact be *indirect*, adding an additional step in the pathways; if we add the microbiome into the equation, everything is a bit less straightforward. Do interventions such as dietary restriction actually regulate lifespan by affecting bacteria, which in turn act on host cells, rather than acting directly on host metabolism? Do compounds like metformin and rapamycin act on the host cells, or microbiota, or both? Do bacteria signal to the host cells to influence their metabolism and signaling pathways? How do the mechanisms that we think about as being mitochondrial in nature relate to bacteria in the gut, since mitochondria are ancient endosymbiotic relatives of bacteria? But, most importantly: *Can we change our aging rates or disease likelihood by changing our microbiome?* If so, what's the underlying mechanism, and how can we induce these changes? I'll try to summarize what we know now, but with the exception of a few key studies, much of what we know about the role of the microbiome and its influence on aging, longevity, and age-related disease is correlative at this point. Unraveling this complex mystery of cause versus effect is one of the major goals of the microbiome and aging field, and it is slowly emerging but is currently far from solved. But a few new studies suggest there is an influence of the gut microbiome on longevity and healthspan.

There are many hypotheses about how changes in the microbiome with age *might* affect host cell health and aging. The list of hypotheses include changes in host metabolism due to changes in production of short-chain fatty acids (SCFAs, which include butyrate, propionate, and acetate); interactions of bacterial metabolites with the insulin signaling pathway; lower levels of production of vitamins and amino acids; loss of DNA repair mechanisms; increased intestinal barrier dysfunction, intestinal stem cell dysfunction, and chronic inflammation; effects on the enteric nervous system; and changes in the host's mitochondrial dynamics and function. Microbial-induced inflammation is at the center of many of the aging microbiome hypotheses, as "inflammaging" is one of the major drivers and biomarkers of aging.

A few of these hypotheses, particularly the effects of various bacterial metabolites produced by the young microbiota, have been tested to see whether they can improve some aspects of aging. SCFAs such as butyrate are produced by the young microbiome and by the "beneficial" bacteria in probiotics, and have been shown to have some protective effects, lending some credence to this hypothesis. Of all the compounds tested thus far, the most consistently positive results in many studies seem to be from butyrate, so this might be our best bet not just in distilling down beneficial bacteria in the gut microbiome but also in determining which of their products might act as a therapeutic.

Butyrate is not the only compound from young bacteria that appears to be beneficial: polyamine (spermidine) supplementation improves mouse lifespan; colanic acid increases worm and fly lifespan; ellagitannins, which are metabolized into the compound urolithin A, improve mitochondrial function; and berberine appears to act through the toll-like receptor LPS-TLR4 signaling pathway to reduce insulin resistance in mice on a high-fat diet. Thus, the benefits of specific bacteria are likely to be due to the compounds that they produce.

Testing Heterochronic Microbiomes
Using Fecal Microbial Transplants

Identifying the functional components of the microbiome requires some deconvolution of cause and effect. The key here is testing whether the different conditions caused the changes in the microbiome, or if instead the microbiome simply reflects differences in the conditions; most studies do not address this question directly, but instead suggest that microbes from the young microbiome produce "good" things, while the old microbiome produces "bad" things. There is a way to test these ideas, though: put the microbiome of an animal of one condition into another animal of a different condition (one that has had its own microbiome wiped out or was raised in a germ-free environment) via fecal microbiome transplant, or FMT. This procedure has already been used under isochronic (same-aged) conditions to test the role of the microbiome in disease. For example, fecal microbial transplants from genetically diabetic mice and from genetically obese mice both recapitulated aspects of the disease (including obesity) in the previously healthy host mice, suggesting that the microbes themselves are responsible at least in part for the phenotypes.[20] Together these data suggest that multiple positive factors (various young microbiome products) and the absence of other negative factors from the old microbiome might together produce a healthy outcome for the host.

Whether changes in the microbiome are a consequence of aging or a cause of aging was largely untested until recently. (I haven't seen the term used in papers, but I'd call this heterochronic microbiome transplantation.) Obviously, testing the role of the aging microbiome in humans ranges from difficult to impossible, although fecal transplants are being tested to determine whether they can reduce the recurrence of C. *difficile* infections, which can be deadly and are rampant in hospitals. To really get at the question, we must turn to model systems once again. In mice, the question of whether the aged gut microbiome is deleterious to young animals seems to be answered: yes, it's bad. And it might even affect the brain: fecal transplants from aged donor mice messed up the ability of young mice to carry out spatial learning and memory tasks,[21] indicating negative effects on the hippocampus. Interestingly, the authors didn't find effects on locomotor activity or anxiety-like behavior, or changes in gut permeability or inflammatory cytokines, suggesting that the hippocampal effects might be temporally distinct from and precede other aging-associated changes. Gene expression analysis on the hippocampus of the mice that received the aged FMT revealed changes in neurotransmission genes and learning-related proteins, correlating with the decreased ability to learn and remember, but *how* this information is transmitted on the gut-brain axis is not immediately obvious. However, proteomics analysis showed that glucose transport proteins that function across the blood-brain barrier were downregulated, indicating one possible mechanism by which metabolism of the brain might be impaired. Microglia, the resident central nervous system immune cells, were changed by the transplant, perhaps suggesting a further point of damage.

Later, another group similarly using transplants of gut microbiota from old to young germ-free mice unexpectedly found that that instead of solely having negative effects, there was at least one positive: the mice that received old gut microbiome showed signs of increased neurogenesis at both 8 and 16 weeks after transplantation.[22] They also found an increase in intestinal and colon size, without signs of dysfunction or intestinal permeability. Gene expression analysis of recipient livers suggested an upregulation of the unfolded protein response (perhaps suggesting a hormetic effect?) and changes in hepatic metabolism, and metagenomic (bacterial) analysis showed an increase in butyrate-producing bacteria (Firmicutes, Lachnospiraceae). Increases in butyrate levels might be responsible for increased activity of brain-derived neurotrophic factor (BDNF), which is associated with neurogenesis. Oral administration of sodium butyrate increased the level of circulating FGF21, the hormone that acts via β-Klotho and appears to keep animals youthful via its activation of AMPK (see chapter 7). While confusing when put in context with the previous work, the simplest model is that butyrate might stimulate some positive effects in mice that are still young enough to receive signals and respond to them. The recipient mice in the latter study were only five to six weeks old, while in the previous study they were already three months old. While this still sounds young, if all of the results are to be believed, then something about the recipient changes during that six-week to three-month window that shifts the response from positive (increased neurogenesis via FGF21 activation) to negative effects. Perhaps the window of FGF21 activation by butyrate shifts during this time. Further analyses with additional and earlier time points for the recipients within the same study would be helpful to reconcile these observations.

The microbiome produces chemical products, and those products change with microbiome composition. Some of those small molecules may have positive effects, like stimulating neurogenesis or properly regulating the insulin pathway, while others may be harmful. In fact, some chemical compounds secreted by the human microbiome resemble clinically used drugs, and may interact with medicines, possibly explaining why shifts in the microbiome can have effects on our health.[23] FMTs might also become a useful treatment for some cases of progeria: FMT of wild-type microbiome to progeria model mice (*LmnaC609G*) increased lifespan significantly ($p = 0.0029$, 13% increase in median lifespan) and other health markers[24]—promising results for the concept of FMT as a therapeutic intervention for at least some progerias.

Can the Young Microbiome Actually Slow Aging?

After reading many papers where groups moved the old microbiome into young animals and found deleterious effects, but didn't ask the million dollar question—*Does the YOUNG microbiome make old animals live longer?*—it was a joy to read the work done by Dario Valenzano's lab (Max Planck Institute for Biology of Ageing in Cologne) using a short-lived vertebrate, the turquoise killifish. Valenzano's lab asked not just whether the makeup of the microbiome changes with age, and whether lab fish retain the diversity of wild fish, but also whether the age-specific gut microbial population can influence the lifespan of the host.[25] After wiping out the gut microbiome of middle-aged fish with antibiotics, they transferred the gut microbiome from young fish to these middle-aged fish, and found that indeed, *the killifish that received the young gut microbiome lived longer than their control counterparts*. In addition to living significantly longer, the fish that received the young gut microbiome also maintained their motor function and maintained microbial diversity into old age. This is a huge deal, since all the other work I told you about involved transplanting old gut microbiota and asking if there are any deleterious effects, and even the neurogenesis result I just described was a somewhat nonintuitive surprise. Transcriptional analysis of the killifish also gives some clues about possible mechanisms that slow decline, including possible roles for cellular proliferation, ribosomal function, hyaluronic acid metabolism, and immune function, although specific pathways were not tested to determine their influence. Nevertheless, these were the first results that really suggested that the young microbiome might not just reflect aging status of the individual, but also slow age-related decline.

As far as I know, no one has yet done the equivalent lifespan experiment treating middle-aged mice with young microbiome, but, excitingly, John Cryan's lab (University College Cork, Ireland) did treat old mice with young microbiome and tested for changes in age-associated behavioral deficits.[26] They found that FMT from young (3–4 months old) mice into aged (19–20 months old) mice appeared to reverse many of the cognitive signs of aging: although locomotor function was not changed, long-term memory (as assessed by Morris water maze) was improved in the old mice receiving young microbiome, as were deficits in environmental interaction (with novel objects) and in anxiety tests. While the microbiomes of the FMT-treated animals were shown to have shifted, and peripheral and hippocampal immunity were altered, the molecular mechanisms and cells driving these changes were not determined. So we do not yet know whether there is a specific metabolite that helps the aging mice slow some of these age-induced cognitive changes or what the signal is between the gut microbiome and brain, but the results are quite promising.

The Microbiota-Gut-Brain Axis and Neurodegenerative Diseases

While we generally think of neurodegenerative diseases as autonomous to the nervous system, interactions between the gut, microbiome, and brain are being examined as a mechanism by which neurological and neurodegenerative diseases might originate, from disorders such as anxiety and depression to aggregation-based diseases like Parkinson's and Alzheimer's disease. At the very least, there seems to be bidirectional signaling between the gut and the brain through the enteric nervous system, the vagus nerve, metabolites (gamma-aminobutyric acid), hormones, neuropeptides (e.g., serotonin), neurotransmitters, and immune signaling molecules produced by the intestine that communicate with the brain. The effect of the microbiome on the enteric nervous system is an active area of research; the vagus nerve can sense gut microbiome metabolites and communicate this information to the central nervous system, which has come to be known as the microbiota-gut-brain axis.[27] How this axis is affected by aging and concomitant changes in the gut microbiome is not yet known. More controversially, it has also been suggested that some aggregation-prone proteins that are associated with neurodegenerative diseases might originate in the gut and travel to the brain, where aggregation causes or contributes to neuronal damage and dysfunction. This theory is known as "Braak's hypothesis," and echoes Metchnikoff's ideas linking the gut microbiome with "senility."

The microbiome of Alzheimer's and Parkinson's disease patients has been found to differ from unaffected controls. A study of patients with dementia showed differences in their gut microbiome, as well, and shifts in the oral microbiome and gingivitis have also been noted in dementia patients.[28] In fact, some specific bacterial species (in the families Prevotellaceae and Ruminococcaceae) have been reported to be more abundant in people with lower AD risk genotypes (APOE3/E3 and APOE2/E3) than in higher AD risk APOE4 carriers; these bacteria are associated with the healthy production of SCFAs, thiamine, and folate. Given these roles, it's not hard to surmise that shifts in the microbiome with age and disease might be deleterious, even accelerating disease, but again, there is no cause and effect tested here, just correlations between the presence of AD risk alleles and bacterial species. However, from the skeptic's viewpoint, one can easily imagine how a neurodegenerative disease might affect diet and hygiene practices, resulting in a difference in the microbiome that is not disease-causing, but indirectly caused by disease. Can cause and effect be more carefully parsed out?

Recently, it has been hypothesized that gut dysfunction and gastrointestinal symptoms (constipation, gut leakiness) that precede frank neurodegenerative disease onset might arise from changes in the gut microbiome, and its signals and metabolites, and in turn might either cause or contribute to neurodegeneration. Dysfunction of the gut microbiome and gastrointestinal problems have been associated with Parkinson's in particular. In 2003, Braak and colleagues proposed that the aggregation-prone protein α-synuclein, which is associated with (but not the only cause of) Parkinson's disease, originates in the gut and travels via the vagus nerve to the brain[29]—but this hypothesis is of course controversial. In 2016, Sarkis Mazmanian's group (Caltech) tested this two-hit theory in a few ways.[30] First, wiping out the microbiome in α-synuclein-overexpressing mice had a positive effect in delaying the onset of motor function defects and α-synuclein pathology, suggesting that the microbiome itself at least contributes to Parkinson's phenotypes of the mice. They also found that microglia are affected by the gut microbiome in these animals, and that treatment with SCFAs (acetate, propionate, and butyrate) activated microglia in affected brain regions; they proposed that the SCFAs produced by bacteria cause inflammation, microglia activation, and α-synuclein aggregation. But the most interesting test they performed involved human patient samples and mice. They transplanted fecal samples from six unmedicated Parkinson's patients and controls into transgenic mice that overexpress α-synuclein, and found that the Parkinson's sample treatment *increased motor dysfunction* of the mice. Although this was a limited study, the data connecting Parkinson's patients' microbiomes with exacerbated α-synuclein-mediated motor deficits is compelling, and perhaps the most direct test of the Braak model to date. Of course, this contrasts with the autism story I shared above, which suggested that the patients' preferences due to ASD caused changes in their gut microbiome, not the other way around.[31] It seems that the microbiome can reflect disease states (e.g., ASD, Hutchinson-Gilford progeria syndrome) and may also influence and promote disease states (Parkinson's)—and tests like fecal transplant experiments are critical for this disambiguation.

What Can We Do to Maintain or Acquire a Healthy Microbiome?

No one is suggesting that we do fecal transplants from young to old people to slow their aging or improve their cognitive function (yet!). But perhaps we could change our diets a bit. Can we eat our way to a healthier microbiome, or try to mimic the effects of a youthful microbiome? Specifically, it would be great if we could increase our gut's levels of beneficial SCFA/butyrate-producing bacteria, like Lachnospiraceae, Clostridiaceae, Bifidobacteriaceae, and Bacteroidaceae, and decrease levels of inflammaging-promoting and frailty-associated bacteria (Erysipelotrichaceae, *Actinomyces*) that produce harmful compounds, like p-cresol, secondary bile acids, and trimethylamine. One approach might be to ingest not just probiotics but also "prebiotics"—various oligosaccharides and resistant starches that are associated with increased levels of "good" bacteria (e.g., *Bifidobacterium* and *Lactobacillus*), which in turn produce SCFAs (butyrate, propionate) that are associated with youthfulness and health.[32] An example of a dietary treatment to improve the microbiome was demonstrated in a recent study of the effect of kefir in the diet on an autism model. Despite the lack of causation by the microbiome in ASD, there was a very modest ($p = 0.051$) amelioration of some mouse ASD model repetitive behavioral defects by kefir[33]—so although the gut microbiome doesn't cause ASD, it might not hurt to follow Metchnikoff's advice and drink a little fermented milk.

Another approach might be to adopt the Mediterranean diet, which places an emphasis on plant-based foods and olive oil, and is one of the factors in the longevity of people living in Blue Zones.[34] This diet has been associated with many positive health outcomes (longer lifespans and reduced cognitive decline and frailty), and at least part of this effect could arise from the types of gut bacteria that thrive on this diet.[35] To test this concept, a group of European researchers embarked on a randomized, multicenter, one-year study across five European countries, the NU-AGE study.[36] By profiling the gut microbiota of more than 600 elderly subjects before and after the subjects followed the Mediterranean diet for 12 months, they observed that the microbiome was indeed remodeled to contain more of the bacterial species (OTUs) that are associated with health, SCFA production, and anti-inflammatory properties (e.g., *Faecalibacterium prausnitzii, Roseburia, Eubacterium*), and fewer of those that are associated with aging, atherosclerosis, diabetes, colorectal cancer, and frailty (e.g., *Ruminococcus, Collinsella, Coprococcus*, etc.). Therefore, the adoption of the Mediterranean diet might have beneficial effects that are direct—that is, acting on our own cells—and others that are indirect, acting through the microbiota that the diet promotes in the gut.

A more extreme approach to remodel the microbiome might be to use a fasting-mimicking diet (FMD). Valter Longo's group (USC) treated a mouse model of inflammatory bowel disease (IBD) with FMD, and found that this diet reduced intestinal inflammation, increased intestinal size via increased stem cell activity, and promoted the growth of beneficial gut microbiota, while ameliorating the IBD disease phenotypes.[37] Some of these phenotypes were due to intestinal inflammation, which was reduced by the FMD. The beneficial bacteria Lactobacillaceae and Bifidobacteriaceae also made a comeback, showing that the FMD remodeled the microbiome. In fact, fecal transplant from FMD-treated mice to IBD model mice also led to improvements in the latter, further supporting the model that at least some of the effects of the FMD were due to the remodeled microbiome. While this study was primarily done in mice, Longo's group also looked at human subjects after FMD cycles and found some suggestions of positive effects as well, but a clinical trial is necessary to determine whether FMD could also affect IBD in patients.

It seems clear that the gut microbiome can be remodeled to promote health, whether from fecal microbiome transfer, an extreme dietary approach like FMD, or a gentler lifestyle and dietary change, such as the Mediterranean diet. Future studies will determine whether specific components, like butyrate, can mimic these diets. As for our attempts to link the microbiome to beneficial longevity outcomes, Metchnikoff would approve, I think—a round of sour milk on the house!

17

Long Life in a Pill?

THE FUTURE OF LONGEVITY: FROM BENCH TO BIOTECH

People assume aging is immutable and that it is a fool's errand to look for drugs that slow the aging process—but they are wrong. . . . People would like to be able to swallow an anti-aging pill that would help them stay healthy and live 10 to 20 years longer.

—DR. RICHARD MILLER (UNIVERSITY OF MICHIGAN),
INTERVIEWED IN *MICHIGAN TODAY*

INVARIABLY, after I finish giving a seminar—no matter what topic I just spoke about, even worm retroviruses and transgenerational inheritance of immune memory—someone will ask me, "So, . . . what can I do to live longer?" If you've been paying attention, by now you could give them a pretty good answer: eat less, exercise regularly, and sleep—and be lucky enough to have the "right" longevity gene alleles and to have access to good health care. Due to inequities in our society, being white and rich greatly improves your chances of living long; being female, childless, and single will add a few years as well.

But you already know that's not the answer they want to hear, especially the eating less and exercising more part. What they *really* want to know is *What drugs can I take to live longer?* A few years ago I might have chuckled at the naïvety of this question, but now it's not so crazy to think that we will be able to take some sort of medicine to extend our healthy lifespans in the foreseeable future. The possibility of increased longevity is in the news all the time these

> I was once invited, along with a few other speakers, to give a talk to a company interested in investing in new longevity biotech. When the speaker before me—one of the well-known and oft-quoted "immortalists" I'll mention below—was asked "What can I do to live longer?" his answer was "Give me money. Give me money and I'll make you live longer." No details, no elaboration. I was flabbergasted at the hubris. I'm sure they did give him money, though. Whether they will live longer, who knows?

days, as discoveries in the field of aging are reported and new "anti-aging" companies are formed almost weekly—mostly based on discoveries first made in academic labs, the kind of work you have been reading about here. The hope is that the many findings made in labs over the past few decades will be translated into useful drugs and treatments. The recent explosion of longevity-related biotech companies—and there are a *lot*!—suggests that there is not only excitement about that possibility but also the financial will—at least among those who fund these companies—to move forward.[1]

This is a great time to make the leap from academic to clinical. The ability to extend lifespan is an ancient desire, but only now do we have a mountain of data from foundational research done in worms, flies, mice, and other model systems, together with cell culture experiments and human genetics data that may allow us to translate these findings into real treatments for people—and now there's plenty of Silicon Valley money to invest in companies that will turn that information into candidate therapies. We really do understand a *lot* more about aging and longevity mechanisms now than we did when worms with doubled lifespans were being first discovered, before the molecular identity of these genes was even known. What's astounding to me is the number of ways, genetically speaking, to extend lifespan and improve healthspan, even though scientists are often found arguing about them as if they are mutually exclusive. Often these mechanisms do boil down to a set of the basics—ways to "turn on" the effects of caloric restriction or stress responses, for example, or to help cells regenerate. I've written about these various mechanisms throughout this book, so you probably already have a good sense of where we can go from here. This is a great time to learn more about the regulation of aging, so that you can make decisions about which approaches seem most promising, and which might be sketchy, or at least over-hyped. In this chapter I'll review the current state of possible therapeutic approaches that have emerged as a result of the aging research we've discussed throughout the book.

There is an excruciating semantic debate in the field about the use of "anti-aging" that I find pointless to air here. Most readers will recognize that when I mention drugs to improve healthspan, it qualifies broadly as an "anti-aging" treatment. However, living to a thousand years is simply silly to discuss. I have to assume that those who bring up these outlandish claims are just trying to be provocative, not serious about solving real problems.

Categories of Candidate Longevity Therapy

If we look back through the previous chapters, we can see that candidate aging treatments fall into a few broad categories.[2] For many years, treatments for specific downstream changes in aging were the main focus, such as reducing protein aggregation and oxidative damage, or more recently increasing autophagy to get rid of damaged cellular components. While those approaches are valuable, systemic treatments that target caloric restriction pathways or insulin signaling might be more effective at slowing aging, because they are most likely to engage multiple, parallel longevity-promoting mechanisms. The appeal of interventions at this level is that they could have broad effects on the whole body, regardless of whether a cell is renewable or post-mitotic—and it's worth noting that the downstream targets of the insulin and FOXO signaling pathway have effects on most of the specific target classes that we have already discussed.

Cells and organs whose functions decline with age might benefit from therapies that enhance stem cell activity and regeneration, and personalized organ growth and replacement. Reducing the runaway inflammation that increases with age could have broad systemic effects, as well. The damaging effects of secretions from dying cells might be mitigated by drugs that seek out and kill these senescent cells. For each of these categories, there are several biotech companies developing or testing drugs to see if they are the key to a longer and better lifespan.*

Old- and New-School Damage Prevention and Cleanup: Antioxidants, Proteostasis, Autophagy, and Mitochondria

Probably the first thing you ever heard about aging research involved antioxidants, which for years were touted as a fix-all. Lots of different molecules, including aspirin, spermidine, vitamin E, quercetin, and the famous red wine

---

* Because of the rapidly changing landscape of longevity biotech, I won't name many specific companies here, but the information I share here is all publicly available.

compound resveratrol, as well as natural compounds, such as those found in berries and in green tea, have been described as antioxidants.[3] Many products in the health-food aisle claim to have antioxidant activity, so the development of new antioxidants isn't a hot biotech area (as far as I know), although there are always new compounds being found that have these properties. The fact that vitamins C and E were reported to block the beneficial effects of exercise is a sobering reminder that we can't always assume that antioxidants are beneficial, though.[4] Other drugs block runaway inflammation that happens with age—like aspirin, or nonsteroidal anti-inflammatory drugs (NSAIDs) like ibuprofen, or the NG52 kinase inhibitor that blocks the COX-2 cascade—although whether these are effective drugs to treat age-related neurodegeneration is less clear.

More recent research in genetics and cell biology has uncovered the components that are critical for proteostasis and autophagy, two mechanisms that clean up damage in the cell. Lysosomes and mitochondria are critical for these processes. Companies that aim to improve proteostasis have focused largely on cardiovascular and neurodegenerative diseases, since it is widely acknowledged that many are diseases of protein aggregation. Autophagy enhancers, such as retinoic acid receptor inhibitors and TFEB activators, are being explored for treating cancers, Alzheimer's, eye diseases, and other age-related indications.[5] Mitochondrial diseases, such as Pearson's syndrome, are the target of companies who aim to transplant healthy mitochondria into patients. Parkinson's disease is an obvious indication to test for mitophagy boosters, since the mitophagy protein genes parkin and PINK1 are predictors for PD.

The trick here is finding a drug that selectively targets and fixes the right pathway without damaging other things at the same time. Right now, these drugs are all in preclinical stages; they are being developed by biotech start-ups that focus specifically on small molecule therapeutics that could tweak these "cleanup" pathways to slow aging. Perhaps the best idea, though, is to work upstream of these specific pathways to coordinate all of these functions via systemic regulators.

Resetting Protein Production

Although we usually think of a stress response as being a good thing, sometimes it can be detrimental. The integrated stress response, or ISR, shuts down protein translation in a cell when it senses danger, such as from a viral infection. But aberrant ISR activity can be deleterious for the brain. Memory defects that are caused by traumatic brain injury seem to be a reversible blockage. Peter Walter's group (UCSF) found that an inhibitor of the ISR, ISRIB, seems to improve brain function by blocking the ISR, so that it can no longer block protein synthesis aberrantly.[6] Excitingly, ISRIB treatment appears to reverse memory defects in animals with TBI, and has potential for the treatment of age-related cognitive decline.[7] ISRIB is now licensed for preclinical drug development.[8]

Targeting the Whole System: Dietary Restriction Mimetics

Only a tiny number of people are like the Biosphere 2 scientist Roy Walford and so interested in delaying aging that they will actually adopt a regimen that is strict enough to elicit the beneficial effects of DR and stick to it long enough to get some results. Let's face it: fasting is a tough task, since we are evolutionarily programmed to survive by craving sugar and fats, and our signaling systems haven't caught up with the food excess that has existed only in the modern era. Diets that are less extreme than DR, like intermittent fasting (IF) and time-restricted feeding are more tolerable, and commercial products (mostly shakes and prepackaged foods) that mimic fasting are available to make this a bit easier. (And if you are too delirious from hunger to tell time, now you can also download an iPhone app that will keep track of your fasting regimen for you.) While perhaps not a long-term solution, fasting-mimicking diet products might help with preparation and recovery from surgery or in tandem with medical treatments. Other products provide specific components (e.g., a caproic acid ketone diester) to mimic ketosis, the metabolic state that the "keto" diet induces, without having to follow a strict diet.

Many of the best-known candidate "anti-aging" drugs fall into the category of CR or DR activators or mimetics because they affect the activity of the key biochemical regulators of DR, the signaling molecules that turn lack of food into cellular changes that are beneficial. Some of these drugs affect the activity of AMP kinase (AMPK), which you'll a remember is the cell's sensor of energy status, so is one of the most important downstream components of DR-mediated lifespan. AMPK activators, such as metformin, 5-aminoimidazole-4-carboxamide ribonucleoside (AICAR), and the mTOR inhibitor rapamycin have shown promising effects on lifespan and healthspan outcomes. Even aspirin has been described as an inhibitor of mTOR and an activator of AMPK. mTOR inhibitors are also used for immunosuppression and cancer and might be useful to test in aging assays. Metformin is the subject of the TAME clinical trial, which I describe below.[9]

Another approach is to find mimetics, or "biosimilars," of rapamycin or metformin, or derivatives of natural compounds, such as epigallocatechin gallate and isoliquiritigenin, that can carry out similarly beneficial functions with fewer adverse effects.[10] In silico approaches are being used to identify small molecules for similar structures and to screen compound libraries for small molecules (or combinations of them) that have resulting gene expression profiles like rapamycin's but without the same predicted adverse effects. And of course, as I mentioned in chapter 7, Hugo Aguilaniu's DR gene candidate screen revealed that reduction of a sphingosine kinase allowed the effects of DR to continue even once *C. elegans* had resumed feeding, suggesting that an inhibitor of this kinase might make a great DR mimetic. While I am not aware of its development yet as a drug, it seems that such an inhibitor would be the perfect pharmaceutical from a financial perspective: while the drug might trick your body into thinking that you are still doing DR, stopping taking it would mean that the effects might stop as well (kind of like the enhanced cognition drug that Bradley Cooper's character was addicted to in *Limitless*).*

Although it seems like we have some good candidate drugs already, it is worth searching for effective drugs that are better tolerated. For example, rapamycin (aka Sirolimus and other brand names) was first used as an immunosuppressant in organ transplants and has been used in treatment of some cancers. Rapamycin has positive effects on lifespan in many organisms but can have unpleasant and sometime dangerous side effects in some people; it might be helpful to know whether lower dosages or shorter treatment courses than those necessary for immunosuppression would still be beneficial while reducing side effects. In fact, the PEARL (Participatory Evaluation of Aging with Rapamycin for Longevity) study trial aims to ask some of these questions by examining the safety and outcomes of different treatment regimens that vary the dosage and frequency of rapamycin treatments.[11] Volunteers aged 50–85 will take rapamycin (or a placebo, as it is a double-blind study) and then undergo tests to measure visceral fat, lean body mass, lipids, liver function, and other important biometrics, as well as adverse events. This trial was first posted in mid-2020 and is expected to be wrapped up at the end of 2023, so we may know relatively soon whether there is an optimal dosage for rapamycin in treatment of age-related metabolic changes.

Although DR mechanisms evolved to slow an individual animal's aging rate for the goal of maintaining reproduction, many of the efforts in this area

---

* *Limitless*, 2011, directed by Neil Burger (Relativity Media).

gamble on the idea that the genetic pathway will work the same way in old, post-reproductive animals as it does in young, reproductive animals. Historically, most genetic-pathway and drug-treatment studies have been done on animals for their whole life, even though post-reproductive lifespan would better model what is likely to happen in humans. (And yeast studies conflate reproduction with lifespan, so the question can't be easily answered in this model.) Of course, this approach is understandable from an experimental design perspective; either the gene in question is being studied in a mutant, or the researchers think that the maximum effect for their drug will be achieved through whole-life treatment. Those are both reasonable rationales for initial studies of a gene or drug. However, whole-life testing might be the wrong approach for two reasons. One, the drug or gene might be completely ineffective post-reproductively, and we need to know that before plowing ahead with drug trials. But, more optimistically, it's also possible that the outcome will be *better* if examined only post-reproductively. If you think back to our discussions of antagonistic pleiotropy, you'll remember that many genes that we think of as "pro-aging" are in fact genes that are necessary early in life but may be deleterious late in life. If you knock down such a gene early in life, you might really mess up your animal's development—but if you wait and knock it down only late in life, you could actually extend lifespan, as shown in the work by Sean Curran and colleagues (USC).[12] And what a gene like that would do mid-reproduction is not clear. Of course, we hope that these pathways are still functional post-reproductively. Because there are already data that mid-life treatment of mice with rapamycin can extend lifespan, the age range being tested for rapamycin in the PEARL trial in humans (50–85 years old) is realistic and has a chance of being effective.[13]

If we are unlucky, however, those pathways will not work once reproduction is over, or will work in only one of the sexes. Starting to take a drug to mimic DR or decreased insulin signaling when you are old enough to finally get interested in prolonging your lifespan—which might very well be after you have stopped being reproductive—might not be effective. I think the possibility that these pathways will stop working in older animals is unlikely, but I would like to see more studies in model organisms that directly ask this question. For example, recently a degradation approach (using a "degron" to target the tagged protein for destruction) was used to test how late the insulin/IGF-1 receptor DAF-2 could be reduced and still extend *C. elegans* lifespan (similar to the Dillin and Kenyon RNAi approach,[14] but without the possibility that the RNAi machinery stops working in mid-life); remarkably, induction of degradation of DAF-2 as late as day 25—when most of the animals have died—can still double the lifespan of the remaining animals.[15] This is great news for all of us who hope to manipulate the insulin pathway late in life! In fact, a 2023 study of 12-month-old mice chronically treated with a cancer drug, an FDA-approved phosphoinositide 3-kinase p110α inhibitor (PI3Ki), Alpelisib (BYL719/Piqray), suggested that pharmacologically blocking the insulin/IGF-1 receptor signaling pathway could extend lifespan (10%) and improve some aspects of healthspan, as we would have predicted from the *C. elegans* work.[16]

I would like to see more mammalian studies carried out in much older animals so that we have a better idea of what might happen when we give a 70+-year-old a drug to slow down age-related declines. I'd also like to have data for both men and women, and know what the clinical study drop-out rate for dietary interventions is for men versus women. My anecdotal evidence suggests that women are less likely to carry through on severe DR diets, which makes sense from a hormonal standpoint—but drugs that mimic DR could be really helpful for both sexes, or even more beneficial for women, even though there are fewer data on them owing to these clinical trial problems. In any case, drugs that target the insulin and DR pathways would be good candidates to affect aging systemically, and thus are likely to have the broadest effects on age-related declines.

Exercise Mimetics

I personally like running, swimming, cycling, and generally doing sports, but even I wouldn't mind being able to skip a few workouts and still stay in shape (or even lose a few pounds and run faster and longer). But if you just hate exercise, an exercise mimetic—a pill to take, instead of sweating on a treadmill[17]—can sound pretty attractive. And for people who are diabetic, bedridden, suffering from degenerative muscle disorders, or frail, an exercise mimetic drug might be a lifesaver, and might also help maintain cognitive function. There is a lot of overlap between the DR and exercise pathways, particularly endurance exercise, since AMPK is a major target of both (as is SIRT1). Endurance exercise activates PCG1α (peroxisome proliferator-activated receptor [PPAR] coactivator 1α), which in turn induces PPARγ, a transcription factor that regulates the remodeling of adipose and skeletal tissues in response to energetic needs, mostly through changes in transcription and mitochondrial biogenesis.[18] These changes increase fatty acid oxidation and improve energy expenditure. Drugs that directly or indirectly activate AMPK by increasing levels of the ribonucleotide ZMP, such as AICAR and compound 14, can have positive effects on blood glucose and body mass in mouse models of obesity.[19] These drugs, as well as compounds like GW1516 (aka GW501516), a synthetic ligand of PPARβ/δ, or activators of REV-ERBα/β or ERRα/γ (other components of the muscle remodeling pathway) can also model the effects of endurance exercise.

Of course, the idea of "exercise in a pill" has not escaped the attention of a group most tuned in to drugs and sports—professional cheaters. As skeptical as one might be of the efficacy of these drugs, athletes—at least the ones willing to cheat—seem to be true believers. These new modern drugs are especially attractive for sports doping because unlike other compounds like synthetic steroids, compounds like AICAR are present naturally in the body, so it's harder to test for them. The World Anti-Doping Agency added AICAR and GW1516 to its lists of banned substances soon after their appearance. In 2012, a doping ring linked to a Spanish cycling team and its doctor, Alberto Beltrán Niño, were arrested for supplying AICAR and other banned substances to athletes.[20] Not surprisingly, the same country that built a whole facility to allow doping while it hosted the Winter Olympics, Russia, is also the home of several athletes who have tested positive for GW1516 (also known as Endurobol), including the 2012 20K racewalking gold medalist Elena Lashmanova and several cyclists.[21] Of course, the downside of exercise mimetics—in addition to the moral problem of cheating—is that ingesting drugs that haven't been tested for safety is dangerous. In fact, several of these performance-enhancing drugs have been linked to increased rapid onset of cancer, so the long-term effects of such drugs need to be fully tested in clinical trials.

Other pathways affect mitochondria in aging cells more directly by promoting the elimination of damaged mitochondria. For example, urolithin A is a microbiome metabolite of ellagitannins (which are naturally present in berries) that induces mitophagy in *C. elegans* and in mammalian cells, improving muscle function in aged rats.[22] Not surprisingly, urolithin A is now being produced and sold as a nutritional supplement. So far, it seems a lot safer than another mitochondrial targeted drug, DNP (2,4-dinitrophenol), which was first used to make bombs in World War I. DNP uncouples oxidative phosphorylation from energy expenditure, resulting in an increased metabolic rate and rapid, uncontrolled weight loss. Munitions workers exposed to DNP had elevated temperatures and lost weight, and many died. This weight loss was noticed (I guess they ignored all the deaths?), and DNP began to be sold as a diet pill. It is extremely toxic, and was eventually banned in the United States, but is still abused by dieters and particularly by body builders who want to eliminate body fat. Its availability online has increased, and the resulting deaths are also rising.[23]

Despite concerns about doping and long-term health effects of some of these drugs, the safe treatment of patients through exercise mimetics seems close to a reality, and while the immediate focus will be on those who need such treatments for genetic disorders, at least some of the exercise mimetics are likely to become effective anti-aging drugs in the not-so-distant future.

"Shiny Objects": The Debated Roles of Resveratrol and Sirtuins in Lifespan Regulation

The red wine polyphenol resveratrol was the darling of the longevity field—and perhaps even more so the popular press—during the early 2000s. Resveratrol is a member of the class of small polyphenol molecules called stilbenes, which are primarily found in the skin of red grapes and other dark berries. The molecule was hyped not only because it seemed like the long-sought fountain of youth had been found but also because it was thought to be the answer to the "French paradox"—the fact that the French eat fatty foods but aren't obese—so red wines got the credit here. (Article writers conveniently forgot about other differences, like smaller portions, more walking, and, most importantly, socialized medicine and a more robust health-care system.) David Sinclair's group first showed that the replicative lifespan of yeast was extended by resveratrol, and later work in worms (see box), flies, and mice showed similar effects. Like the compounds mentioned above, resveratrol was thought to function through the DR signaling pathway. While it is debated,[24] it was most commonly thought that resveratrol had its positive effects through its activation of a protein called Sir2 in yeast, and through sirtuins in other animals. As discussed in chapter 7, sirtuins are a large family of proteins with protein deacetylase or mono-ADP-ribosyltransferase activity, and some are histone deacetylases. Different SIRTs are found in different sites in the cell (nucleus, cytoplasm, or mitochondria) and have different substrate targets, molecular functions, and cellular roles (e.g., metabolism). Leonard (Lenny) Guarente's group (MIT) originally showed that activation of Sir2 increases replicative lifespan of yeast, so the connection with resveratrol was natural. In 2007, Sinclair and Christoph Westphal's resveratrol derivatives company Sirtris Pharmaceuticals went public, and in 2008 it was bought by GlaxoSmithKline for a whopping $720 million.

Resveratrol and sirtuin-activating compounds (STACs) were the darlings of the longevity world for several years but fell out of favor for a few reasons. Although many different groups found that resveratrol treatment of a range of organisms had lifespan and metabolic effects, whether it did so by activating sirtuins specifically, or through some non-sirtuin-dependent mechanism, was not known.[25] Sinclair's group had used an in vitro fluorescent readout of histone deacetylase activity (the cleverly dubbed *Fluor-de-Lys* assay) to show that resveratrol and its derivatives (STACs) activated SIRT1. Resveratrol's efficacy in specifically activating sirtuins was first debated in yeast, as in vitro studies suggested that the reported activation of sirtuins had actually been mediated by the fluorescent moiety used in the *Fluor-de-Lys* in vitro assays, rather than by resveratrol.[26] Moreover, some groups could not replicate the beneficial effects of resveratrol or sirtuin overexpression.[27] There was also evidence that CR-induced longevity did not require Sir2 (suggesting that there are multiple downstream pathways for CR), and a yeast strain with no Sir2 could extend replicative lifespan. The conclusions about resveratrol became quite muddied. Now it's thought that only mice fed high-fat diets benefit from resveratrol, which might sound discouraging if you're a skinny person who wants to live long, but it might be good news for those with pre-diabetes or metabolic dysfunction. Although activation of most of the various sirtuins in the cell seems beneficial, despite their varied sites of action and functions, how one could get specific activation of only the beneficial sirtuin by resveratrol was not addressed in the early work, and the "SIRT1 cult" (according to Charles Brenner [City of Hope National Medical Center], whose lab first found that nicotinamide riboside [NR] is an "unanticipated" NAD$^+$ precursor and milk nutrient)[28] rolled on. But GSK stopped developing Sirtris's compound SRT501, and later shut down Sirtris altogether.[29]

In 2013, Sinclair's group published a study in *Science* in which they used a point mutant of SIRT1 to show that resveratrol's effects in vivo do, in fact, require SIRT1, and are possibly mediated by its substrates, a mechanism they called "assisted allosteric activation." That is, their interpretation of the data is that resveratrol does bind to SIRT1, but does so with the help of the protein that it will actually deacetylate, like FOXO. Whether resveratrol's reputation has been fully rehabilitated yet is not clear, as the Sirtris/GSK fiasco seems to be much better known than this follow-up paper, and the story continues to serve either as a cautionary tale (or inspiration, depending on your perspective) in the "anti-aging" biotech field.

I got caught in the crossfire here. I was a postdoc in Cynthia Kenyon's lab when I read Sinclair's first paper on the effects of resveratrol on yeast lifespan, so I decided to ask whether it did the same in worms. I ordered some resveratrol from Sigma, guessed at a concentration that might get past the worm's thick cuticle (10 times the amount used in yeast sounded about right), found some spare (slightly old) bacterial feeding plates to squirt the resveratrol dissolved in ethanol or ethanol control onto, and set up a bunch of worms to test for lifespan. A few weeks into this experiment I had to go away for a few days for a job interview, so I asked labmates to count the dying worms. When I came back and totaled the results, I was amazed—resveratrol-treated worms showed a 40% lifespan extension! I was able to replicate these results, although the effects were slightly smaller—I had made fresh bacterial plates, so maybe the bacteria were metabolizing the resveratrol. Next I put resveratrol on some mutants so that I could test the genetic pathways. It seemed that worm *sir-2* was required for the effect, largely confirming Sinclair's findings. I think my postdoc advisor, Cynthia Kenyon, was not excited about getting caught up in this mess (and she also had a company with Leonard Guarente, Elixir, one of the first companies focused on slowing aging), so I stopped working on it. Later at a Cold Spring Harbor meeting, I was trying to get down to the restrooms at a break just after Sinclair and Kaeberlein had had a heated exchange about whether resveratrol worked or not—and I got trapped between them at the top of the stairs, explaining what I had found. I never published the results on resveratrol. Later, in my own lab, we did find that a low-copy SIR-2 transgenic strain did live longer,[a] despite claims to the contrary,[b] adding fuel to another debate. It was only a 15% increase in lifespan, but was consistent and significant—but paltry compared to *daf-2*'s doubling of lifespan, which is why we could never get too excited about *sir-2*.

The Importance of NAD$^+$

Regardless of the effects of resveratrol or sirtuins on longevity, it is generally agreed that the biochemical basis of many pro-longevity factors in the cell is the coenzyme nicotinamide adenine dinucleotide, or NAD$^+$. This cofactor and its precursors, such as nicotinamide (vitamin B3), are extremely important for human health in general. For example, in the United States through the early 1900s, pellagra outbreaks were common in places where corn was ingested without adopting maize nixtamalization (prepping corn with alkali) practices previously perfected by Native Americans.* Without nixtamalization, no niacin is made available from the processed corn. Later it was discovered that pellagra is the result of a niacin deficiency caused by stripping off the niacin-containing germ hulls, which made shipping out of the Midwest cheaper.[30] This Jim Crow-era cheap corn-for-cotton trade resulted in a 40-year-long pellagra outbreak in the South that ended only when bulked-up soldiers were needed in the US Army during World War II, and vitamin supplements, including niacin, were added to foods.[31]

NAD^+ is has become the focus of the aging field because it is required for so many of the important functions of the cell: mitochondrial enzymes need NAD^+ for its oxidizing potential (it is converted to its reduced form, NADH, as it oxidizes) to generate ATP in the electron transport chain, and various proteins and sirtuins (NAD^+-dependent protein deacetylases) require NAD^+ for their function.[32] NR is a precursor to NAD^+, and is used by the cell to generate NAD^+; this is why you'll see it sold as a supplement (see Nutraceuticals and Cosmeticeuticals section below) to boost cellular levels of NAD^+, instead of NAD^+ alone, which may not be as easily absorbed and used by the body. NMN (nicotinamide mononucleotide) is also being tested as a treatment for aging and age-related female infertility, with the idea that it might reduce oocyte dysfunction.[33]

The popular press's unbridled enthusiasm for resveratrol and sirtuins has been a mixed bag for the longevity field, overall. These stories dominated the press's reporting on aging for years—in fact, at points it seemed hard to escape stories about resveratrol, Sir2, specific yeast researchers, and, bizarrely, whatever these researchers personally ingested (as if that were a scientifically sound approach at all). On the one hand, these news stories introduced a large lay population to the wonders of aging research, which is great. Several longevity

---

* Sarah Kendall Taber describes the fascinating history of the disastrous health consequences of the corn-for-cotton trade combined with ignorance of Native American practices. The work is as yet unpublished, but she does tweet about it (see Dr Sarah Taber [@SarahTaber_bww], 2021, "Southern US footways include nixtamalization!," Twitter, July 10, 2021, 2:13 a.m., https://twitter.com/SarahTaber_bww/status/1413667784052158466). See also Extra Credits, 2018, "Pellagra—a Medical Mystery—Extra History," YouTube, July 26, 2018, https://www.youtube.com/watch?v=reYKBgdrZsM.

companies were launched from this work, which is at least nice for those founders and investors, and if they are successful, that could help slow aging and have a huge impact on human health. On the other hand, resveratrol and sirtuins being sold as the "fountain of youth" makes one's skeptic spidey-sense tingle, especially since it was well known that other pathways, particularly the insulin pathway, had greater effects on longevity in most animals but were not acknowledged in these reports, and resveratrol and sirtuins were being touted as the sole mechanism to extend lifespan. The subsequent failures to replicate resveratrol's effects, or to get much out of the sirtuin work, likely made the whole field seem less trustworthy. Maybe this was a natural reaction to the overconfidence of the sirtuin and resveratrol crowd, but they might be vindicated in the end—we will have to see how this plays out in the long run.

Blood Factors and Other Systemic Regulators of Longevity

Like unhappy families, aging cells each fail in their own way. Cellular processes need to be coordinated for an animal to live longer so it can successfully reproduce; these pathways are linked systemically, upstream of both proliferating and non-proliferating cells. Caloric restriction isn't the only way to live longer, as I've covered extensively in this book. We can think of aging therapies the same way: unless the therapies combat both proliferating and non-proliferating cell aging, ultimately they will fail to extend lifespan, even if they successfully treat one kind of cellular aging, because the other kind will eventually lead to organ failure.

Systemic factors include of course the insulin pathway, the growth factor hormone FGF21, and factors that are found in blood, such as Klotho and GDF11. Several biotech ventures, such as Elevian, Rejuvenate Bio, and Alkahest, are aimed at developing these factors as aging therapeutics, either through single-factor treatment, or via pooled plasma molecules.* Because there is some

---

* Don't confuse these sound experimental approaches that aim to understand specific blood factors and downstream mechanisms with frankly suspect companies like Ambrosia that market $8,000 "young plasma" transfusions; after an FDA warning, Ambrosia shut down (Brodwin, E., 2019, "The Founder of a Startup that Charged $8,000 to Fill Your Veins with Young Blood Says He's Shuttered the Company and Started a New One," *Insider*, August 14, 2019, https://www.businessinsider.com/young-blood-transfusions-ambrosia-shut-down-2019-6), but its founder, Jesse Karmazin, went on to found yet another plasma start-up, Ivy Plasma.

overlap with those being developed as therapeutics for diabetes, metabolic disorders, and neurodegenerative diseases (particularly Alzheimer's disease) it is possible that the new systemic therapies will also benefit those experiencing normal age-related decline. (Another idea is to remove "old" blood factors, but that is considerably more complicated.) My favorite blood plasma story is from Saul Villeda's lab (UCSF), where they found that factors from exercising mice can improve the cognitive function of sedentary mice. The effect was somewhat indirect, since the plasma factor is an enzyme from liver (Gpld1) that alters signaling downstream of GPI-anchored substrate cleavage.[34] While this might be complicated to administer, it is very likely that other exercise-induced blood factors would make great drugs eventually.

Senolytherapeutics: Targeting Senescent Cells for Destruction

Senescent cells are dangerous not only because they cause loss of that cell's function but also because of toxic secretions from the dying cell that injure neighboring cells, an effect known as the senescence-associated secretory phenotype (SASP).[35] Some of these cells refuse to die—they have dialed up the pathways to block apoptosis (programmed cell death) that should be activated when cells are sick. Drugs that target these senescent cells and eliminate them, called "senolytics," could be helpful in reducing age-related inflammation, cancers, diabetes, atherosclerosis, and other age-related illnesses—which gives pharma several indications other than lifespan to measure senolytic success (see below). First-generation senolytics, such as navitoclax and dasatinib, were developed through hypothesis-driven approaches, while other compounds have been identified through unbiased, high-throughput screening. Others activate the immune-system-mediated clearance of senescent cells. Senomorphic drugs suppress senescent phenotypes and block SASP without killing the cell.[36] Yet another approach is to deplete the deleterious signaling molecules themselves. Current companies developing senolytics as clinical therapeutics have had successful phase I trials with subsequent phase II trial failures (e.g., for treatment of osteoarthritis). However, recently senolytic treatments for some eye diseases are looking hopeful even in early phase II. Despite initial failures, drugs targeting senolytic cells may be some of the first anti-aging drugs to emerge onto the market from recent academic longevity research.

Rejuvenation, Reprogramming, and Regeneration: Gene Therapy and Beyond

Other therapeutic approaches focus on regenerating and replacing cells we lose with age, particularly those that normally proliferate but lose the ability to divide with age. Stem cells seem to offer the possibility of immortality, and this area uses the most sci-fi tech of all the longevity biotech companies: in vitro organ growth, induced pluripotent stem cells (iPSCs)—taking a patient's own cells and turning them into stem cells to study and replace tissues—postpartum placenta cell harvesting, tissue allografting, epigenetic reprogramming, parabiosis, and genetic engineering: a veritable *Who's Who* of dystopian science fiction approaches.

One of the first things you'll find if you are looking into aging research is an almost religious belief among some that telomeres are the fix-all for every aging ailment. Telomeres were the main focus of excitement in the longevity biotech field years ago. Telomere rescue approaches—for example, by expressing telomerase (hTERT) in aging cells—might prevent cell replication exhaustion exemplified by the Hayflick limit. Short telomeres lead to cell senescence, while cells that express more telomerase have a longer life—at least in vitro. In fact, a cohort of Ashkenazi Jewish centenarians was found to have longer telomeres and were enriched for a variant of hTERT, supporting this notion.[37] However, a major caveat to any scheme to boost telomerase activity is the possibility of increasing the risk of cancer, since making cells immortal is exactly what cancer does. Malignant tumors express higher levels of hTERT, and melanoma immunotherapies even target hTERT,[38] which tells you exactly how bad it is to have runaway telomerase function in your cells. It is likely that centenarians benefit from multiple cellular advantages that not only maintain cells longer but also prevent cancer development as a result of those changes. Whether "telomerase activating molecules" are useful or even active is hard to know, in the absence of the kind of testing that the FDA would do—if aging were an FDA indication. (Normal sleep and exercise both seem to help maintain telomere length in a healthy way, so figuring out exactly how this happens without promoting cancer is the next direction.) Boosting hTERT is not a mainstream approach at this time, but a company called Libella Gene Therapeutics was charging patients $1 million to take part in a "pay-to-play clinical trial" for telomere boosting genetic therapy; the trial was not authorized by the FDA—in fact, it was being carried out at a clinic near Bogotá, Columbia.[39] Needless to say, this approach seems quite ethically fraught.

At least one person *really* believed in the telomere-boosting approach: Elizabeth Parrish, a 44-year-old housewife and part-time employee turned longevity-biotech-start-up CEO, flew to Columbia and subjected herself to dozens of these unapproved "anti-aging" injections of a myostatin inhibitor and telomerase as "patient zero," and has claimed that her white blood cell telomeres are 9% longer now than before the treatment (i.e., within the normal range of variability (10%) for telomere length measurements, so unremarkable)—but the data haven't been published or subjected to scrutiny or follow-up by scientists.[a]

Gene therapy hasn't been slowed down because of "a lot of academics sitting on the fence bickering," as Elizabeth Parrish claims, but rather because of the serious risks that must be carefully assessed and controlled. It's definitely not something one should just show up at a clinic and have done one afternoon with no oversight or follow-up, particularly since some adverse effects, like leukemia, won't be obvious right away in the same way that an acute immune reaction might be.

Gene therapy was the great medical hope of the 1990s but screeched to a halt in 1999 after the death of 18-year-old Jesse Gelsinger, who had a massive, lethal immune response to the adenovirus vector that was being used to deliver the gene to cure his metabolic disorder, partial ornithine transcarbamolyase deficiency.[40] Additionally, several boys who had been treated for severe combined immunodeficiency later developed leukemia, apparently caused by the unexpected integration of the corrective gene into the genome at a site that activated genes that caused leukemia.[41] I mention these issues to underscore the fact that although gene therapy has great promise, it's no joke. During a hiatus it took to review its oversight procedures, the gene therapy field has fixed many of its problems, both technical and legal, and has lately experienced a comeback. With a different viral vector (adeno-associated virus) the method has been successfully used to treat forms of blindness and spinal muscular atrophy, and many more treatments—particularly for debilitating childhood diseases—are currently underway.

Importantly for our interests here, there has been a gene therapy trial going on since 2019 at Weill Medical College for Alzheimer's Disease: patients over 50 with two copies of the deleterious APOE4 allele exhibiting mild cognitive impairment will have cognitive and PET scan assessments, and receive the beneficial APOE2 delivered to their central nervous system.[42] In addition to safety and dosage information, cognitive status, CSF markers, and amyloid plaques will be measured. Gene therapy for Huntington's disease (phase III since 2017) and several different trials for Parkinson's disease are also ongoing.[43] There is hope that gene therapy may be used to treat at least some aspects of aging. The possible development of cancer is a major concern for most approaches that aim to increase telomere length or increase regeneration in general, so telomere length might not ever be the right target, in contrast to the genetic lesions linked to specific age-related diseases. Immunotherapy has been successful in the treatment of several cancers; whether there are good targets in aging for this approach, such as by targeting senolytic cells, is unclear but a possibility.

The rapid development of CRISPR gene editing approaches will add specificity that might help avoid some of the dangerous effects of earlier gene therapy approaches.[44] In fact, David Liu's group reported that they had used in vivo base editing to repair the lamin A mutation that causes Hutchinson-Gilford progeria in mice and in fibroblasts derived from HGPS patients;[45] the hope is that this approach, with appropriate monitoring for immune responses, can be moved into children to prevent this devastating rapid-aging disease soon. The recent success of antisense oligonucleotide treatment (e.g., for Batten's disease)[46] suggests that small RNA manipulation could also become a powerful form of personalized therapy.

Regeneration treatments are likely to be targeted to specific cell and tissue types, since the goal is to renew a specific cell type that has been lost to aging or disease. For example, a recent study in which three of the four "Yamanaka" stem cell factors (the Oct3/4, Sox2, and Klf4 transcription factors—OSK) were activated in retinal ganglion cells (peripheral neurons) successfully restored sight in a mouse model of age-related degeneration;[47] these results suggest that a regeneration approach could become a treatment for age-related decline, perhaps even in neurons of the central nervous system. Obviously, this approach is many steps away from clinical application; in addition to the genetic engineering required to express these Yamanaka factors in cells and putting them back into the patient, the ability to cross the blood-brain barrier needs to be solved before this can be used to treat loss of neurons in the brain. But for the treatment of age-related ophthalmic disease, this approach already has a lot of immediate potential (and in fact seems to be moving into clinical trials), but as with any of these kinds of treatments, would need to address any concerns about the introduction of new, genetically altered cells into a patient.

The next level of regeneration is organ regeneration and replacement. This might be more straightforward than the other approaches I've described since the medical field has decades of organ transplant and wound healing expertise. Growing new organs from one's own differentiated iPSCs should reduce immune complications, and is the focus of several legitimate companies—but, again, beware of clinics claiming to offer stem-cell therapies that seem too good to be true.

The Search for New Drugs

Do we really need more drugs to treat aging? Probably. Right now the drugs that are being developed for market are largely the ones that showed some mechanism in an academic lab, which is sufficient for understanding the molecular and genetic pathways involved but may not address real-world problems in human therapeutics: drugs must be safe, bioavailable, and effective—and for cognitive issues, will need to cross the blood-brain barrier, which is tricky. We are lucky that a few old drugs like metformin and rapamycin have been used for decades already, so repurposing them to treat age-related diseases will be more straightforward. Nevertheless, even getting them to the clinic to treat aging is requiring a ton of effort and money (see section on the TAME trial, below), so one hopes that there will be a smoother path paved by these initial aging trials.

There are already several drugs that are at least associated with slowing aging; many are assembled in the DrugAge database, which contains a list of hundreds of geroprotectors.[48] This database shows lifespan results for drugs that have been tested in several model organisms. I've mentioned the Intervention Testing Program (ITP), an NIH-funded, multi-lab program that is aimed at testing good candidate drugs for their effects on life and healthspan in male and female mice (and it has a *C. elegans* counterpart, the CITP).[49] The candidates are submitted by scientists, who provide a rationale for testing, along with suggestions for dosage and delivery methods. So far there have been positive results from aspirin, rapamycin, 17α-estradiol, acarbose, NDGA, protandim, and glycine, and there is a long list of candidates that are still being tested.[50] Some of the large libraries of small molecules that have been generated in academic and biotech labs have been bought by larger longevity companies, which are now screening them for age-related phenotypes. But there might be better small molecules or entirely different drug classes that are safer, function at lower doses, and cause fewer off-target effects. Companies that use high-throughput screening or artificial intelligence (AI) approaches to find new molecules that target different aging pathways may identify these new candidate drugs.

To begin to address "translatability" of geroprotectors from treatments in model systems to treatment of humans, Janssens and Houtkooper recently probed the list of drugs in the DrugAge database for possible side effects reported in humans in the SEP-L1000 database, summing the total side effects listed for each compound (which included terms ranging from the serious—pain, dizziness, malaise, and hypertension—to uncomfortable, like "flatulence").[51] Using this information, they identified possible geroprotectors with fewer side effects and with a greater chance of increasing lifespan. The best candidates after this assessment were spermidine, clofibrate, D-glucosamine, and gallic acid, as they had the biggest predicted lifespan effects with relatively mild side effects, like "rash" and "nausea"—rather than "cardiac failure" or "death" (yikes!). This type of prioritization based on previous clinical information could be very valuable in moving drugs from the lab to the clinic more rapidly.

Strategies to Test, Develop, and Sell Longevity Therapies

The problem with developing effective drugs to treat aging immediately becomes obvious: how would you know if your drug is actually slowing your aging and making you live longer? It's easy to know whether your molecule of choice works in worms, flies, or mice, since lifespan assays can be carried out very quickly and with the proper numbers to assess statistics—but this is entirely different in humans, owing to decades-long time scales, variability, and ethical issues. How would you run a clinical trial to test a longevity treatment? And even if you figured out how to do this, the FDA does not currently recognize aging as an indication, so it's not possible to get FDA approval for a longevity drug. Biotech companies are taking several different approaches to tackle this problem from multiple directions, with the goal of turning the discoveries we've made in academic labs into therapeutics. One approach is to use age-related diseases as a proxy for aging (for example, osteoarthritis and age-related macular degeneration). Another is to bypass the FDA altogether by developing products—both compounds and diagnostics—that can be sold without FDA approval. Other companies focus on developing and screening new drug libraries or using AI approaches to discover new compounds that directly target longevity in model systems, but will eventually need to address the FDA problem. And yet another option is to use our own house pets— dogs—to assess changes in healthspan with age and the effects of candidate longevity drugs on healthspan.

Testing "Anti-aging" Drugs by Instead Tackling Age-Related Diseases

One important principle behind the promise of longevity biotech is that slowing aging should help slow down the onset of many of the aging-related diseases (heart disease, cognitive decline, cancer, neurodegenerative disorders, arthritis, etc.) that make the prospect of aging so unappealing. The goal of most of the geroscience field is not to live as long as possible but instead to reduce the occurrence of age-related disease and the accompanying loss of quality of life. This might take many forms, since there are so many things that go wrong with age, ranging from the inconvenient to the debilitating. While living longer and maintaining independence, cognitive function, and motility longer with age are obvious goals of a better lifespan, reducing cancer, arthritis, cardiovascular disease, metabolic disorders, deafness, blindness, and various other degenerative and neurodegenerative disorders is also critical. To be clear, there are also companies focused on each of those individual diseases, particularly cancer and neurodegeneration. For some longevity companies, the view is that rather than playing whack-a-mole by trying to treat each of these individual diseases, slowing aging may slow the onset of all of them. However, since aging itself is not (currently) recognized as a disease by the FDA, and would be too slow to test anyway, testing drugs for their effects on lifespan is both difficult and perhaps pointless. (The TAME trial of metformin's effects might help change this, though—see TAME section, below.) Using age-related diseases as a proxy for aging can actually be helpful from a clinical testing standpoint: if you can show that a drug slows an aging-related disease and is safe, it might become FDA approved. If a drug also happens to extend lifespan or slow the onset of other aging-related disorders, it would then be considered a beneficial side effect. The treatment of cardiovascular disease illustrates this approach, since the widespread use of statins has already helped extend lifespan. One could make a similar argument for diabetes drugs, as metabolic disorders rise with age, and anti-diabetes treatments are some of the best candidates for systemic longevity drugs.

Testing the effects of candidate drugs on specific age-related diseases as a proxy for aging is the strategy that some companies have already adopted.* For example, Unity Biotechnology is testing senolytics—drugs that target senescent cells—for their effects on age-related diseases, rather than testing for effects on lifespan. Their first trials testing senolytics for reducing osteoarthritis initially looked promising, but then failed in phase II trials. The reasons for this failure are unknown—did they pick the wrong disease, molecular target, dosage, small molecule, or site of administration? These points can be debated, but apparently the company has already moved on: Unity appears to have shifted to a different disease target, age-related eye diseases (specifically, wet age-related macular degeneration and diabetic macular edema). Their drug UBX1325 passed phase I trials and looked promising in early phase II trials, so we should know in a few years how it fares. They have other drugs in the pipeline to test for effects on other age-related diseases, including cognitive decline and neurodegeneration. While it's disappointing that the osteoarthritis trials failed, at least it lays out a plan for how a company might start to tackle aging from a pharmaceutical standpoint. (It is also a bit sobering, though, given the clearly overly optimistic declarations by some of "curing aging" from the early 2000s.)

Other companies are focusing on indications such as cardiovascular disease, rather than focusing directly on longevity, while many others are using neurodegeneration or cancer as their target indications.[52] Obesity is another indication that can be tested; for example, derivatives of mitochondrially derived peptides, such as the small peptide MOTS-c hormone that is encoded by the mitochondrial genome, are being developed as possible therapeutics, since these muscle-derived peptides might act in a systemic manner like insulin. Excitingly, recent work in aged mice suggests that intermittent MOTS-c treatment improves many healthspan parameters[53]—but, of course, this still needs to be tested in humans. Muscle aging is another area that can be easily measured clinically, since improvements in muscle mass, strength, and repair should all have very obvious results. BioAge is testing the activation of a receptor of a secreted peptide hormone, apelin receptor J, and activation of the hypoxia pathway (via the inhibition of HIF-1-PR [hypoxia induced factor prolyl hydroxylase]) for muscle-related reductions

---

* I have no stake in any of the companies I mention here.

in frailty, which could improve the quality of life for the elderly. Another indication that can be tested is runaway inflammation in the elderly, or "inflammaging," and there are reliable markers for these measurements. A reduction in inflammation might be especially important for the reduction of Covid and perhaps even long Covid symptoms that we do not yet fully understand.

Despite the fact that they are measuring specific aging diseases, all of these companies are considered longevity or "anti-aging" companies. What will be interesting is if any find their drugs not only treat the disease they are testing but also extend lifespan.

Diagnostics

The ability to measure one's "biological age" from biomarkers and therefore predict one's future lifespan is a growing market. It is unclear how accurate these measurements are, and it is important to note that they are being marketed without FDA approval. The earliest products in this category arose from the discovery of telomeres and the enzyme to lengthen them, telomerase; once it was found that cells lost their ability to replicate once they reached the Hayflick limit imposed by shortening telomeres, it immediately suggested that telomere length might reflect one's true cellular age; one can buy tests that purport to reveal telomere length, and therefore something about your biological age. Other biomarker tests include blood panels—much like your doctor might order if you are sick—to measure the number of blood and immune cells and levels of specific sugars, lipids, enzymes, protein glycation, NAD^+, and a few other metabolites; although there's nothing new about these components, the tests are also being sold as metrics of aging.

More recently, several companies have begun to sell kits that allow you to measure your biological age based on DNA methylation marks, the epigenetic clocks we discussed in chapter 15. There are several different clocks, so using one that best reports on "biological" age from the tissue you are providing (blood or saliva) is probably your best bet. Cognitive tests and facial imaging round out the list of assessments that used to be done solely in academic studies or clinics but now are being offered to assess one's biological age.[54] Three-dimensional facial imaging is particularly interesting because it correlates well with blood biomarkers of aging and can reflect rates of aging. Being able to assess your biological age before you start taking any treatment might be a good "control," and then monitoring whether it changes with treatment if you can afford to do so.[55] However, it should be noted that these are not FDA approved, so one is putting a lot of trust in the companies selling the products; finding a way to actually get FDA approval (or not) would give consumers more confidence.

Nutraceuticals and Cosmeticeuticals

One way around the FDA approval problem altogether, as well as the long time frame needed to assess the effects of a drug on aging, is to use the label "nutraceuticals" instead of "pharmaceuticals." Nutraceutical products are largely unregulated and do not require approval by the FDA. For this reason, they can be sold without proof of efficacy or safety. Most vitamins and dietary supplements you are familiar with fall into this category and have been sold this way for decades, and are also subject to the concerns I listed above. More recently, compounds discovered in aging research labs have been developed into nutraceuticals. For example, Elysium's Basis is one of several NAD^+-based anti-aging supplements that are being sold; Basis includes nicotinamide riboside to generate NAD^+, and pterostilbene, an antioxidant and cousin of resveratrol, the famous red wine component.[56] Elysium sells a few other products, including a B vitamin/fatty acid compound that is marketed for brain health, and a diagnostic for aging that is based on Morgan Levine's DNA methylation biological aging clock (discussed in chapter 15). In theory, one could use the clock in tandem with the pills to track whether they have any influence on your aging rate, but I do not know if there is sufficient resolution to do so. While all of these products are based on solid, published science, the products themselves are not FDA approved—so they might be great and actually work to slow aging, or they might completely ineffective (so it's just money down the toilet, indirectly but literally)—we just don't know yet.

Even if you don't care about living longer, you might want to look better. The most obvious signs of aging, wrinkles and hair loss, might be reversible, and these outward signs of aging are the target of companies that focus on approaches like mitochondrial restoration. Like nutraceuticals, cosmeticeuticals—cosmetic products that are intended to have a therapeutic effect through penetration of the epidermis—similarly do not require FDA approval. While this sounds superficial in every way, such a treatment might also help patients recover their hair after chemotherapy treatments, so vanity is not the only reason to develop these products.

The Road to FDA Approval: Targeting Aging with Metformin (TAME)

Another approach is to change the goalposts by getting the FDA to recognize aging as a disease, so that drugs aimed at aging itself can be clinically tested and approved—and long-term, developing a path toward FDA approval is hugely important with this growing market. Dr. Nir Barzilai (Albert Einstein College of Medicine) has led an effort to prove that aging is a treatable indication, using the same clinical trial process that is used to test other diseases. Metformin is a good place to start, because it is already FDA approved for the treatment of diabetes, polycystic ovary syndrome, and other blood sugar disorders; it's generally well-tolerated and safe; it's cheap; and it's been well studied in many organisms for its role in aging. Enough geroscientists with medical degrees are convinced by the data that they are taking it themselves, particularly if they have been diagnosed with pre-diabetes. Surprisingly, the exact molecular mechanism of metformin is not yet understood; it has been variously described as an antioxidant, an AMPK activator, a mitochondrial complex I inhibitor, and a microbiome modifier.[57] Nevertheless, metformin's effects on aging have been documented in many animals and it seems to play a positive role in several different cellular processes. A small, six-week clinical trial ($n = 14$) of patients over 70 was positive, suggesting that a larger trial is necessary to determine whether metformin could be broadly helpful in treating aging.[58] The TAME trial will enroll over 3000 people aged 65–79 from across the United States at different sites, where metformin's effects on several age-related diseases (dementia, heart disease, cancer, etc.) will be tracked. If successful, this large-scale approach should pave the way for other candidate anti-aging drugs to be tested in FDA clinical trials. In addition to age-related diseases, a companion study called TAME BIO will track biomarkers—that is, components of blood, urine, serum, CSF, and so on—in the TAME participants. This companion study may shed light on which biomarkers best reflect aging progression, which could become extremely useful as a proxy for aging—likely better than the ones that are currently being used in commercial diagnostics, since those were originally chosen for other diseases. The use of biomarkers will speed the assessment of progress, since the end result (delayed death) will not be necessary to track changes with age and those that might be slowed with drug treatment. Although there is great excitement about the possibility that metformin might slow aging, it is also important to ask whether there is any benefit at all to healthy, low-risk people. While it is clear that prediabetic and people with type 2 diabetes benefit from metformin, recent studies on healthy people don't show clear benefits, and it may have no effects on all-cause, cancer, or cardiovascular mortality rates.[59] Additionally, in a 12-week placebo-controlled study of aerobic exercise training in older adults (~62 years), metformin blunted the positive effects of exercise, and another study showed that metformin may lower testosterone.[60] Therefore, metformin might be beneficial to the overweight, prediabetic, or diabetic population, but less helpful to those who are already at low risk for metabolic disease.

Man's Best Friend and/or Test Subject

Perhaps the most endearing of ongoing clinical trial approaches are those that use pet dogs. Dogs can be as genetically heterogeneous as people are, live in shared environments with people (so can be sentinel species, like coal mine canaries), and could possibly help us understand whether longevity drugs can extend the lives of longer-lived species. This is the approach that Dan Promislow and Matt Kaeberlein (University of Washington) have taken with the Dog Aging Project, a longevity study of 10,000 dogs over 10 years that first began as a crowdfunded, open science project.[61] In general, people are very fond of their dogs and want them to live long and healthy lives, so there are lots of volunteers for the study—and since owners know their dogs' behaviors well, there is even the possibility of monitoring their neurological function using the Canine Cognitive Dysfunction Scale.[62] Each dog's genome will be sequenced, so genes that associate not only with long life (like the lower IGF-1 of long-lived small dogs) but also other healthspan factors can be studied. A subset of the dogs will be enrolled in a double-blind study of rapamycin to see if it has beneficial healthspan effects, the TRIAD (Test of Rapamycin in Aging Dogs) trial.[63] A few companies with similar strategies (e.g., FidoGen and Loyal, with a healthspan study and drug trials) have started, as well. Rejuvenate Bio is using gene therapy to treat specific ailments in dogs (mitral valve disease) by boosting levels of longevity factors, such as FGF21 and Klotho.[64] Presumably these results will then be applied to people who are also suffering from heart disease. But at the very least some of these drugs might help our pets live a bit longer.

Conclusions: Immortalists versus Realists:
Beware the Hype

"Treating arguments and proposals that are not backed up by scientific evidence as though they were scientific ideas carries the risk of making them impressive to laypersons, whose main way of distinguishing among hypotheses is to take note of those that are promoted in public media or presented to them by advocates whose style they like. A conference devoted to public transport systems would not be tempted to include a debate on teleportation as an approach to reduce traffic congestion; neither would an editor assembling a special issue on food shortages in the developing world solicit an essay on Aladdin's lamp.

. . . Science—unlike fantasy—works and leads to discoveries that serve as the foundation for material progress."[65]

Scientists are a weird bunch, simultaneously skeptical (the best are critical of their own work) and yet ever optimistic that the work will lead to a treatment that will make a difference in someone's life. It is one thing to declare that immortality is nigh, and quite another to determine exactly *how* to make people live even a little bit longer than they do now, much less decades—particularly since there can be deleterious effects from the alteration of some of the pathways we have discussed. There's a small faction of these "immortalists" that accuses these academic scientists of thinking too small and suggests that they are somehow the problem in aging research, the reason we don't already have "anti-aging" therapies. Of course, it's easy to grab headlines with statements

It's funny to me that in interviews, several non-scientist venture capitalists claim that other people called them "nuts" for working on aging (I have never heard anyone say that, but OK), and then they declare themselves visionaries. But most of the companies are using approaches that are based on published work from academic labs, which doesn't strike me as particularly visionary or crazy, just the next obvious direction for development of the drugs and approaches that academic scientists in the aging field have had as a goal all along. In fact, our mandate from the NIH is to help improve human health, so the eventual development of drugs that results from NIH-funded research is not only natural, it's expected.

that *the person who will live a thousand years has already been born*—whether you have evidence for it or not.[66] What is harder is to actually prove that it can be done. To suggest that those who say that the prospect of near immortality is currently unrealistic are somehow "pro-aging" puts responsible scientists in the awkward position of having to explain why it might take time to translate their findings safely to the clinic, while also being rightfully excited about the therapeutic potential of their discoveries. There are a few loud voices making outlandish statements about living forever, and "biohackers" using dubious methods (there's a weird Venn diagram of cryptocurrency enthusiasts with this "biohacking longevity" group)—but these usually are not the scientists who are actually doing the hard work of discovering the genetic and cellular mechanisms that will make living longer, or at least better, a real possibility.[67] The latter seem boring in comparison with the hype of the former, so you're less likely to encounter much of this work in a TED talk or in *Wired* magazine, but discoveries by the realists—the scientific findings you've been reading about in this book—are most likely to improve the quality of your life, within your lifetime.

———

So how should I answer that person who asks me "What can I do to live longer?" Getting to a longer life is unlikely to be because of one magic bullet. While this is not great news for anyone who wants to claim to be able to "fix" aging forever, the fact that there are many ways—possibly even additive pathways—to get to extend life- and healthspan is encouraging. Maybe a cocktail of the best drugs in each of these categories will get us closer to that point. There is no FDA-approved drug to slow aging or age-related decline *yet*, but several are on the horizon. There are hundreds of candidate drugs being tested for different age-related indications by dozens of different companies at this point; if you're an investor, the trick is to guess which one is going to be successful first. Most realistically, a company might get a hit with one of the age-related diseases, and that drug might be useful for many indications, including aging in general. But even if they are "only" successful in treating some of the more debilitating aspects of aging, such as cancer, diabetes, cognitive decline, or muscle frailty, it will be a win for a large part of the population. Similarly, there are no FDA-approved diagnostics yet—everything being sold now is a nutraceutical or non-FDA-approved diagnostic—but some of the ones out there might work, so trying a noninvasive one might be interesting, especially if it can be used to track your anti-aging treatments accurately and in an unbiased manner. A functioning aging diagnostic might also be useful to indicate when one would start to benefit from an anti-aging treatment.

Longevity Drugs in a Just and Sustainable World

Lives remaining: Zero

—ALEXIS OHANIAN (REDDIT FOUNDER AND HUSBAND
OF SERENA WILLIAMS), ON *DESUS & MERO*

Real treatments to slow aging would help alleviate late-in-life frailty and morbidity, declines in cognitive function, and loss of independence with age. In a just world, these treatments would be distributed regardless of income, not restricted to the wealthiest among us. While anti-aging treatments would be beneficial for individuals, the reduction of the need for long-term and invasive treatments for debilitated aging patients would relieve burden on our medical systems, a problem that is likely to be exacerbated by the long-term effects of Covid that are now becoming apparent. Most optimistically, perhaps the possibility of a long life would convince those in positions of power to finally make real efforts to combat climate change (not just colonization of Mars) to make sure there is a world worth living a long life in. Looking to the future, I'm more hopeful than ever that the work that has been done in academic labs over the past 25 years has led us to the point where clinical approaches can be safely developed to treat some age-related disorders—and I'm excited to see whether the next decade gets us closer to helping everyone live a higher quality of life with age.

NOTES

Notes to Introduction

1. Kenyon, C., Chang, J., Gensch, E., Rudner, A., and Tabtiang, R. A., 1993, "*C. elegans* Mutant that Lives Twice as Long as Wild Type," *Nature* 366:461–64.

2. Murphy, C. T., et al., 2003, "Genes that Act Downstream of DAF-16 to Influence the Lifespan of *Caenorhabditis elegans*," *Nature* 424:277–83.

3. Fire, A., et al., 1998, "Potent and Specific Genetic Interference by Double-Stranded RNA in *Caenorhabditis elegans*," *Nature* 391:806–11, https://doi.org/10.1038/35888; Matzke, M. A., and Matzke, A., 1995, "How and Why Do Plants Inactivate Homologous (Trans)genes?," *Plant Physiology* 107:679–685, https://doi.org/10.1104/pp.107.3.679.

4. Fraser, A. G., et al., 2000, "Functional Genomic Analysis of *C. elegans* Chromosome I by Systematic RNA Interference," *Nature* 408:325–30, https://doi.org/10.1038/35042517.

5. On CRISPR: Doudna, J. A., and Charpentier, E., 2014, "Genome Editing: The New Frontier of Genome Engineering with CRISPR-Cas9," *Science* 346:1258096, https://doi.org/10.1126/science.1258096.

Chapter 1

1. Centers for Disease Control and Prevention, National Center for Health Statistics, 2004, *Health, United States, 2004: With Chartbook on Trends in the Health of Americans* (Atlanta, GA: CDC).

2. Levine, M. E., and Crimmins, E. M., 2014, "Evidence of Accelerated Aging among African Americans and Its Implications for Mortality," *Social Science and Medicine* 118:27–32, https://doi.org/10.1016/j.socscimed.2014.07.022.

3. Flaskerud, J. H., 2012, "Coping and Health Status: John Henryism," *Issues in Mental Health Nursing* 33:712–15, https://doi.org/10.3109/01612840.2012.673695.

4. "Interview with Elizabeth Arias: Decline in Life Expectancy in 2020," CDC National Center for Health Statistics, February 19, 2021, https://www.cdc.gov/nchs/pressroom/podcasts/2021/20210219/20210219.htm.

5. Seeman, T. E., et al., 2004, "Cumulative Biological Risk and Socio-economic Differences in Mortality: MacArthur Studies of Successful Aging," *Social Science and Medicine* 58:1985–97, https://doi.org/10.1016/S0277-9536(03)00402-7.

6. See Goldsmith, J., 2017, "America's Health and the 2016 Election: An Unexpected Connection," *The Health Care Blog,* January 4, 2017, https://thehealthcareblog.com/blog/2017/01/04/americas-health-and-the-2016-election-an-unexpected-connection/.

7. World Health Organization, 2015, *Trends in Maternal Mortality: 1990 to 2015,* WHO/RHR/15.23, https://apps.who.int/iris/bitstream/handle/10665/193994/WHO_RHR_15.23_eng.pdf.

8. US Department of Health and Human Services, Health Resources and Services Administration, 2019, *HSRA Maternal Mortality Summit: Promising Global Practices to Improve Maternal Health Outcomes,* technical report, February 15, 2019, https://www.hrsa.gov/sites/default/files/hrsa/maternal-health/maternal-mortality-technical-report.pdf.

9. Davis, M. F., and de Londras, F., 2021, "Most Democracies Are Expanding Abortion Access. The U.S. Is Retracting It," wbur, *Cognoscenti,* October 21, 2021, https://www.wbur.org/cognoscenti/2021/10/21/supreme-court-abortion-dobbs-texas-martha-f-davis-fiona-de-londras; Latt, S. M., Milner, A., and Kavanagh, A., 2019, "Abortion Laws Reform May Reduce Maternal Mortality: An Ecological Study in 162 Countries," *BMC Women's Health* 19, https://doi.org/10.1186/s12905-018-0705-y.

10. MacDorman, M. F., Declercq, E., and Thoma, M. E., 2017, "Trends in Maternal Mortality by Sociodemographic Characteristics and Cause of Death in 27 States and the District of Columbia," *Obstetrics and Gynecology* 129:811–18, https://doi.org/10.1097/aog.0000000000001968.

11. Dong, X., Milholland, B., and Vijg, J., 2016, "Evidence for a Limit to Human Lifespan," *Nature* 538:257–59, https://doi.org/10.1038/nature19793.

12. Lenart, A., and Vaupel, J. W., 2017, "Questionable Evidence for a Limit to Human Lifespan," *Nature* 546:e13–e14, https://doi.org/10.1038/nature22790; Van Santen, Hester, 2016, "Nature Article Is Wrong about 115 Year Limit on Human Lifespan," NRC, October 7, 2016, https://www.nrc.nl/nieuws/2016/10/07/human-lifespan-limited-to-115-years-a1525476. See chapter 2 for a claim alleging that Calment's age was faked.

13. Modig, K., Andersson, T., Vaupel, J., Rau, R., and Ahlbom, A., 2017, "How Long Do Centenarians Survive? Life Expectancy and Maximum Lifespan," *Journal of Internal Medicine* 282:156–63, https://doi.org/10.1111/joim.12627.

14. Crimmins, E. M., 2015, "Lifespan and Healthspan: Past, Present, and Promise," *Gerontologist* 55:901–11, https://doi.org/10.1093/geront/gnv130.

15. Olshansky, S. J., 2015, "Articulating the Case for the Longevity Dividend," in *Aging: The Longevity Dividend,* edited by S. J. Olshansky, G. M. Martin, and J. L. Kirkland (New York: Cold Spring Harbor Laboratory Press), 191.

16. Goldman, D., 2015, "The Economic Promise of Delayed Aging," in *Aging: The Longevity Dividend,* 197.

17. Fries, J. F., Bruce, B., and Chakravarty, E., 2011, "Compression of Morbidity 1980–2011: A Focused Review of Paradigms and Progress," *Journal of Aging Research* 2011:261702, https://doi.org/10.4061/2011/261702.

18. Milman, S., and Barzilai, N., 2015, "Dissecting the Mechanisms Underlying Unusually Successful Human Health Span and Life Span," in *Aging: The Longevity Dividend,* 93; Sierra, F., 2015, "The Emergence of Geroscience as an Interdisciplinary Approach to the Enhancement of Health Span and Life Span," in *Aging: The Longevity Dividend,* 1.

Chapter 2

1. Goudeau, J., and Aguilaniu, H., 2010, "Carbonylated Proteins Are Eliminated during Reproduction in *C. elegans*," *Aging Cell* 9:991–1003, https://doi.org/10.1111/j.1474-9726.2010 .00625.x.

2. Metchnikoff, E., *The Prolongation of Life: Optimistic Studies*, trans. P. C. Mitchell (Putnam, NY, 1908), part 3, "Diseases that Shorten Life," 149.

3. Austad, S. N., and Fischer, K. E., 1991, "Mammalian Aging, Metabolism, and Ecology: Evidence from the Bats and Marsupials," *Journal of Gerontology* 46:B47–53, https://doi.org/10 .1093/geronj/46.2.B47.

4. Williams, G. C., 1957, "Pleiotropy, Natural Selection and the Evolution of Senescence," *Evolution* 11: 398–411.

5. George Williams: "Natural selection will always be in greatest opposition to the decline of the most senescence-prone system," and, therefore, "senescence should always be a generalized deterioration, and never due largely to changes in a single system. . . . This conclusion banishes the fountain of youth to the limbo of scientific impossibilities where other human aspirations, like the perpetual motion machine and Laplace's 'superman' have already been placed by other theoretical considerations. Such conclusions are always disappointing, but they have the desirable consequence of channeling research in directions that are likely to be fruitful."

6. For increased lifespan in worms: Klass, M. R., 1983, "A Method for the Isolation of Longevity Mutants in the Nematode *Caenorhabditis elegans* and Initial Results," *Mechanisms of Ageing and Development* 22:279–86, https://doi.org/10.1016/0047-6374(83)90082-9; Johnson, T. E., 1990, "Increased Life-Span of *age-1* Mutants in *Caenorhabditis elegans* and Lower Gompertz Rate of Aging," *Science* 249:908–12, https://doi.org/10.1126/science .2392681; Kenyon, C., Chang, J., Gensch, E., Rudner, A., and Tabtiang, R., 1993, "A *C. elegans* Mutant that Lives Twice as Long as Wild Type," *Nature* 366:461–64.

7. Tacutu, R., et al., 2012, "Prediction of *C. elegans* Longevity Genes by Human and Worm Longevity Networks," *PLOS One* 7:e48282, https://doi.org/10.1371/journal.pone.0048282; Curran, S. P., and Ruvkun, G., 2007, "Lifespan Regulation by Evolutionarily Conserved Genes Essential for Viability," *PLOS Genetics* 3:e56, https://doi.org/10.1371/journal.pgen .0030056.

8. David, D. C., et al., 2010, "Widespread Protein Aggregation as an Inherent Part of Aging in *C. elegans*," *PLOS Biology* 8:e1000450, https://doi.org/10.1371/journal.pbio.1000450.

9. Blagosklonny, M. V., 2021, "The Hyperfunction Theory of Aging: Three Common Misconceptions," *Oncoscience* 8:103–7, https://doi.org/10.18632/oncoscience.545.

10. Garigan, D., et al., 2002, "Genetic Analysis of Tissue Aging in *Caenorhabditis elegans*: A Role for Heat-Shock Factor and Bacterial Proliferation," *Genetics* 161:1101–12; Herndon, L. A., et al., 2002, "Stochastic and Genetic Factors Influence Tissue-Specific Decline in Ageing *C. elegans*," *Nature* 419:808–14.

11. Murphy, C. T., et al., 2003, "Genes that Act Downstream of DAF-16 to Influence the Lifespan of *Caenorhabditis elegans*," *Nature* 424:277–83.

12. Sornda, T., et al., 2019, "Production of YP170 Vitellogenins Promotes Intestinal Senescence in *Caenorhabditis elegans*," *Journals of Gerontology, Series A, Biological Sciences and Medical Sciences* 74:1180–88, https://doi.org/10.1093/gerona/glz067.

13. Honda, Y., and Honda, S., 2002, "Oxidative Stress and Life Span Determination in the Nematode *Caenorhabditis elegans*," *Annals of the New York Academy of Sciences* 959:466–74.

14. Van Raamsdonk, J., and Hekimi, S., 2012, "Superoxide Dismutase Is Dispensable for Normal Animal Lifespan," *Proceedings of the National Academy of Sciences of the USA* 109:5785–90, https://doi.org/10.1073/pnas.1116158109.

15. Ungvari, Z., et al., 2011, "Extreme Longevity Is Associated with Increased Resistance to Oxidative Stress in *Arctica islandica*, the Longest-Living Non-Colonial Animal," *Journals of Gerontology, Series A, Biological Sciences and Medical Sciences* 66:741–50, https://doi.org/10.1093/gerona/glr044.

16. Andziak, B., O'Connor, T. P., and Buffenstein, R., 2005, "Antioxidants Do Not Explain the Disparate Longevity between Mice and the Longest-Living Rodent, the Naked Mole-Rat," *Mechanisms of Ageing and Development* 126:1206–12, https://doi.org/10.1016/j.mad.2005.06.009.

17. Selman, C., et al., 2006, "Life-Long Vitamin C Supplementation in Combination with Cold Exposure Does Not Affect Oxidative Damage or Lifespan in Mice, but Decreases Expression of Antioxidant Protection Genes," *Mechanisms of Ageing and Development* 127:897–904, https://doi.org/10.1016/j.mad.2006.09.008.

18. Ristow, M., et al., 2009, "Antioxidants Prevent Health-Promoting Effects of Physical Exercise in Humans," *Proceedings of the National Academy of Sciences of the USA* 106:8665–70, https://doi.org/10.1073/pnas.0903485106.

19. Cypser, J. R., and Johnson, T. E., 2002, "Multiple Stressors in *Caenorhabditis elegans* Induce Stress Hormesis and Extended Longevity," *Journals of Gerontology, Series A, Biological Sciences and Medical Sciences* 57:B109–14; Lithgow, G. J., 2001, "Hormesis: A New Hope for Ageing Studies or a Poor Second to Genetics?," *Human and Experimental Toxicology* 20:301–3, discussion 319–20; Olsen, A., Vantipalli, M. C., and Lithgow, G. J., 2006, "Lifespan Extension of *Caenorhabditis elegans* following Repeated Mild Hormetic Heat Treatments," *Biogerontology* 7:221–30, https://doi.org/10.1007/s10522-006-9018-x.

20. Kirkwood, T. B., 1987, "Immortality of the Germ-Line versus Disposability of the Soma," *Basic Life Sciences* 42:209–18.

21. Min, K. J., Lee, C. K., and Park, H. N., 2012, "The Lifespan of Korean Eunuchs," *Current Biology* 22:R792–93, https://doi.org/10.1016/j.cub.2012.06.036.

22. Le Bourg, E., 2015, "No Ground for Advocating that Korean Eunuchs Lived Longer than Intact Men," *Gerontology* 62:69–70, https://doi.org/10.1159/000435854.

23. Hsin, H., and Kenyon, C., 1999, "Signals from the Reproductive System Regulate the Lifespan of *C. elegans*," *Nature* 399:362–66.

24. Shi, C., Booth, L. N., and Murphy, C. T., 2019, "Insulin-like Peptides and the mTOR-TFEB Pathway Protect *Caenorhabditis elegans* Hermaphrodites from Mating-Induced Death," *eLife* 8, https://doi.org/10.7554/eLife.46413; Shi, C., and Murphy, C. T., 2014, "Mating Induces Shrinking and Death in *Caenorhabditis* Mothers," *Science* 343:536–40, https://doi.org/10.1126/science.1242958; Shi, C., Runnels, A. M., and Murphy, C. T., 2017, "Mating and Male Pheromone Kill *Caenorhabditis* Males through Distinct Mechanisms," *eLife* 6, https://doi.org/10.7554/eLife.23493.

25. Dillin, A., Crawford, D. K., and Kenyon, C., 2002, "Timing Requirements for Insulin/IGF-1 Signaling in *C. elegans*," *Science* 298:830–34.

26. Venz, R., Pekec, T., Katic, I., Ciosk, R., and Ewald, C. Y., 2021, "End-of-Life Targeted Auxin-Mediated Degradation of DAF-2 Insulin/IGF-1 Receptor Promotes Longevity Free from Growth-Related Pathologies," *eLife* 10:e71335, https://doi.org/10.7554/eLife.71335.

27. Sohrabi, S., Cota, V., and Murphy, C. T., 2023, "CeLab, a Microfluidic Platform for the Study of Life History Traits, Reveals Metformin and SGK-1 Regulation of Longevity and Reproductive Span," *bioRxiv*, 2023.01.09.523184, https://doi.org/10.1101/2023.01.09.523184.

28. Perls, T. T., Alpert, L., and Fretts, R. C., 1997, "Middle-Aged Mothers Live Longer," *Nature* 389:133, https://doi.org/10.1038/38148.

29. An accessible review of many of these theories can be found at "Biological Aging Theories" (Programmed-Aging.org, last revised 2015, http://www.programmed-aging.org/theories/).

30. Mitteldorf, J., and Pepper, J., 2009, "Senescence as an Adaptation to Limit the Spread of Disease," *Journal of Theoretical Biology* 260:186–95, https://doi.org/10.1016/j.jtbi.2009.05.013.

31. Mitteldorf, J., and Pepper, J. W., 2007, "How Can Evolutionary Theory Accommodate Recent Empirical Results on Organismal Senescence?," *Theory in Biosciences* 126:3–8, https://doi.org/10.1007/s12064-007-0001-0.

32. Brunet-Rossinni, A. K., and Austad, S. N., 2004, "Ageing Studies on Bats: A Review," *Biogerontology* 5, 211–22, https://doi.org/10.1023/B:BGEN.0000038022.65024.d8.

33. Turbill, C., Bieber, C., and Ruf, T., 2011, "Hibernation Is Associated with Increased Survival and the Evolution of Slow Life Histories among Mammals," *Proceedings of the Royal Society B: Biological Sciences* 278:3355–63, https://doi.org/10.1098/rspb.2011.0190.

34. Godfrey-Smith, P. 2016, "Octopuses and the Puzzle of Aging," *New York Times*, December 2, 2016.

35. Reznick, D., Bryant, M., and Holmes, D., 2006, "The Evolution of Senescence and Post-reproductive Lifespan in Guppies (*Poecilia reticulata*)," *PLOS Biology* 4:e7, https://doi.org/10.1371/journal.pbio.0040007.

36. Carroll, S. B., 2016, *The Serengeti Rules: The Quest to Discover How Life Works and Why It Matters* (Princeton, NJ: Princeton University Press).

37. Hawkes, K., 2003, "Grandmothers and the Evolution of Human Longevity," *American Journal of Human Biology* 15:380–400, https://doi.org/10.1002/ajhb.10156.

38. Maliha, G., and Murphy, C. T., 2016, "A Simple Offspring-to-Mother Size Ratio Predicts Post-reproductive Lifespan," *bioRxiv*, 048835, https://doi.org/10.1101/048835.

39. Francis, N., Gregg, T., Owen, R., Ebert, T., and Bodnar, A., 2006, "Lack of Age-Associated Telomere Shortening in Long- and Short-Lived Species of Sea Urchins," *FEBS Letters* 580:4713–17, https://doi.org/10.1016/j.febslet.2006.07.049.

40. Giaimo, S., and Baudisch, A., 2015, "The Effect of Post-Reproductive Lifespan on the Fixation Probability of Beneficial Mutations," *PLOS One* 10:e0133820, https://doi.org/10.1371/journal.pone.0133820.

41. Mao, K., et al., 2018, "Late-Life Targeting of the IGF-1 Receptor Improves Healthspan and Lifespan in Female Mice," *Nature Communications* 9:2394, https://doi.org/10.1038/s41467-018-04805-5.

42. Laron, Z., 2002, "Effects of Growth Hormone and Insulin-like Growth Factor 1 Deficiency on Ageing and Longevity," *Novartis Foundation Symposium* 242:125–37; discussion 137–42.

43. Suh, Y., et al., 2008, "Functionally Significant Insulin-like Growth Factor I Receptor Mutations in Centenarians," *Proceedings of the National Academy of Sciences of the USA* 105:3438–42.

44. Dobzhansky, T., 1973, "Nothing in Biology Makes Sense Except in the Light of Evolution," *American Biology Teacher* 35:125–29, https://doi.org/10.2307/4444260.

Chapter 3

1. Zak, N., 2019, "Evidence that Jeanne Calment Died in 1934—Not 1997," *Rejuvenation Research* 22:3–12, https://doi.org/10.1089/rej.2018.2167.

2. Dong, X., Milholland, B., and Vijg, J., 2016, "Evidence for a Limit to Human Lifespan," *Nature* 538:257–59, https://doi.org/10.1038/nature19793.

3. Lenart, A., and Vaupel, J. W., 2017, "Questionable Evidence for a Limit to Human Lifespan," *Nature* 546:E13–14, https://doi.org/10.1038/nature22790.

4. Rozing, M. P., Kirkwood, T.B.L., and Westendorp, R.G.J., 2017, "Is There Evidence for a Limit to Human Lifespan?," *Nature* 546:E11–12, https://doi.org/10.1038/nature22788.

5. Barbi, E., Lagona, F., Marsili, M., Vaupel, J. W., and Wachter, K. W., 2018, "The Plateau of Human Mortality: Demography of Longevity Pioneers," *Science* 360:1459–61, https://doi.org/10.1126/science.aat3119.

6. Ismail, K., et al., 2016, "Compression of Morbidity Is Observed across Cohorts with Exceptional Longevity," *Journal of the American Geriatrics Society* 64, 1583–91, https://doi.org/10.1111/jgs.14222.

7. Fries, J. F., 1980, "Aging, Natural Death, and the Compression of Morbidity," New England Journal of Medicine 303:130–35, https://doi.org/10.1056/NEJM198007173030304.

8. Buettner, D., and Skemp, S., 2016, "Blue Zones: Lessons From the World's Longest Lived," *American Journal of Lifestyle Medicine* 10:318–21, https://doi.org/10.1177/1559827616637066; Poulain, M., et al., 2004, "Identification of a Geographic Area Characterized by Extreme Longevity in the Sardinia Island: The AKEA Study," *Experimental Gerontology* 39:1423–29, https://doi.org/10.1016/j.exger.2004.06.016.

9. Buettner, D., 2012, *The Blue Zones: 9 Lessons for Living Longer from the People Who've Lived the Longest*, 2nd ed. (Washington, DC: National Geographic).

10. Herskind, A. M., et al., 1996, "The Heritability of Human Longevity: A Population-Based Study of 2872 Danish Twin Pairs Born 1870–1900," *Human Genetics* 97:319–23; Yashin, A. I., Iachine, I. A., Harris, J. R. ,1999, "Half of Variation in Susceptibility to Mortality is Genetic: Findings from Swedish Twin Survival Data," *Behavior Genetics* 29:11–19; Christensen, K., Johnson, T. E., and Vaupel, J. W., 2006, "The Quest for Genetic Determinants of Human Longevity: Challenges and Insights," *Nature Reviews Genetics* 7:436–48; Willcox, B. J., Willcox, D. C., and Suzuki, M., 2017, "Demographic, Phenotypic, and Genetic Characteristics of Centenarians in Okinawa and Japan: Part 1, Centenarians in Okinawa," *Mechanisms of Ageing and Development* 165:75–79, https://doi.org/10.1016/j.mad.2016.11.001.

11. Deelen, J., et al., 2019, "A Meta-analysis of Genome-wide Association Studies Identifies Multiple Longevity Genes," *Nature Communications* 10:3669, https://doi.org/10.1038/s41467-019-11558-2.

12. Milman, S. and Barzilai, N., 2015, "Dissecting the Mechanisms Underlying Unusually Successful Human Health Span and Life Span," in *Aging: The Longevity Dividend*, edited by S. J. Olshansky, G. M. Martin, and J. L. Kirkland (New York: Cold Spring Harbor Laboratory Press), 93.

13. Sebastiani, P., and Perls, T. T., 2012, "The Genetics of Extreme Longevity: Lessons from the New England Centenarian Study," *Frontiers in Genetics* 3:277, https://doi.org/10.3389 /fgene.2012.00277.

14. Joshi, P. K., et al., 2017, "Genome-wide Meta-analysis Associates HLA-DQA1/DRB1 and LPA and Lifestyle Factors with Human Longevity," *Nature Communications* 8:910, https:// doi.org/10.1038/s41467-017-00934-5.

15. Nygaard, M., Thinggaard, M., Christensen, K., and Christiansen, L., 2017, "Investigation of the 5q33.3 Longevity Locus and Age-Related Phenotypes," *Aging (Albany NY)* 9:247–55, https://doi.org/10.18632/aging.101156.

16. Newman, A. B., et al., 2011, "Health and Function of Participants in the Long Life Family Study: A Comparison with Other Cohorts," *Aging (Albany NY)* 3:63–76, https://doi .org/10.18632/aging.100242; Koropatnick, T. A., et al., 2008, "A Prospective Study of High-Density Lipoprotein Cholesterol, Cholesteryl Ester Transfer Protein Gene Variants, and Healthy Aging in Very Old Japanese-American Men," *Journals of Gerontology, Series A, Biological Sciences and Medical Sciences* 63:1235–40, https://doi.org/10.1093/gerona/63.11.1235.

17. Sanders, A. E., et al., 2010, "Association of a Functional Polymorphism in the Cholesteryl Ester Transfer Protein (CETP) Gene with Memory Decline and Incidence of Dementia," *JAMA* 303:150–58, https://doi.org/10.1001/jama.2009.1988.

18. On *C. elegans*: Henderson, S. T., and Johnson, T. E., 2001, "*daf-16* Integrates Developmental and Environmental Inputs to Mediate Aging in the Nematode *Caenorhabditis elegans*," *Current Biology* 11:1975–80; Lin, K., Dorman, J. B., Rodan, A., and Kenyon, C., 1997, "*daf-16*: An HNF-3/Forkhead Family Member that Can Function to Double the Life-Span of *Caenorhabditis elegans*," *Science* 278:1319–22; Ogg, S., et al.,1997, "The Fork Head Transcription Factor DAF-16 Transduces Insulin-like Metabolic and Longevity Signals in *C. elegans*," *Nature* 389:994–99. On flies: Hwangbo, D. S., Gersham, B., Tu, M. P., Palmer, M., and Tatar, M., 2004, "*Drosophila* dFOXO Controls Lifespan and Regulates Insulin Signalling in Brain and Fat Body," *Nature* 429:562–66; Giannakou, M. E., et al., 2004, "Long-Lived *Drosophila* with Overexpressed dFOXO in Adult Fat Body," *Science* 305:361, https://doi.org /10.1126/science.1098219. On mice: Shimokawa, I., et al., 2015, "The Life-Extending Effect of Dietary Restriction Requires Foxo3 in Mice," *Aging Cell* 14:707–9, https://doi.org/10.1111 /acel.12340; Yamaza, H., et al., 2010, "FoxO1 Is Involved in the Antineoplastic Effect of Calorie Restriction," *Aging Cell* 9:372–82, https://doi.org/10.1111/j.1474-9726.2010.00563 .x; Paik, J. H., et al., 2007, "FoxOs Are Lineage-Restricted Redundant Tumor Suppressors and Regulate Endothelial Cell Homeostasis," *Cell* 128:309–23, https://doi.org/10.1016/j.cell .2006.12.029.

19. Willcox, B. J., et al., 2008, "FOXO3A Genotype Is Strongly Associated with Human Longevity," *Proceedings of the National Academy of Sciences of the USA* 105:13987–92, https:// doi.org/10.1073/pnas.0801030105; Anselmi, C. V., et al., 2009, "Association of the FOXO3A Locus with Extreme Longevity in a Southern Italian Centenarian Study," *Rejuvenation Research* 12:95–104, https://doi.org/10.1089/rej.2008.0827; Flachsbart, F., et al., 2009, "Association of

FOXO3A Variation with Human Longevity Confirmed in German Centenarians," *Proceedings of the National Academy of Sciences of the USA* 106:2700–2705, https://doi.org/10.1073/pnas .0809594106.

20. Suh, Y., et al., 2008, "Functionally Significant Insulin-like Growth Factor I Receptor Mutations in Centenarians," *Proceedings of the National Academy of Sciences of the USA* 105:3438–42.

21. Teumer, A., et al., 2016, "Genomewide Meta-analysis Identifies Loci Associated with IGF-I and IGFBP-3 Levels with Impact on Age-Related Traits," *Aging Cell* 15:811–24, https:// doi.org/10.1111/acel.12490.

22. Joshi et al. 2017.

23. Lopez-Otin, C., Blasco, M. A., Partridge, L., Serrano, M., and Kroemer, G., 2013, "The Hallmarks of Aging," *Cell* 153:1194–217, https://doi.org/10.1016/j.cell.2013.05.039.

24. Joshi et al. 2017.

25. Andriani, G. A., et al., 2016, "Whole Chromosome Instability Induces Senescence and Promotes SASP," *Scientific Reports* 6:35218, https://doi.org/10.1038/srep35218.

26. Zeng, Y., et al., 2016, "Novel Loci and Pathways Significantly Associated with Longevity," *Scientific Reports* 6:21243, https://doi.org/10.1038/srep21243.

27. Zeng, Y., et al., 2016, "Interaction between the FOXO1A-209 Genotype and Tea Drinking Is Significantly Associated with Reduced Mortality at Advanced Ages," *Rejuvenation Research* 19, 195–203, https://doi.org/10.1089/rej.2015.1737.

28. Harmon, A., 2017, "The Secret to Long Life? It May Lurk in the DNA of the Oldest among Us," *New York Times*, November 13, 2017, https://www.nytimes.com/2017/11/13 /health/supercentenarians-genetics-longevity.html.

29. Betterhumans Inc., "Collaborate with Betterhumans," Betterhumans Supercentenarian Research Study, 2022, https://supercentenarianstudy.com/collaborate.html.

30. Garagnani, P., et al., 2021, "Whole-Genome Sequencing Analysis of Semi-supercentenarians," *eLife* 10:e57849, https://doi.org/10.7554/eLife.57849.

31. Teumer et al. 2016; Johnson, S. C., Dong, X., Vijg, J., and Suh, Y., 2015, "Genetic Evidence for Common Pathways in Human Age-Related Diseases," *Aging Cell* 14:809–17, https:// doi.org/10.1111/acel.12362.

32. Fortney, K., et al., 2015, "Genome-wide Scan Informed by Age-Related Disease Identifies Loci for Exceptional Human Longevity," *PLOS Genetics* 11:e1005728, https://doi.org/10 .1371/journal.pgen.1005728.

33. Ruby, J. G., et al., 2018, "Estimates of the Heritability of Human Longevity Are Substantially Inflated Due to Assortative Mating," *Genetics* 210:1109–24, https://doi.org/10.1534 /genetics.118.301613.

Chapter 4

1. Austad, S. N., 2005, "Diverse Aging Rates in Metazoans: Targets for Functional Genomics," *Mechanisms of Ageing and Development* 126:43–49, https://doi.org/10.1016/j.mad.2004.09.022.

2. See Barber, E., 2013, "Scientists Discover World's Oldest Clam, Killing It in the Process," *Christian Science Monitor*, November 15, 2013, https://www.csmonitor.com/Science/2013 /1115/Scientists-discover-world-s-oldest-clam-killing-it-in-the-process.

3. Nielsen, J., et al., 2016, "Eye Lens Radiocarbon Reveals Centuries of Longevity in the Greenland Shark (*Somniosus microcephalus*)," *Science* 353:702–4, https://doi.org/10.1126/science.aaf1703.

4. Atwal, S., 2022, "190-Year-Old Jonathan Becomes World's Oldest Tortoise Ever," Guinness World Records, January 12, 2022, https://www.guinnessworldrecords.com/news/2022/1/190-year-old-jonathan-becomes-worlds-oldest-tortoise-ever-688683.

5. Keane, M., et al., 2015, "Insights into the Evolution of Longevity from the Bowhead Whale Genome," *Cell Reports* 10:112–22, https://doi.org/10.1016/j.celrep.2014.12.008.

6. Laron, Z., 2002, "Effects of Growth Hormone and Insulin-like Growth Factor 1 Deficiency on Ageing and Longevity," *Novartis Foundation Symposium* 242:125–37; discussion 137–42.

7. Rosenbloom, A. L., Aguirre, J. G., Rosenfeld, R. G., and Fielder, P. J., 1990, "The Little Women of Loja: Growth Hormone–Receptor Deficiency in an Inbred Population of Southern Ecuador," *New England Journal of Medicine* 323:1367–74, https://doi.org/10.1056/nejm199011153232002.

8. Austad, S. N., and Fischer, K. E., 1991, "Mammalian Aging, Metabolism, and Ecology: Evidence from the Bats and Marsupials," *Journal of Gerontology* 46:B47–53, https://doi.org/10.1093/geronj/46.2.B47.

9. Turbill, C., Smith, S., Deimel, C., and Ruf, T., 2012, "Daily Torpor Is Associated with Telomere Length Change over Winter in Djungarian Hamsters," *Biology Letters* 8:304–7, https://doi.org/10.1098/rsbl.2011.0758.

10. Foley, N. M., et al., 2018, "Growing Old, Yet Staying Young: The Role of Telomeres in Bats' Exceptional Longevity," *Science Advances* 4:eaao0926, https://doi.org/10.1126/sciadv.aao0926; Zhang, G., et al., 2013, "Comparative Analysis of Bat Genomes Provides Insight into the Evolution of Flight and Immunity," *Science* 339:456–60, https://doi.org/10.1126/science.1230835.

11. Seim, I., et al., 2013, "Genome Analysis Reveals Insights into Physiology and Longevity of the Brandt's Bat *Myotis brandtii*," *Nature Communications* 4:2212, https://doi.org/10.1038/ncomms3212.

12. Lewis, K. N., Andziak, B., Yang, T., and Buffenstein, R., 2013, "The Naked Mole-Rat Response to Oxidative Stress: Just Deal with It," *Antioxidants and Redox Signaling* 19:1388–99, https://doi.org/10.1089/ars.2012.4911.

13. Buffenstein, R., 2008, "Negligible Senescence in the Longest Living Rodent, the Naked Mole-Rat: Insights from a Successfully Aging Species," *Journal of Comparative Physiology B* 178:439–45, https://doi.org/10.1007/s00360-007-0237-5.

14. Zhao, Y., et al., 2018, "Naked Mole Rats Can Undergo Developmental, Oncogene-Induced and DNA Damage-Induced Cellular Senescence," *Proceedings of the National Academy of Sciences of the USA* 115:1801–6, https://doi.org/10.1073/pnas.1721160115.

15. Lee, B. P., Smith, M., Buffenstein, R., and Harries, L. W., 2020, "Negligible Senescence in Naked Mole Rats May Be a Consequence of Well-Maintained Splicing Regulation," *Geroscience* 42:633–651, https://doi.org/10.1007/s11357-019-00150-7.

16. Lewis, K. N., et al., 2016, "Unraveling the Message: Insights into Comparative Genomics of the Naked Mole-Rat," *Mammalian Genome* 27:259–78, https://doi.org/10.1007/s00335-016-9648-5.

17. Kim, E. B., et al., 2011, "Genome Sequencing Reveals Insights into Physiology and Longevity of the Naked Mole Rat," *Nature* 479:223–27, https://doi.org/10.1038/nature 10533.

18. Tian, X., et al., 2013, "High-Molecular-Mass Hyaluronan Mediates the Cancer Resistance of the Naked Mole Rat," *Nature* 499:346–49, https://doi.org/10.1038/nature12234.

19. On regulation of the insulin signaling pathway: Ihle, K. E., Baker, N. A., and Amdam, G. V., 2014, "Insulin-like Peptide Response to Nutritional Input in Honey Bee Workers," *Journal of Insect Physiology* 69:49–55, https://doi.org/10.1016/j.jinsphys.2014.05.026.

20. Hutchinson, E. W., Shaw, A. J., and Rose, M. R., 1991, "Quantitative Genetics of Postponed Aging in *Drosophila* melanogaster: II, Analysis of Selected Lines," *Genetics* 127:729–37, https://doi.org/10.1093/genetics/127.4.729; Leroi, A. M., Chen, W. R., and Rose, M. R., 1994, "Long-Term Laboratory Evolution of a Genetic Life-History Trade-off in *Drosophila melanogaster*: 2, Stability of Genetic Correlations," *Evolution* 48:1258–68, https://doi.org/10.1111/j.1558–5646.1994.tb05310.x.

21. Chapman, T., Liddle, L. F., Kalb, J. M., Wolfner, M. F., and Partridge, L., 1995, "Cost of Mating in *Drosophila melanogaster* Females Is Mediated by Male Accessory Gland Products," *Nature* 373:241–44, https://doi.org/10.1038/373241a0.

22. Piper, M.D.W., et al., 2017, "Matching Dietary Amino Acid Balance to the In Silico-Translated Exome Optimizes Growth and Reproduction without Cost to Lifespan," *Cell Metabolism* 25:610–21, https://doi.org/10.1016/j.cmet.2017.02.005.

23. Klass, M. R., 1983, "A Method for the Isolation of Longevity Mutants in the Nematode *Caenorhabditis elegans* and Initial Results," *Mechanisms of Ageing and Development* 22:279–86, https://doi.org/10.1016/0047-6374(83)90082-9.

24. Friedman, D. B., and Johnson, T. E., 1988, "A Mutation in the *age-1* Gene in *Caenorhabditis elegans* Lengthens Life and Reduces Hermaphrodite Fertility," *Genetics* 118:75–86.

25. Morris, J. Z., Tissenbaum, H. A., and Ruvkun, G., 1996, "A Phosphatidylinositol-3-OH Kinase Family Member Regulating Longevity and Diapause in *Caenorhabditis elegans*," *Nature* 382:536–39.

26. Kenyon, C., Chang, J., Gensch, E., Rudner, A., and Tabtiang, R., 1993, "A *C. elegans* Mutant that Lives Twice as Long as Wild Type," *Nature* 366:461–64.

27. Kimura, K. D., Tissenbaum, H. A., Liu, Y., and Ruvkun, G., 1997, "*daf-2*, an Insulin Receptor-like Gene that Regulates Longevity and Diapause in *Caenorhabditis elegans*," *Science* 277:942–46.

28. Ogg, S., et al., 1997, "The Fork Head Transcription Factor DAF-16 Transduces Insulin-like Metabolic and Longevity Signals in *C. elegans*," *Nature* 389:994–99; Lin, K., Dorman, J. B., Rodan, A., and Kenyon, C., 1997, "*daf-16*: An HNF-3/Forkhead Family Member that Can Function to Double the Life-Span of *Caenorhabditis elegans*," *Science* 278:1319–22.

29. Ogg, S., and Ruvkun, G., 1998, "The *C. elegans* PTEN Homolog, DAF-18, Acts in the Insulin Receptor-like Metabolic Signaling Pathway," *Molecular Cell* 2:887–93.

30. Paradis, S., Ailion, M., Toker, A., Thomas, J. H., and Ruvkun, G., 1999, "A PDK1 Homolog Is Necessary and Sufficient to Transduce AGE-1 PI3 Kinase Signals that Regulate Diapause in *Caenorhabditis elegans*," *Genes and Development* 13:1438–52; Paradis, S., and Ruvkun, G., 1998, "*Caenorhabditis elegans* Akt/PKB Transduces Insulin Receptor-like Signals from AGE-1 PI3 Kinase to the DAF-16 Transcription Factor," *Genes and Development* 12:2488–98.

31. Arrested Development narrator: It isn't.

32. Dillin, A., Crawford, D. K., and Kenyon, C., 2002, "Timing Requirements for Insulin/IGF-1 Signaling in *C. elegans*," *Science* 298:830–34.

33. Lakowski, B., and Hekimi, S., 1996, "Determination of Life-Span in *Caenorhabditis elegans* by Four Clock Genes," *Science* 272:1010–13; Lakowski, B., and Hekimi, S., 1998, "The Genetics of Caloric Restriction in *Caenorhabditis elegans*," *Proceedings of the National Academy of Sciences of the USA* 95:13091–96.

34. Tatar, M., et al., 2001, "A Mutant *Drosophila* Insulin Receptor Homolog that Extends Life-Span and Impairs Neuroendocrine Function," *Science* 292:107–10, https://doi.org/10.1126/science.1057987; Clancy, D. J., et al., 2001, "Extension of Life-Span by Loss of CHICO, a *Drosophila* Insulin Receptor Substrate Protein," *Science* 292:104–6, https://doi.org/10.1126/science.1057991.

35. Hwangbo, D. S., Gersham, B., Tu, M. P., Palmer, M., and Tatar, M., 2004, "*Drosophila* dFOXO Controls Lifespan and Regulates Insulin Signalling in Brain and Fat Body," *Nature* 429:562–66.

36. Fire, A., et al., 1998, "Potent and Specific Genetic Interference by Double-Stranded RNA in *Caenorhabditis elegans*," *Nature* 391:806–11, https://doi.org/10.1038/35888.

37. Fraser, A. G., et al., 2000, "Functional Genomic Analysis of *C. elegans* Chromosome I by Systematic RNA Interference," *Nature* 408:325–30, https://doi.org/10.1038/35042517.

38. Hansen, M., Hsu, A. L., Dillin, A., and Kenyon, C., 2005, "New Genes Tied to Endocrine, Metabolic, and Dietary Regulation of Lifespan from a *Caenorhabditis elegans* Genomic RNAi Screen," *PLOS Genetics* 1:119–28, https://doi.org/10.1371/journal.pgen.0010017; Timmons, L., Tabara, H., Mello, C. C., and Fire, A. Z., 2003, "Inducible Systemic RNA Silencing in *Caenorhabditis elegans*," *Molecular Biology of the Cell* 14:2972–83, https://doi.org/10.1091/mbc.e03-01-0858.

39. Stroustrup, N., et al., 2013, "The *Caenorhabditis elegans* Lifespan Machine," *Nature Methods* 10:665–70, https://doi.org/10.1038/nmeth.2475; Churgin, M. A., et al., 2017, "Longitudinal Imaging of *Caenorhabditis elegans* in a Microfabricated Device Reveals Variation in Behavioral Decline during Aging," *eLife* 6:e26652, https://doi.org/10.7554/eLife.26652; Zhang, W. B., et al., 2016, "Extended Twilight among Isogenic *C. elegans* Causes a Disproportionate Scaling between Lifespan and Health," *Cell Systems* 3:333–45.e334, https://doi.org/10.1016/j.cels.2016.09.003.

40. Kaletsky, R., et al., 2018, "Transcriptome Analysis of Adult *Caenorhabditis elegans* Cells Reveals Tissue-Specific Gene and Isoform Expression," *PLOS Genetics* 14:e1007559, https://doi.org/10.1371/journal.pgen.1007559; Yao, V., et al., 2018, "An Integrative Tissue-Network Approach to Identify and Test Human Disease Genes," *Nature Biotechnology*, https://doi.org/10.1038/nbt.4246.

41. Mattison, J. A., et al., 2017, "Caloric Restriction Improves Health and Survival of Rhesus Monkeys," *Nature Communications* 8:14063, https://doi.org/10.1038/ncomms14063.

42. Salmon, A. B., 2016, "Moving Toward 'Common' Use of the Marmoset as a Non-human Primate Aging Model," *Pathobiology of Aging and Age-Related Diseases* 6:32758, https://doi.org/10.3402/pba.v6.32758.

43. Hansen, M., Hsu, A. L., Dillin, A., and Kenyon, C., 2005, "New Genes Tied to Endocrine, Metabolic, and Dietary Regulation of Lifespan from a *Caenorhabditis elegans* Genomic RNAi Screen," *PLOS Genetics* 1:119–28, https://doi.org/10.1371/journal.pgen.0010017; Dillin, A., et al., 2002, "Rates of Behavior and Aging Specified by Mitochondrial Function during

Development," *Science* 298:2398–2401, https://doi.org/10.1126/science.1077780; Lee, S. S., et al., 2003, "A Systematic RNAi Screen Identifies a Critical Role for Mitochondria in *C. elegans* Longevity," *Nature Genetics* 33:40–48, https://doi.org/10.1038/ng1056.

44. Powers, R. W., 3rd, Kaeberlein, M., Caldwell, S. D., Kennedy, B. K., and Fields, S., 2006, "Extension of Chronological Life Span in Yeast by Decreased TOR Pathway Signaling," *Genes and Development* 20:174–84, https://doi.org/10.1101/gad.1381406.

45. Shi, C., et al., 2020, "Allocation of Gene Products to Daughter Cells Is Determined by the Age of the Mother in Single *Escherichia coli* Cells," *Proceedings of the Royal Society B: Biological Sciences* 287:20200569, https://doi.org/10.1098/rspb.2020.0569; Aguilaniu, H., Gustafsson, L., Rigoulet, M., and Nyström, T., 2003, "Asymmetric Inheritance of Oxidatively Damaged Proteins during Cytokinesis," *Science* 299:1751–53, https://doi.org/10.1126/science.1080418.

46. Yang, J., et al., 2015, "Systematic Analysis of Asymmetric Partitioning of Yeast Proteome between Mother and Daughter Cells Reveals "Aging Factors" and Mechanism of Lifespan Asymmetry," *Proceedings of the National Academy of Sciences of the USA* 112:11977–82, https://doi.org/10.1073/pnas.1506054112.

47. Henderson, K. A., Hughes, A. L., and Gottschling, D. E., 2014, "Mother-Daughter Asymmetry of pH Underlies Aging and Rejuvenation in Yeast," *eLife* 3:e03504, https://doi.org/10.7554/eLife.03504.

48. Valenzano, D. R., Sharp, S., and Brunet, A., 2011, "Transposon-Mediated Transgenesis in the Short-Lived African Killifish *Nothobranchius furzeri*, a Vertebrate Model for Aging," *G3: Genes|Genomes|Genetics* 1:531–38, https://doi.org/10.1534/g3.111.001271.

49. Valenzano, D. R., et al., 2015, "The African Turquoise Killifish Genome Provides Insights into Evolution and Genetic Architecture of Lifespan," *Cell* 163:1539–54, https://doi.org/10.1016/j.cell.2015.11.008; Harel, I., Valenzano, D. R., and Brunet, A., 2016, "Efficient Genome Engineering Approaches for the Short-Lived African Turquoise Killifish," *Nature Protocols* 11:2010–28, https://doi.org/10.1038/nprot.2016.103; Harel, I., et al., 2015, "A Platform for Rapid Exploration of Aging and Diseases in a Naturally Short-Lived Vertebrate," *Cell* 160:1013–26, https://doi.org/10.1016/j.cell.2015.01.038.

50. McKay, A., et al., 2021, "An Automated Feeding System for the African Killifish Reveals Effects of Dietary Restriction on Lifespan and Allows Scalable Assessment of Associative Learning," *bioRxiv* 2021.03.30:437790, https://doi.org/10.1101/2021.03.30.437790.

51. Harel, I., Valenzano, D. R., and Brunet 2016.

52. Reichard, M., Polačik, M., and Sedláček, O., 2009, "Distribution, Colour Polymorphism and Habitat Use of the African Killifish *Nothobranchius furzeri*, the Vertebrate with the Shortest Life Span," *Journal of Fish Biology* 74:198–212.

53. Smith, P., et al., 2017, "Regulation of Life Span by the Gut Microbiota in the Short-Lived African Turquoise Killifish," *eLife* 6:27014, https://doi.org/10.7554/eLife.27014.

Chapter 5

1. Rowe, J. W., Andres, R., Tobin, J. D., Norris, A. H., and Shock, N. W., 1976, "The Effect of Age on Creatinine Clearance in Men: A Cross-sectional and Longitudinal Study," *Journal of Gerontology* 31:155–63, https://doi.org/10.1093/geronj/31.2.155.

2. Bennett, D. A., Shannon, K. M., Beckett, L. A., and Wilson, R. S., 1999, "Dimensionality of Parkinsonian Signs in Aging and Alzheimer's Disease," *Journals of Gerontology, Series A,*

Biological Sciences and Medical Sciences 54:M191–96, https://doi.org/10.1093/gerona/54.4 .m191.

3. Ferrucci, L., et al., 2000, "Subsystems Contributing to the Decline in Ability to Walk: Bridging the Gap between Epidemiology and Geriatric Practice in the InCHIANTI Study," *Journal of the American Geriatrics Society* 48:1618–25, https://doi.org/10.1111/j.1532–5415.2000.tb03873.x.

4. Yashin, A. I., et al., 2010, "'Predicting' Parental Longevity from Offspring Endophenotypes: Data from the Long Life Family Study (LLFS)," *Mechanisms of Ageing and Development* 131:215–22, https://doi.org/10.1016/j.mad.2010.02.001.

5. Sebastiani, P., et al., 2009, "A Family Longevity Selection Score: Ranking Sibships by Their Longevity, Size, and Availability for Study," *American Journal of Epidemiology* 170:1555–62, https://doi.org/10.1093/aje/kwp309.

6. Sebastiani, P., et al., 2017, "Biomarker Signatures of Aging," *Aging Cell* 16:329–38, https://doi.org/10.1111/acel.12557.

7. Ismail, K., et al., 2016, "Compression of Morbidity Is Observed across Cohorts with Exceptional Longevity," *Journal of the American Geriatrics Society* 64:1583–91, https://doi.org/10.1111/jgs.14222.

8. Ko, S.-u., Hausdorff, J. M., and Ferrucci, L., 2010, "Age-Associated Differences in the Gait Pattern Changes of Older Adults during Fast-Speed and Fatigue Conditions: Results from the Baltimore Longitudinal Study of Ageing," *Age and Ageing* 39:688–94, https://doi.org/10.1093/ageing/afq113.

9. Guralnik, J. M., et al., 1994, "A Short Physical Performance Battery Assessing Lower Extremity Function: Association with Self-Reported Disability and Prediction of Mortality and Nursing Home Admission," *Journal of Gerontology* 49:M85–94, https://doi.org/10.1093/geronj/49.2.m85.

10. Owusu, C., Margevicius, S., Schluchter, M., Koroukian, S. M., and Berger, N. A., 2017, "Short Physical Performance Battery, Usual Gait Speed, Grip Strength and Vulnerable Elders Survey Each Predict Functional Decline among Older Women with Breast Cancer," *Journal of Geriatric Oncology* 8:356–62, https://doi.org/10.1016/j.jgo.2017.07.004.

11. Vasunilashorn, S., et al., 2009, "Use of the Short Physical Performance Battery Score to Predict Loss of Ability to Walk 400 Meters: Analysis from the InCHIANTI Study," *Journals of Gerontology, Series A, Biological Sciences and Medical Sciences* 64:223–29, https://doi.org/10.1093/gerona/gln022.

12. Sebastiani et al. 2017.

13. Re, D. E., Tskhay, K. O., Tong, M.-O., Wilson, J. P., Zhong, C.-B., and Rule, N. O., 2015, "Facing Fate: Estimates of Longevity from Facial Appearance and Their Underlying Cues," *Archives of Scientific Psychology* 3:30–36, https://doi.org/10.1037/arc0000015.

14. Ganel, T., and Goodale, M. A., 2021, "The Effect of Smiling on the Perceived Age of Male and Female Faces across the Lifespan," *Scientific Reports* 11:23020, https://doi.org/10.1038/s41598-021-02380-2.

15. Chen, W., et al., 2015, "Three-Dimensional Human Facial Morphologies as Robust Aging Markers," *Cell Research* 25:574–87, https://doi.org/10.1038/cr.2015.36.

16. Garigan, D., et al., 2002, "Genetic Analysis of Tissue Aging in *Caenorhabditis elegans*: A Role for Heat-Shock Factor and Bacterial Proliferation," *Genetics* 161:1101–12, https://doi.org/10.1093/genetics/161.3.1101; Herndon, L. A., et al., 2002, "Stochastic and Genetic Factors Influence Tissue-Specific Decline in Ageing *C. elegans*," *Nature* 419:808–14.

17. Tank, E. M., Rodgers, K. E., and Kenyon, C., 2011, "Spontaneous Age-Related Neurite Branching in *Caenorhabditis elegans*," *Journal of Neuroscience* 31:9279–88, https://doi.org/10.1523/JNEUROSCI.6606-10.2011; Toth, M. L., et al., 2012, "Neurite Sprouting and Synapse Deterioration in the Aging *Caenorhabditis elegans* Nervous System," *Journal of Neuroscience* 32:8778–90, https://doi.org/10.1523/JNEUROSCI.1494-11.2012.

18. Zhang, W. B., et al., 2016, "Extended Twilight among Isogenic *C. elegans* Causes a Disproportionate Scaling between Lifespan and Health," *Cell Systems* 3:333–45.e334, https://doi.org/10.1016/j.cels.2016.09.003.

19. Gerstbrein, B., Stamatas, G., Kollias, N., and Driscoll, M., 2005, "In Vivo Spectrofluorimetry Reveals Endogenous Biomarkers that Report Healthspan and Dietary Restriction in *Caenorhabditis elegans*," *Aging Cell* 4:127–37, https://doi.org/10.1111/j.1474–9726.2005.00153.x; Pincus, Z., Mazer, T. C., and Slack, F. J., 2016, "Autofluorescence as a Measure of Senescence in *C. elegans*: Look to Red, Not Blue or Green," *Aging (Albany NY)* 8:889–98, https://doi.org/10.18632/aging.100936.

20. Rea, S. L., Wu, D., Cypser, J. R., Vaupel, J. W., and Johnson, T. E., 2005, "A Stress-Sensitive Reporter Predicts Longevity in Isogenic Populations of *Caenorhabditis elegans*," *Nature Genetics* 37:894–98.

21. Zhao, Y., et al., 2017, "Two Forms of Death in Ageing *Caenorhabditis elegans*," *Nature Communications* 8:15458, https://doi.org/10.1038/ncomms15458.

22. Garigan et al. 2002.

23. Stroustrup, N., Ulmschneider, B. E., Nash, Z. M., López-Moyado, I. F., Apfeld, J., and Fontana, W., 2013, "The *Caenorhabditis elegans* Lifespan Machine," *Nature Methods* 10:665–70, https://doi.org/10.1038/nmeth.22475.

24. Shi, C., and Murphy, C. T., 2013, "Mating Induces Shrinking and Death in Caenorhabditis Mothers," *Science* 343:536–40, https://doi.org/10.1126/science.1242958; Rahimi, M., Sohrabi, S., and Murphy, C. T., 2022, "Novel Elasticity Measurements Reveal *C. elegans* Cuticle Stiffens with Age and in a Long-Lived Mutant," *Biophysical Journal* 121:515–24, https://doi.org/10.1016/j.bpj.2022.01.013.

25. Stroustrup et al. 2013.

26. Huang, C., Xiong, C., and Kornfeld, K., 2004, "Measurements of Age-Related Changes of Physiological Processes that Predict Lifespan of *Caenorhabditis elegans*," *Proceedings of the National Academy of Sciences of the USA* 101:8084–89.

27. Bansal, A., Zhu, L., Yen, K., and Tissenbaum, H. A., 2015, "Uncoupling Lifespan and Healthspan in *Caenorhabditis elegans* Longevity Mutants," *Proceedings of the National Academy of Sciences of the USA* 112:E277–86, https://doi.org/10.1073/pnas.1412192112.

28. Kauffman, A. L., Ashraf, J. M., Corces-Zimmerman, M. R., Landis, J. N., and Murphy, C. T., 2010, "Insulin Signaling and Dietary Restriction Differentially Influence the Decline of Learning and Memory with Age," *PLOS Biology* 8:e1000372, https://doi.org/10.1371/journal.pbio.1000372.

29. Luo, S., Shaw, W. M., Ashraf, J., and Murphy, C. T., 2009, "TGF-Beta Sma/Mab Signaling Mutations Uncouple Reproductive Aging from Somatic Aging," *PLOS Genetics* 5:e1000789, https://doi.org/10.1371/journal.pgen.1000789.

30. Hahm, J. H., et al., 2015, "*C. elegans* Maximum Velocity Correlates with Healthspan and Is Maintained in Worms with an Insulin Receptor Mutation," *Nature Communications* 6:8919, https://doi.org/10.1038/ncomms9919.

31. Hahm et al. 2015.

32. Yang, J., et al., 2019, "Association between Push-Up Exercise Capacity and Future Cardiovascular Events among Active Adult Men," *JAMA Network Open* 2:e188341, https://doi.org/10.1001/jamanetworkopen.2018.8341.

33. Ryan, D. A., et al., 2014, "Sex, Age, and Hunger Regulate Behavioral Prioritization through Dynamic Modulation of Chemoreceptor Expression," *Current Biology* 24:2509–17, https://doi.org/10.1016/j.cub.2014.09.032.

34. Hsu, A. L., Feng, Z., Hsieh, M. Y., and Xu, X. Z., 2009, "Identification by Machine Vision of the Rate of Motor Activity Decline as a Lifespan Predictor in *C. elegans*," *Neurobiology of Aging* 30:1498–1503, https://doi.org/10.1016/j.neurobiolaging.2007.12.007; Wu, C. Y., et al., 2018, "Enhancing GABAergic Transmission Improves Locomotion in a *Caenorhabditis elegans* Model of Spinal Muscular Atrophy," *eNeuro* 5:0289-18.2018, https://doi.org/10.1523/ENEURO.0289-18.2018.

35. Li, L. B., et al., 2016, "The Neuronal Kinesin UNC-104/KIF1A Is a Key Regulator of Synaptic Aging and Insulin Signaling-Regulated Memory," *Current Biology* 26:605–15, https://doi.org/10.1016/j.cub.2015.12.068.

36. Tank, Rodgers, and Kenyon 2011; Toth et al. 2012.

37. Kauffman et al. 2010.

38. Lakhina, V., et al., 2019, "ZIP-5/bZIP Transcription Factor Regulation of Folate Metabolism Is Critical for Aging Axon Regeneration," *bioRxiv*, https://doi.org/10.1101/727719; Byrne, A. B., et al., 2014, "Insulin/IGF1 Signaling Inhibits Age-Dependent Axon Regeneration," *Neuron* 81:561–73, https://doi.org/10.1016/j.neuron.2013.11.019.

39. Kaletsky, R., et al., 2016, "The *C. elegans* Adult Neuronal IIS/FOXO Transcriptome Reveals Adult Phenotype Regulators," *Nature* 529:92–96, https://doi.org/10.1038/nature16483.

40. Jacobson, J., et al., 2010, "Biomarkers of Aging in *Drosophila*," *Aging Cell* 9:466–77, https://doi.org/10.1111/j.1474-9726.2010.00573.x.

41. Mair, W., Goymer, P., Pletcher, S. D., and Partridge, L., 2003, "Demography of Dietary Restriction and Death in *Drosophila*," *Science* 301:1731–33.

42. Rera, M., Clark, R. I., and Walker, D. W., 2012, "Intestinal Barrier Dysfunction Links Metabolic and Inflammatory Markers of Aging to Death in *Drosophila*," *Proceedings of the National Academy of Sciences of the USA* 109:21528–33, https://doi.org/10.1073/pnas.1215849110.

43. Bier, E., and Bodmer, R., 2004, "*Drosophila*, an Emerging Model for Cardiac Disease," *Gene* 342:1–11, https://doi.org/10.1016/j.gene.2004.07.018.

44. Chaudhuri, J., et al., 2018, "The Role of Advanced Glycation End Products in Aging and Metabolic Diseases: Bridging Association and Causality," *Cell Metabolism* 28:337–52, https://doi.org/10.1016/j.cmet.2018.08.014.

45. Berman, G. J., Choi, D. M., Bialek, W., and Shaevitz, J. W., 2014, "Mapping the Stereotyped Behaviour of Freely Moving Fruit Flies," *Journal of the Royal Society Interface* 11:0672, https://doi.org/10.1098/rsif.2014.0672.

46. Palliyaguru, D. L., et al., 2021, "Study of Longitudinal Aging in Mice: Presentation of Experimental Techniques," *Journals of Gerontology, Series A, Biological Sciences and Medical Sciences* 76:552–60, https://doi.org/10.1093/gerona/glaa285.

47. Tiku, V., et al., 2017, "Small Nucleoli Are a Cellular Hallmark of Longevity," *Nature Communications* 8:16083, https://doi.org/10.1038/ncomms16083.

Chapter 6

1. Klass, M. R., 1983, "A Method for the Isolation of Longevity Mutants in the Nematode *Caenorhabditis elegans* and Initial Results," *Mechanisms of Ageing and Development* 22:279–86, https://doi.org/10.1016/0047-6374(83)90082-9; Klass, M. R., 1977, "Aging in the Nematode *Caenorhabditis elegans*: Major Biological and Environmental Factors Influencing Life Span," *Mechanisms of Ageing and Development* 6:413–29, https://doi.org/10.1016/0047-6374(77)90043-4.

2. Friedman, D. B., and Johnson, T. E., 1988, "Three Mutants that Extend both Mean and Maximum Life Span of the Nematode, *Caenorhabditis elegans*, Define the *age-1* Gene," *Journal of Gerontology* 43:B102–9, https://doi.org/10.1093/geronj/43.4.b102; Johnson, T. E., 1990, "Increased Life-Span of *age-1* Mutants in *Caenorhabditis elegans* and Lower Gompertz Rate of Aging," *Science* 249:908–12, https://doi.org/10.1126/science.2392681.

3. Gottlieb, S., and Ruvkun, G., 1994, "*daf-2, daf-16* and *daf-23*: Genetically Interacting Genes Controlling Dauer Formation in *Caenorhabditis elegans*," *Genetics* 137:107–20; Larsen, P. L., Albert, P. S., and Riddle, D. L., 1995, "Genes that Regulate Both Development and Longevity in *Caenorhabditis elegans*," *Genetics* 139:1567–83; Malone, E. A., Inoue, T., and Thomas, J. H., 1996, "Genetic Analysis of the Roles of *daf-28* and *age-1* in Regulating *Caenorhabditis elegans* Dauer Formation," *Genetics* 143:1193–205.

4. Kimura, K. D., Tissenbaum, H. A., Liu, Y., and Ruvkun, G., 1997, "*daf-2*, an Insulin Receptor-like Gene that Regulates Longevity and Diapause in *Caenorhabditis elegans*," *Science* 277:942–46.

5. Wolkow, C. A., Muñoz, M. J., Riddle, D. L., and Ruvkun, G., 2002, "Insulin Receptor Substrate and p55 Orthologous Adaptor Proteins Function in the *Caenorhabditis elegans* *daf-2*/Insulin-like Signaling Pathway," *Journal of Biological Chemistry* 277:49591–97, https://doi.org/10.1074/jbc.M207866200; Ogg, S., and Ruvkun, G., 1998, "The *C. elegans* PTEN Homolog, DAF-18, Acts in the Insulin Receptor-like Metabolic Signaling Pathway," *Molecular Cell* 2:887–93; Paradis, S., Ailion, M., Toker, A., Thomas, J. H., and Ruvkun, G., 1999, "A PDK1 Homolog Is Necessary and Sufficient to Transduce AGE-1 PI3 Kinase Signals that Regulate Diapause in *Caenorhabditis elegans*," *Genes and Development* 13:1438–52; Paradis, S., and Ruvkun, G., 1998, "*Caenorhabditis elegans* Akt/PKB Transduces Insulin Receptor-like Signals from AGE-1 PI3 Kinase to the DAF-16 Transcription Factor," *Genes and Development* 12:2488–98.

6. Ogg, S., et al., 1997, "The Fork Head Transcription Factor DAF-16 Transduces Insulin-like Metabolic and Longevity Signals in *C. elegans*," *Nature* 389:994–99; Lin, K., Dorman, J. B., Rodan, A., and Kenyon, C., 1997, "*daf-16*: An HNF-3/Forkhead Family Member that Can Function to Double the Life-Span of *Caenorhabditis elegans*," *Science* 278:1319–22.

7. Lee, R. Y., Hench, J., and Ruvkun, G., 2001, "Regulation of *C. elegans* DAF-16 and Its Human Ortholog FKHRL1 by the *daf-2* Insulin-like Signaling Pathway," *Current Biology* 11:1950–57.

8. Pierce, S. B., et al., 2001, "Regulation of DAF-2 Receptor Signaling by Human Insulin and *ins-1*, a Member of the Unusually Large and Diverse *C. elegans* Insulin Gene Family," *Genes and Development* 15:672–86.

9. Wang, Y., et al., 2006, "*C. elegans* 14-3-3 Proteins Regulate Life Span and Interact with SIR-2.1 and DAF-16/FOXO," *Mechanisms of Ageing and Development* 127:741–47.

10. Dillin, A., Crawford, D. K., and Kenyon, C., 2002, "Timing Requirements for Insulin/IGF-1 Signaling in *C. elegans*," *Science* 298:830–34.

11. Venz, R., Pekec, T., Katic, I., Ciosk, R., and Ewald, C. Y., 2021, "End-of-Life Targeted Auxin-Mediated Degradation of DAF-2 Insulin/IGF-1 Receptor Promotes Longevity Free from Growth-Related Pathologies," *eLife* 10:e71335, https://doi.org/10.7554/eLife.71335.

12. Clancy, D. J., et al., 2001, "Extension of Life-Span by Loss of CHICO, a *Drosophila* Insulin Receptor Substrate Protein," *Science* 292:104–6, https://doi.org/10.1126/science .1057991; Tatar, M., et al., 2001, "A Mutant *Drosophila* Insulin Receptor Homolog that Extends Life-Span and Impairs Neuroendocrine Function," *Science* 292:107–10, https://doi.org/10.1126 /science.1057987.

13. Lamb, M. J., 1968, "Temperature and Lifespan in *Drosophila*," *Nature* 220:808–9, https://doi.org/10.1038/220808a0; Aigaki, T., and Ohba, S., 1984, "Effect of Mating Status on *Drosophila virilis* Lifespan," *Experimental Gerontology* 19:267–78, https://doi.org/10.1016/0531 -5565(84)90022-6; Chapman, T., Hutchings, J., and Partridge, L., 1993, "No Reduction in the Cost of Mating for *Drosophila melanogaster* Females Mating with Spermless Males," *Proceedings of the Royal Society B: Biological Sciences* 253:211–17, https://doi.org/10.1098/rspb.1993.0105; Chapman, T., Herndon, L. A., Heifetz, Y., Partridge, L., and Wolfner, M. F., 2001, "The Acp26Aa Seminal Fluid Protein Is a Modulator of Early Egg Hatchability in *Drosophila melanogaster*," *Proceedings of the Royal Society B: Biological Sciences* 268:1647–54, https://doi.org/10.1098 /rspb.2001.1684; Rose, M. R., 1989, "Genetics of Increased Lifespan in *Drosophila*," *Bioessays* 11:132–35, https://doi.org/10.1002/bies.950110505.

14. Clancy et al. 2001.

15. Tatar et al. 2001.

16. Hwangbo, D. S., Gersham, B., Tu, M. P., Palmer, M., and Tatar, M., 2004, "*Drosophila* dFOXO Controls Lifespan and Regulates Insulin Signalling in Brain and Fat Body," *Nature* 429:562–66.

17. Libina, N., Berman, J. R., and Kenyon, C., 2003, "Tissue-Specific Activities of *C. elegans* DAF-16 in the Regulation of Lifespan," *Cell* 115:489–502.

18. Broughton, S. J., et al., 2010, "DILP-Producing Median Neurosecretory Cells in the *Drosophila* Brain Mediate the Response of Lifespan to Nutrition," *Aging Cell* 9:336–46, https:// doi.org/10.1111/j.1474-9726.2010.00558.x.

19. Holzenberger, M., et al., 2003, "IGF-1 Receptor Regulates Lifespan and Resistance to Oxidative Stress in Mice," *Nature* 421:182–87; Hsieh, C. C., DeFord, J. H., Flurkey, K., Harrison, D. E., and Papaconstantinou, J., 2002, "Effects of the Pit1 Mutation on the Insulin Signaling Pathway: Implications on the Longevity of the Long-Lived Snell Dwarf Mouse," *Mechanisms of Ageing and Development* 123:1245–55; Hsieh, C. C., DeFord, J. H., Flurkey, K., Harrison, D. E., and Papaconstantinou, J., 2002, "Implications for the Insulin Signaling Pathway in Snell Dwarf Mouse Longevity: A Similarity with the *C. elegans* Longevity Paradigm," *Mechanisms of Ageing and Development* 123:1229–44.

20. Holzenberger et al. 2003.

21. Bluher, M., Kahn, B. B., and Kahn, C. R., 2003, "Extended Longevity in Mice Lacking the Insulin Receptor in Adipose Tissue," *Science* 299:572–74.

22. Burks, D. J., et al., 2000, "IRS-2 Pathways Integrate Female Reproduction and Energy Homeostasis," *Nature* 407:377–82, https://doi.org/10.1038/35030105.

23. Suh, Y., et al., 2008, "Functionally Significant Insulin-Like Growth Factor I Receptor Mutations in Centenarians," *Proceedings of the National Academy of Sciences of the USA* 105:3438–42.

24. Hedges, C. P., Shetty, B., Broome, S. C., MacRae, C., Koutsifeli, P., Buckels, E. J., et al., 2023, "Dietary Supplementation of Clinically Utilized PI3K p110α Inhibitor Extends the Lifespan of Male and Female Mice," *Nature Aging*, https://doi.org/10.1038/s43587-022-00349-y.

25. Murphy, C. T., et al., 2003, "Genes that Act Downstream of DAF-16 to Influence the Lifespan of *Caenorhabditis elegans*," *Nature* 424:277–83.

26. Tepper, R. G., Ashraf, J., Kaletsky, R., Kleemann, G., Murphy, C. T., and Bussemaker, H. J., 2013, "PQM-1 Complements DAF-16 as a Key Transcriptional Regulator of DAF-2-Mediated Development and Longevity," *Cell* 154:676–90, https://doi.org/10.1016/j.cell.2013.07.006.

27. Furuyama, T., Nakazawa, T., Nakano, I., and Mori, N., 2000, "Identification of the Differential Distribution Patterns of mRNAs and Consensus Binding Sequences for Mouse DAF-16 Homologues," *Biochemical Journal* 349:629–34.

28. Lee, S. S., Kennedy, S., Tolonen, A. C., and Ruvkun, G., 2003, "DAF-16 Target Genes that Control *C. elegans* Life-Span and Metabolism," *Science* 300:644–47.

29. McElwee, J. J., Schuster, E., Blanc, E., Thomas, J. H., and Gems, D., 2004, "Shared Transcriptional Signature in *Caenorhabditis elegans* Dauer Larvae and Long-Lived *daf-2* Mutants Implicates Detoxification System in Longevity Assurance," *Journal of Biological Chemistry* 279:44533–43.

30. Tepper et al. 2013. The *pqm-1* gene had emerged once from a paraquat sensitivity screen in the mid-1990s, but had not been studied after that.

31. Hsu, A. L., Murphy, C. T., and Kenyon, C., 2003, "Regulation of Aging and Age-Related Disease by DAF-16 and Heat-Shock Factor," *Science* 300:1142–45; Tullet, J. M., et al., 2008, "Direct Inhibition of the Longevity-Promoting Factor SKN-1 by Insulin-like Signaling in *C. elegans*," *Cell* 132:1025–38, https://doi.org/10.1016/j.cell.2008.01.030; Lapierre, L. R., et al., 2013, "The TFEB Orthologue HLH-30 Regulates Autophagy and Modulates Longevity in *Caenorhabditis elegans*," *Nature Communications* 4:2267, https://doi.org/10.1038/ncomms3267; Riedel, C. G., et al., 2013, "DAF-16 Employs the Chromatin Remodeller SWI/SNF to Promote Stress Resistance and Longevity," *Nature Cell Biology* 15:491–501, https://doi.org/10.1038/ncb2720.

32. Soukas, A. A., Kane, E. A., Carr, C. E., Melo, J. A., and Ruvkun, G., 2009, "Rictor/ TORC2 Regulates Fat Metabolism, Feeding, Growth, and Life Span in *Caenorhabditis elegans*," *Genes and Development* 23:496–511, https://doi.org/10.1101/gad.1775409; Oh, S. W., et al., 2005, "JNK Regulates Lifespan in *Caenorhabditis elegans* by Modulating Nuclear Translocation of Forkhead Transcription Factor/DAF-16," *Proceedings of the National Academy of Sciences of the USA* 102:4494–99.

33. Berdichevsky, A., Viswanathan, M., Horvitz, H. R., and Guarente, L., 2006, "*C. elegans* SIR-2.1 Interacts with 14-3-3 Proteins to Activate DAF-16 and Extend Life Span," *Cell* 125:1165–77.

34. Tepper et al. 2013.

35. Perez, M. F., Francesconi, M., Hidalgo-Carcedo, C., and Lehner, B., 2017, "Maternal Age Generates Phenotypic Variation in *Caenorhabditis elegans*," *Nature* 552:106–9, https://doi.org/10.1038/nature25012.

Notes to text box: "Why Isn't Decreased *daf-2* Activity Deleterious, like Diabetes?" (Chapter 6, page 94)

a. Honjoh, S., Yamamoto, T., Uno, M., and Nishida, E., 2009, "Signalling through RHEB-1 Mediates Intermittent Fasting-Induced Longevity in *C. elegans*," *Nature* 457:726–30, https://doi .org/10.1038/nature07583.

b. Lee, S. J., Murphy, C. T., and Kenyon, C., 2009, "Glucose Shortens the Life Span of *C. elegans* by Downregulating DAF-16/FOXO Activity and Aquaporin Gene Expression," *Cell Metabolism* 10:379–91, https://doi.org/10.1016/j.cmet.2009.10.003.

Chapter 7

1. Most, J., Tosti, V., Redman, L. M., and Fontana, L., 2017, "Calorie Restriction in Humans: An Update," *Ageing Research Reviews* 39:36–45, https://doi.org/10.1016/j.arr.2016.08.005.

2. Hindhede, M., 1920, "The Effect of Food Restriction during War on Mortality in Copenhagen," *JAMA* 74:381–82; Strom, A., and Jensen, R. A., 1951, "Mortality from Circulatory Diseases in Norway 1940–1945," *Lancet* 1:126–29, https://doi.org/10.1016/s0140 -6736(51)91210-x.

3. Most et al. 2017.

4. Osborne, T. B., Mendel, L. B., and Ferry, E. L., 1917, "The Effect of Retardation of Growth upon the Breeding Period and Duration of Life of Rats," *Science* 45:294–95, https://doi.org/10 .1126/science.45.1160.294.

5. Hansel, W., 2016, "Clive McCay: A Man Before His Time," *Endocrinology and Metabolic Syndrome* 5:3, https://doi.org/10.4172/2161-1017.1000236.

6. Kopec, S., 1928, "On the Influence of Intermittent Starvation on the Longevity of the Imaginal Stage of *Drosophila melanogaster*," *Journal of Experimental Biology* 5:204–11.

7. Greer, E. L., and Brunet, A., 2009, "Different Dietary Restriction Regimens Extend Lifespan by Both Independent and Overlapping Genetic Pathways in *C. elegans*," *Aging Cell* 8:113– 27, https://doi.org/10.1111/j.1474-9726.2009.00459.x.

8. Bonkowski, M. S., et al., 2009, "Disruption of Growth Hormone Receptor Prevents Calorie Restriction from Improving Insulin Action and Longevity," *PLOS One* 4:e4567, https://doi.org/10.1371/journal.pone.0004567; Bonkowski, M. S., Rocha, J. S., Masternak, M. M., Al Regaiey, K. A., and Bartke, A., 2006, "Targeted Disruption of Growth Hormone Receptor Interferes with the Beneficial Actions of Calorie Restriction," *Proceedings of the National Academy of Sciences of the USA* 103:7901–5, https://doi.org/10.1073/pnas.0600161103.

9. Mattison, J. A., et al., 2017, "Caloric Restriction Improves Health and Survival of Rhesus Monkeys," *Nature Communications* 8:14063, https://doi.org/10.1038/ncomms14063.

10. Mattison et al. 2017.

11. Mair, W., Goymer, P., Pletcher, S. D., and Partridge, L., 2003, "Demography of Dietary Restriction and Death in *Drosophila*," *Science* 301:1731–33.

12. Padamsey, Z., Katsanevaki, D., Dupuy, N., and Rochefort, N. L, 2022,." Neocortex Saves Energy by Reducing Coding Precision during Food Scarcity," *Neuron* 110:280–96.e210, https://doi.org/10.1016/j.neuron.2021.10.024.

13. Kauffman, A. L., Ashraf, J. M., Corces-Zimmerman, M. R., Landis, J. N., and Murphy, C. T., 2010, "Insulin Signaling and Dietary Restriction Differentially Influence the Decline of

Learning and Memory with Age," *PLOS Biology* 8:e1000372, https://doi.org/10.1371/journal.pbio.1000372.

14. *Drosophila*: Plaçais, P. Y., and Preat, T., 2013, "To Favor Survival under Food Shortage, the Brain Disables Costly Memory," *Science* 339:440–42, https://doi.org/10.1126/science.1226018. Blowflies: Longden, K. D., Muzzu, T., Cook, D. J., Schultz, S. R., and Krapp, H. G., 2014, "Nutritional State Modulates the Neural Processing of Visual Motion," *Current Biology* 24:890–95, https://doi.org/10.1016/j.cub.2014.03.005.

15. Skorupa, D. A., Dervisefendic, A., Zwiener, J., and Pletcher, S. D., 2008, "Dietary Composition Specifies Consumption, Obesity, and Lifespan in *Drosophila melanogaster*," *Aging Cell* 7:478–90, https://doi.org/10.1111/j.1474-9726.2008.00400.x.

16. Babygirija, R., and Lamming, D. W., 2021, "The Regulation of Healthspan and Lifespan by Dietary Amino Acids," *Translational Medicine of Aging* 5:17–30, https://doi.org/10.1016/j.tma.2021.05.001.

17. Lopez-Torres, M., and Barja, G., 2008, "Lowered Methionine Ingestion as Responsible for the Decrease in Rodent Mitochondrial Oxidative Stress in Protein and Dietary Restriction: Possible Implications for Humans," *Biochimica et Biophysica Acta* 1780:1337–47, https://doi.org/10.1016/j.bbagen.2008.01.007; Pamplona, R., and Barja, G., 2006, "Mitochondrial Oxidative Stress, Aging and Caloric Restriction: The Protein and Methionine Connection," *Biochimica et Biophysica Acta* 1757:496–508, https://doi.org/10.1016/j.bbabio.2006.01.009.

18. Grandison, R. C., Piper, M. D., and Partridge, L., 2009, "Amino-Acid Imbalance Explains Extension of Lifespan by Dietary Restriction in *Drosophila*," *Nature* 462:1061–64, https://doi.org/10.1038/nature08619; Zhou, G., et al., 2020, "Methionine Increases Yolk Production to Offset the Negative Effect of Caloric Restriction on Reproduction without Affecting Longevity in *C. elegans*," *Aging (Albany NY)* 12:2680–97, https://doi.org/10.18632/aging.102770.

19. Richardson, N. E., et al., 2021, "Lifelong Restriction of Dietary Branched-Chain Amino Acids has Sex-Specific Benefits for Frailty and Lifespan in Mice," *Nature Aging* 1:73–86, https://doi.org/10.1038/s43587-020-00006-2.

20. Lagiou, P., et al., 2007, "Low Carbohydrate-High Protein Diet and Mortality in a Cohort of Swedish Women," *Journal of Internal Medicine* 261:366–74, https://doi.org/10.1111/j.1365-2796.2007.01774.x.

21. MacArthur, M. R., et al., 2021, "Total Protein, Not Amino Acid Composition, Differs in Plant-Based versus Omnivorous Dietary Patterns and Determines Metabolic Health Effects in Mice," *Cell Metabolism* 33:1808–19.E2, https://doi.org/10.1016/j.cmet.2021.06.011.

22. McElwee, J. J., et al., 2007, "Evolutionary Conservation of Regulated Longevity Assurance Mechanisms," *Genome Biology* 8:R132, https://doi.org/10.1186/gb-2007-8-7-r132.

23. De Cabo, R., and Mattson, M. P., 2019, "Effects of Intermittent Fasting on Health, Aging, and Disease," *New England Journal of Medicine* 381:2541–51, https://doi.org/10.1056/NEJMra1905136; Heilbronn, L. K., Smith, S. R., Martin, C. K., Anton, S. D., and Ravussin, E., 2005, "Alternate-Day Fasting in Nonobese Subjects: Effects on Body Weight, Body Composition, and Energy Metabolism," *American Journal of Clinical Nutrition* 81:69–73, https://doi.org/10.1093/ajcn/81.1.69; Patterson, R. E., and Sears, D. D., 2017 "Metabolic Effects of Intermittent Fasting," *Annual Review of Nutrition* 37:371–93, https://doi.org/10.1146/annurev-nutr-071816-064634; Sutton, E. F., et al., 2018, "Early Time-Restricted Feeding Improves Insulin Sensitivity, Blood Pressure, and Oxidative Stress Even without Weight Loss

in Men with Prediabetes," *Cell Metabolism* 27:1212–21.e1213, https://doi.org/10.1016/j
.cmet.2018.04.010.

24. Longo, V. D., and Panda, S., 2016, "Fasting, Circadian Rhythms, and Time-Restricted
Feeding in Healthy Lifespan," *Cell Metabolism* 23:1048–59, https://doi.org/10.1016/j.cmet
.2016.06.001; Mattson, M. P., et al., 2014, "Meal Frequency and Timing in Health and Disease,"
Proceedings of the National Academy of Sciences of the USA 111:16647–53, https://doi.org/10
.1073/pnas.1413965111. Satchin Panda has a whole book about this topic, but we can summarize
by saying that the coordination of caloric intake, metabolism, light exposure, and sleep is
important.

25. Leung, G.K.W., Huggins, C. E., Ware, R. S., and Bonham, M. P., 2020, "Time of Day
Difference in Postprandial Glucose and Insulin Responses: Systematic Review and Meta-
analysis of Acute Postprandial Studies," *Chronobiology International* 37:311–26, https://doi.org
/10.1080/07420528.2019.1683856.

26. Brandhorst, S., et al., 2015, "A Periodic Diet that Mimics Fasting Promotes Multi-system
Regeneration, Enhanced Cognitive Performance, and Healthspan," *Cell Metabolism* 22:86–99,
https://doi.org/10.1016/j.cmet.2015.05.012; Fanti, M., Mishra, A., Longo, V. D., and Brandhorst,
S., 2021, "Time-Restricted Eating, Intermittent Fasting, and Fasting-Mimicking Diets in Weight
Loss," *Current Obesity Reports* 10:70–80, https://doi.org/10.1007/s13679-021-00424-2.

27. Ulgherait, M., et al., 2021, "Circadian Autophagy Drives iTRF-Mediated Longevity,"
Nature 598:353–58, https://doi.org/10.1038/s41586-021-03934-0.

28. Lowe, D. A., et al., 2020, "Effects of Time-Restricted Eating on Weight Loss and Other
Metabolic Parameters in Women and Men with Overweight and Obesity: The TREAT Randomized Clinical Trial," *JAMA Internal Medicine* 180:1491–99, https://doi.org/10.1001
/jamainternmed.2020.4153.

29. Most et al. 2017.

30. Holloszy, J. O., and Fontana, L., 2007, "Caloric Restriction in Humans," *Experimental
Gerontology* 42:709–12, https://doi.org/10.1016/j.exger.2007.03.009.

31. Most et al. 2017; Kalm, L. M., and Semba, R. D., 2005, "They Starved so that Others Be
Better Fed: Remembering Ancel Keys and the Minnesota Experiment," *Journal of Nutrition*
135:1347–52, https://doi.org/10.1093/jn/135.6.1347.

32. Velthuis-te Wierik, E. J., van den Berg, H., Schaafsma, G., Hendriks, H. F., and Brouwer,
A., 1994, "Energy Restriction, a Useful Intervention to Retard Human Ageing? Results of a
Feasibility Study," *European Journal of Clinical Nutrition* 48:138–48.

33. Hall, K. D., and Kahan, S., 2018, "Maintenance of Lost Weight and Long-Term Manage-
ment of Obesity," *Medical Clinics of North America* 102:183–97, https://doi.org/10.1016/j.mcna
.2017.08.012.

34. Walford, R. L., Mock, D., Verdery, R., and MacCallum, T., 2002, "Calorie Restriction in
Biosphere 2: Alterations in Physiologic, Hematologic, Hormonal, and Biochemical Parameters
in Humans Restricted for a 2-Year Period," *Journals of Gerontology, Series A, Biological Sciences
and Medical Sciences* 57:B211–24, https://doi.org/10.1093/gerona/57.6.b211.

35. Lassinger, B. K., Kwak, C., Walford, R. L., and Jankovic, J., 2004, "Atypical Parkinsonism
and Motor Neuron Syndrome in a Biosphere 2 Participant: A Possible Complication of Chronic
Hypoxia and Carbon Monoxide Toxicity?," *Movement Disorders* 19:465–69, https://doi.org/10
.1002/mds.20076.

36. Hamadeh, M. J., Rodriguez, M. C., Kaczor, J. J., and Tarnopolsky, M. A., 2005, "Caloric Restriction Transiently Improves Motor Performance but Hastens Clinical Onset of Disease in the Cu/Zn-Superoxide Dismutase Mutant G93A Mouse," *Muscle and Nerve* 31:214–20, https://doi.org/10.1002/mus.20255.

37. Klass, M. R., 1983, "A Method for the Isolation of Longevity Mutants in the Nematode *Caenorhabditis elegans* and Initial Results," *Mechanisms of Ageing and Development* 22:279–86, https://doi.org/10.1016/0047-6374(83)90082-9.

38. Lakowski, B., and Hekimi, S., 1996, "Determination of Life-Span in *Caenorhabditis elegans* by Four Clock Genes," *Science* 272:1010–13.

39. Kauffman et al. 2010.

40. Lakowski and Hekimi 1996.

41. Houthoofd, K., Braeckman, B. P., Johnson, T. E., and Vanfleteren, J. R., 2003, "Life Extension via Dietary Restriction Is Independent of the Ins/IGF-1 Signalling Pathway in *Caenorhabditis elegans*," *Experimental Gerontology* 38:947–54, https://doi.org/10.1016/s0531-5565(03)00161-x.

42. Howitz, K. T., et al., 2003, "Small Molecule Activators of Sirtuins Extend *Saccharomyces cerevisiae* Lifespan," *Nature* 425:191–96, https://doi.org/10.1038/nature01960.

43. Kaeberlein, M., McVey, M., and Guarente, L., 1999, "The SIR2/3/4 Complex and SIR2 Alone Promote Longevity in *Saccharomyces cerevisiae* by Two Different Mechanisms," *Genes and Development* 13:2570–80, https://doi.org/10.1101/gad.13.19.2570.

44. Howitz et al. 2003.

45. Rine, J., and Herskowitz, I., 1987, "Four Genes Responsible for a Position Effect on Expression from HML and HMR in *Saccharomyces cerevisiae*," *Genetics* 116:9–22, https://doi.org/10.1093/genetics/116.1.9.

46. Hirschey, M. D., Shimazu, T., Huang, J. Y., and Verdin, E., 2009, "Acetylation of Mitochondrial Proteins," *Methods in Enzymology* 457:137–47, https://doi.org/10.1016/s0076-6879(09)05008-3.

47. Berdichevsky, A., Viswanathan, M., Horvitz, H. R., and Guarente, L., 2006, "*C. elegans* SIR-2.1 Interacts with 14-3-3 Proteins to Activate DAF-16 and Extend Life Span," *Cell* 125:1165–77.

48. Sinclair, D. A., and Guarente, L., 1997, "Extrachromosomal rDNA Circles: A Cause of Aging in Yeast," *Cell* 91:1033–42, https://doi.org/10.1016/s0092-8674(00)80493-6.

49. Aguilaniu, H., Gustafsson, L., Rigoulet, M., and Nyström, T., 2003, "Asymmetric Inheritance of Oxidatively Damaged Proteins during Cytokinesis," *Science* 299:1751–53, https://doi.org/10.1126/science.1080418.

50. Goudeau, J., and Aguilaniu, H., 2010, "Carbonylated Proteins Are Eliminated during Reproduction in *C. elegans*," *Aging Cell* 9:991–1003, https://doi.org/10.1111/j.1474-9726.2010.00625.x.

51. Longo, V. D., and Fabrizio, P., 2012, "Chronological Aging in *Saccharomyces cerevisiae*," in *Aging Research in Yeast*, edited by M. Breitenbach, S. Jazwinski, and P. Laun, Subcellular Biochemistry 57:101–21 (Dordrecht: Springer), https://doi.org/10.1007/978-94-007-2561-4_5.

52. Powers, R. W., 3rd, Kaeberlein, M., Caldwell, S. D., Kennedy, B. K., and Fields, S., 2006, "Extension of Chronological Life Span in Yeast by Decreased TOR Pathway Signaling," *Genes and Development* 20:174–84, https://doi.org/10.1101/gad.1381406.

53. McCormick, M. A., et al., 2015, "A Comprehensive Analysis of Replicative Lifespan in 4,698 Single-Gene Deletion Strains Uncovers Conserved Mechanisms of Aging," *Cell Metabolism* 22:895–906, https://doi.org/10.1016/j.cmet.2015.09.008.

54. Dillin, A., et al., 2002, "Rates of Behavior and Aging Specified by Mitochondrial Function during Development," *Science* 298:2398–2401, https://doi.org/10.1126/science.1077780; Hansen, M., Hsu, A. L., Dillin, A., and Kenyon, C., 2005, "New Genes Tied to Endocrine, Metabolic, and Dietary Regulation of Lifespan from a *Caenorhabditis elegans* Genomic RNAi Screen," *PLOS Genetics* 1:119–28, https://doi.org/10.1371/journal.pgen.0010017; Hansen, M., et al., 2007, "Lifespan Extension by Conditions that Inhibit Translation in *Caenorhabditis elegans*," *Aging Cell* 6:95–110, https://doi.org/10.1111/j.1474-9726.2006.00267.x; Lee, S. S., et al., 2003, "A Systematic RNAi Screen Identifies a Critical Role for Mitochondria in *C. elegans* Longevity," *Nature Genetics* 33:40–48, https://doi.org/10.1038/ng1056.

55. Lindstrom, D. L., and Gottschling, D. E., 2009, "The Mother Enrichment Program: A Genetic System for Facile Replicative Life Span Analysis in *Saccharomyces cerevisiae*," *Genetics* 183:413–22, 411si–413si, https://doi.org/10.1534/genetics.109.106229; Yu, R., Jo, M. C., and Dang, W., 2020, "Measuring the Replicative Lifespan of *Saccharomyces cerevisiae* Using the HYAA Microfluidic Platform," *Methods in Molecular Biology* 2144:1–6, https://doi.org/10.1007/978-1-0716-0592-9_1.

56. Howitz et al. 2003.

57. Wood, J. G., et al., 2004, "Sirtuin Activators Mimic Caloric Restriction and Delay Ageing in Metazoans," *Nature* 430:686–89, https://doi.org/10.1038/nature02789; Baur, J. A., et al., 2006, "Resveratrol Improves Health and Survival of Mice on a High-Calorie Diet," *Nature* 444:337–42, https://doi.org/10.1038/nature05354; Rascón, B., Hubbard, B. P., Sinclair, D. A., and Amdam, G. V., 2012, "The Lifespan Extension Effects of Resveratrol Are Conserved in the Honey Bee and May Be Driven by a Mechanism Related to Caloric Restriction," *Aging (Albany NY)* 4:499–508, https://doi.org/10.18632/aging.100474; Song, J., et al., 2021, "Resveratrol Elongates the Lifespan and Improves Antioxidant Activity in the Silkworm *Bombyx mori*," *Journal of Pharmaceutical Analysis* 11:374–82, https://doi.org/10.1016/j.jpha.2020.06.005; Valenzano, D. R., et al., 2006, "Resveratrol Prolongs Lifespan and Retards the Onset of Age-Related Markers in a Short-Lived Vertebrate," *Current Biology* 16:296–300, https://doi.org/10.1016/j.cub.2005.12.038; Liu, T., et al., 2015, "Resveratrol Attenuates Oxidative Stress and Extends Life Span in the Annual Fish *Nothobranchius guentheri*," *Rejuvenation Research* 18:225–33, https://doi.org/10.1089/rej.2014.1618.

58. Kaeberlein, M., et al., 2005a, "Substrate-Specific Activation of Sirtuins by Resveratrol," *Journal of Biological Chemistry* 280:17038–45, https://doi.org/10.1074/jbc.M500655200.

59. Staats, S., et al., 2018, "Dietary Resveratrol Does Not Affect Life Span, Body Composition, Stress Response, and Longevity-Related Gene Expression in *Drosophila melanogaster*," *International Journal of Molecular Sciences* 19:223, https://doi.org/10.3390/ijms19010223; Bass, T. M., Weinkove, D., Houthoofd, K., Gems, D., and Partridge, L., 2007, "Effects of Resveratrol on Lifespan in *Drosophila melanogaster* and *Caenorhabditis elegans*," *Mechanisms of Ageing and Development* 128:546–52, https://doi.org/10.1016/j.mad.2007.07.007; Pallauf, K., Rimbach, G., Rupp, P. M., Chin, D., and Wolf, I. M., 2016, "Resveratrol and Lifespan in Model Organisms," *Current Medicinal Chemistry* 23:4639–80, https://doi.org/10.2174/0929867323666161024151233.

60. Fernández, A. F., and Fraga, M. F., 2011, "The Effects of the Dietary Polyphenol Resveratrol on Human Healthy Aging and Lifespan," *Epigenetics* 6:870–74, https://doi.org/10.4161/epi.6.7.16499; Bhullar, K. S., and Hubbard, B. P., 2015, "Lifespan and Healthspan Extension by Resveratrol," *Biochimica et Biophysica Acta* 1852:1209–18, https://doi.org/10.1016/j.bbadis.2015.01.012.

61. Hirschey et al. 2009.

62. Greer and Brunet 2009.

63. Arriola Apelo, S. I., and Lamming, D. W., 2016, "Rapamycin: An InhibiTOR of Aging Emerges from the Soil of Easter Island," *Journals of Gerontology, Series A, Biological Sciences and Medical Sciences* 71:841–49, https://doi.org/10.1093/gerona/glw090.

64. Lamming, D. W., 2016, "Inhibition of the Mechanistic Target of Rapamycin (mTOR)-Rapamycin and Beyond," *Cold Spring Harbor Perspectives in Medicine* 6:a025924, https://doi.org/10.1101/cshperspect.a025924; Lamming, D. W., Ye, L., Sabatini, D. M., and Baur, J. A., 2013, "Rapalogs and mTOR Inhibitors as Anti-aging Therapeutics," *Journal of Clinical Investigation* 123:980–89, https://doi.org/10.1172/JCI64099.

65. Vellai, T., et al., 2003, "Influence of TOR Kinase on Lifespan in *C. elegans*," *Nature* 426:620, https://doi.org/10.1038/426620a.

66. Kapahi, P., et al., 2004, "Regulation of Lifespan in *Drosophila* by Modulation of Genes in the TOR Signaling Pathway," *Current Biology* 14:885–90, https://doi.org/10.1016/j.cub.2004.03.059.

67. Powers et al. 2006; Kaeberlein, M., et al., 2005b, "Regulation of Yeast Replicative Life Span by TOR and Sch9 in Response to Nutrients," *Science* 310:1193–96, https://doi.org/10.1126/science.1115535; Hansen et al. 2005.

68. Sheaffer, K. L., Updike, D. L., and Mango, S. E., 2008, "The Target of Rapamycin Pathway Antagonizes pha-4/FoxA to Control Development and Aging," *Current Biology* 18:1355–64, https://doi.org/10.1016/j.cub.2008.07.097.

69. Apfeld, J., O'Connor, G., McDonagh, T., DiStefano, P. S., and Curtis, R., 2004, "The AMP-Activated Protein Kinase AAK-2 Links Energy Levels and Insulin-like Signals to Lifespan in *C. elegans*," *Genes and Development* 18:3004–9, https://doi.org/10.1101/gad.1255404.

70. Soukas, A. A., Kane, E. A., Carr, C. E., Melo, J. A., and Ruvkun, G., 2009, "Rictor/TORC2 Regulates Fat Metabolism, Feeding, Growth, and Life Span in *Caenorhabditis elegans*," *Genes and Development* 23:496–511, https://doi.org/10.1101/gad.1775409.

71. Greer, E. L., et al., 2007, "An AMPK-FOXO Pathway Mediates Longevity Induced by a Novel Method of Dietary Restriction in *C. elegans*," *Current Biology* 17:1646–56, https://doi.org/10.1016/j.cub.2007.08.047.

72. Greer and Brunet 2009.

73. Honjoh, S., Yamamoto, T., Uno, M., and Nishida, E., 2009, "Signalling through RHEB-1 Mediates Intermittent Fasting-Induced Longevity in *C. elegans*," *Nature* 457:726–30, https://doi.org/10.1038/nature07583.

74. Murphy, C. T., et al., 2003, "Genes that Act Downstream of DAF-16 to Influence the Lifespan of *Caenorhabditis elegans*," *Nature* 424:277–83; Murphy, C. T., Lee, S. J., and Kenyon, C., 2007, "Tissue Entrainment by Feedback Regulation of Insulin Gene Expression in the Endoderm of *Caenorhabditis elegans*," *Proceedings of the National Academy of Sciences of the USA* 104:19046–50.

75. Lee, C. K., Klopp, R. G., Weindruch, R., and Prolla, T. A., 1999, "Gene Expression Profile of Aging and Its Retardation by Caloric Restriction," *Science* 285:1390–93.

76. Pletcher, S. D., et al., 2002, "Genome-wide Transcript Profiles in Aging and Calorically Restricted *Drosophila melanogaster*," *Current Biology* 12:712–23.

77. McCarroll, S. A., et al., 2004, "Comparing Genomic Expression Patterns across Species Identifies Shared Transcriptional Profile in Aging," *Nature Genetics* 36:197–204.

78. Pandit, A., Jain, V., Kumar, N., and Mukhopadhyay, A., 2014, "PHA-4/FOXA-Regulated MicroRNA Feed Forward Loops during *Caenorhabditis elegans* Dietary Restriction," *Aging (Albany NY)* 6:835–55, https://doi.org/10.18632/aging.100697.

79. Hou, L., et al., 2016, "A Systems Approach to Reverse Engineer Lifespan Extension by Dietary Restriction," *Cell Metabolism* 23:529–40, https://doi.org/10.1016/j.cmet.2016.02.002.

80. Bishop, N. A., and Guarente, L., 2007, "Two Neurons Mediate Diet-Restriction-Induced Longevity in *C. elegans*," *Nature* 447:545–49, https://doi.org/10.1038/nature05904.

81. Burkewitz, K., et al., 2015, "Neuronal CRTC-1 Governs Systemic Mitochondrial Metabolism and Lifespan via a Catecholamine Signal," *Cell* 160:842–55, https://doi.org/10.1016/j.cell.2015.02.004.

82. Owen, B. M., et al., 2013, "FGF21 Contributes to Neuroendocrine Control of Female Reproduction," *Nature Medicine* 19:1153–56, https://doi.org/10.1038/nm.3250.

83. Bookout, A. L., et al., 2013, "FGF21 Regulates Metabolism and Circadian Behavior by Acting on the Nervous System," *Nature Medicine* 19:1147–52, https://doi.org/10.1038/nm.3249; Talukdar, S., et al., 2016, "FGF21 Regulates Sweet and Alcohol Preference," *Cell Metabolism* 23:344–49, https://doi.org/10.1016/j.cmet.2015.12.008.

84. Zhang, Y., et al., 2012, "Liver LXRalpha Expression is Crucial for Whole Body Cholesterol Homeostasis and Reverse Cholesterol Transport in Mice," *Journal of Clinical Investigation* 122:1688–99, https://doi.org/10.1172/JCI59817.

85. Hansen et al. 2007; Kaeberlein et al. 2005b; Hamilton, B., et al., 2005, "A Systematic RNAi Screen for Longevity Genes in *C. elegans*," *Genes and Development* 19:1544–55.

86. Wu, J., et al., 2018, "PHA-4/FoxA Senses Nucleolar Stress to Regulate Lipid Accumulation in *Caenorhabditis elegans*," *Nature Communications* 9:1195, https://doi.org/10.1038/s41467-018-03531-2.

87. Apfeld et al. 2004.

88. Tiku, V., et al., 2017, "Small Nucleoli Are a Cellular Hallmark of Longevity," *Nature Communications* 8:16083, https://doi.org/10.1038/ncomms16083.

89. Mair et al. 2003.

90. Mitchell, S. J., et al., 2019, "Daily Fasting Improves Health and Survival in Male Mice Independent of Diet Composition and Calories," *Cell Metabolism* 29:221–28.e223, https://doi.org/10.1016/j.cmet.2018.08.011.

91. Ulgherait et al. 2021.

92. Lowe et al. 2020.

93. Padamsey et al. 2022.

Notes to text box: "Don't We Already Know What's Bad for Us? Why It's Important to Know When Researchers Have Conflicts of Interest" (Chapter 7, pages 114–115)

a. Piercy, D., 2017, "How Women in the 1950s Stayed Fit and Trim," Life with Dee, March 8, 2017, https://lifewithdee.com/how-women-in-the-1950s-stayed-fit-and-trim/.

b. Kearns, C. E., Schmidt, L. A., and Glantz, S. A., 2016, "Sugar Industry and Coronary Heart Disease Research: A Historical Analysis of Internal Industry Documents," *JAMA Internal Medicine* 176:1680–85, https://doi.org/10.1001/jamainternmed.2016.5394.

c. Glantz was in the news more recently for less admirable deeds, having been accused of sexual harassment and racially insensitive behavior and retaliation; a settlement was reached in 2018.

Chapter 8

1. Taylor, R. C., and Dillin, A., 2011, "Aging as an Event of Proteostasis Collapse," *Cold Spring Harbor Perspectives in Biology* 3:a004440, https://doi.org/10.1101/cshperspect.a004440.

2. Murphy, C. T., et al., 2003, "Genes that Act Downstream of DAF-16 to Influence the Lifespan of *Caenorhabditis elegans*," *Nature* 424:277–83.

3. Pride, H., et al., 2015, "Long-Lived Species Have Improved Proteostasis Compared to Phylogenetically-Related Shorter-Lived Species," *Biochemical and Biophysical Research Communications* 457:669–75, https://doi.org/10.1016/j.bbrc.2015.01.046.

4. Hansen, M., et al., 2007, "Lifespan Extension by Conditions that Inhibit Translation in *Caenorhabditis elegans*," *Aging Cell* 6:95–110, https://doi.org/10.1111/j.1474–9726.2006.00267.x.

5. Savas, J. N., Toyama, B. H., Xu, T., Yates, J. R., and Hetzer, M. W., 2012, "Extremely Long-Lived Nuclear Pore Proteins in the Rat Brain," *Science* 335:942, https://doi.org/10.1126/science.1217421.

6. Tiku, V., et al., 2017, "Small Nucleoli Are a Cellular Hallmark of Longevity," *Nature Communications* 8:16083, https://doi.org/10.1038/ncomms16083.

7. Sousa, R., 2014, "Structural Mechanisms of Chaperone Mediated Protein Disaggregation," *Frontiers in Molecular Biosciences* 1, https://doi.org/10.3389/fmolb.2014.00012.

8. Hsu, A. L., Murphy, C. T., and Kenyon, C., 2003, "Regulation of Aging and Age-Related Disease by DAF-16 and Heat-Shock Factor," *Science* 300:1142–45.

9. Sengupta, U., Nilson, A. N., and Kayed, R., 2016, "The Role of Amyloid-β Oligomers in Toxicity, Propagation, and Immunotherapy," *eBioMedicine* 6:42–49, https://doi.org/10.1016/j.ebiom.2016.03.035.

10. Mehta, D., Jackson, R., Paul, G., Shi, J., and Sabbagh, M., 2017, "Why Do Trials for Alzheimer's Disease Drugs Keep Failing? A Discontinued Drug Perspective for 2010–2015," *Expert Opinion on Investigational Drugs* 26:735–39, https://doi.org/10.1080/13543784.2017.1323868.

11. Sengupta, Nilson, and Kayed 2016.

12. Mole, B., 2021, "Second Lab Worker with Deadly Prion Disease Prompts Research Pause in France," *Ars Technica*, July 29, 2021, https://arstechnica.com/science/2021/07/second-lab-worker-with-deadly-prion-disease-prompts-research-pause-in-france/.

13. Si, K., Lindquist, S., and Kandel, E. R., 2003, "A Neuronal Isoform of the Aplysia CPEB Has Prion-like Properties," *Cell* 115:879–91, https://doi.org/10.1016/S0092-8674(03)01020-1.

14. Shorter, J., and Lindquist, S., 2005, "Prions as Adaptive Conduits of Memory and Inheritance," *Nature Reviews Genetics* 6:435–50, https://doi.org/10.1038/nrg1616.

15. Chaudhuri, J., et al., 2018, "The Role of Advanced Glycation End Products in Aging and Metabolic Diseases: Bridging Association and Causality," *Cell Metabolism* 28:337–52, https://doi.org/10.1016/j.cmet.2018.08.014.

16. Lee, S. J., Murphy, C. T., and Kenyon, C., 2009, "Glucose Shortens the Life Span of *C. elegans* by Downregulating DAF-16/FOXO Activity and Aquaporin Gene Expression," *Cell Metabolism* 10:379–91, https://doi.org/10.1016/j.cmet.2009.10.003.

17. Chaudhuri, J., et al., 2016, "A *Caenorhabditis elegans* Model Elucidates a Conserved Role for TRPA1-Nrf Signaling in Reactive Alpha-Dicarbonyl Detoxification," *Current Biology* 26:3014–25, https://doi.org/10.1016/j.cub.2016.09.024.

18. Chaudhuri et al. 2018.

19. Chaudhuri et al. 2018.

20. Aguilaniu, H., Gustafsson, L., Rigoulet, M., and Nyström, T., 2003, "Asymmetric Inheritance of Oxidatively Damaged Proteins during Cytokinesis," *Science* 299:1751–53, https://doi.org/10.1126/science.1080418.

21. Goudeau, J., and Aguilaniu, H., 2010, "Carbonylated Proteins Are Eliminated during Reproduction in *C. elegans*," *Aging Cell* 9:991–1003, https://doi.org/10.1111/j.1474-9726.2010.00625.x; Bohnert, K. A., and Kenyon, C., 2017, "A Lysosomal Switch Triggers Proteostasis Renewal in the Immortal *C. elegans* Germ Lineage," *Nature* 551:629–33, https://doi.org/10.1038/nature24620.

22. Walther, D. M., et al., 2015, "Widespread Proteome Remodeling and Aggregation in Aging *C. elegans*," *Cell* 161:919–32, https://doi.org/10.1016/j.cell.2015.03.032; Greer, E. L., and Brunet, A., 2005, "FOXO Transcription Factors at the Interface between Longevity and Tumor Suppression," *Oncogene* 24:7410–25.

23. Tian, Y., Merkwirth, C., and Dillin, A., 2016, "Mitochondrial UPR: A Double-Edged Sword," *Trends in Cell Biology* 26:563–65, https://doi.org/10.1016/j.tcb.2016.06.006.

24. Nargund, A. M., Pellegrino, M. W., Fiorese, C. J., Baker, B. M., and Haynes, C. M., 2012, "Mitochondrial Import Efficiency of ATFS-1 Regulates Mitochondrial UPR Activation," *Science* 337:587–90, https://doi.org/10.1126/science.1223560.

25. Li, T. Y., et al., 2021, "The Transcriptional Coactivator CBP/p300 Is an Evolutionarily Conserved Node that Promotes Longevity in Response to Mitochondrial Stress," *Nature Aging* 1:165–78, https://doi.org/10.1038/s43587-020-00025-z; Merkwirth, C., et al., 2016, "Two Conserved Histone Demethylases Regulate Mitochondrial Stress-Induced Longevity," *Cell* 165:1209–23, https://doi.org/10.1016/j.cell.2016.04.012.

26. Merkwirth et al. 2016.

27. Zhang, Q., et al., 2021, "The Memory of Neuronal Mitochondrial Stress Is Inherited Transgenerationally via Elevated Mitochondrial DNA Levels," *Nature Cell Biology* 23:870–80, https://doi.org/10.1038/s41556-021-00724-8.

28. Lee, S. K., 2021, "Endoplasmic Reticulum Homeostasis and Stress Responses in *Caenorhabditis elegans*," in *Celllular Biology of the Endoplasmic Reticulum*, edited by L. B. Agellon and M. Michalak, Progress in Molecular and Subcellular Biology 59:279–303 (Cham: Springer), https://doi.org/10.1007/978-3-030-67696-4_13; Pintado, C., Macias, S., Dominguez-Martin, H., Castano, A., and Ruano, D., 2017, "Neuroinflammation Alters Cellular Proteostasis by Producing Endoplasmic Reticulum Stress, Autophagy Activation and Disrupting ERAD Activation," *Scientific Reports* 7:8100, https://doi.org/10.1038/s41598-017-08722-3; Sekiya, M., et al., 2017,

"EDEM Function in ERAD Protects against Chronic ER Proteinopathy and Age-Related Physiological Decline in *Drosophila*," *Developmental Cell* 41:652–64.e5, https://doi.org/10.1016/j.devcel.2017.05.019.

29. Joaquin Navajas Acedo (MadScientist, @MadS100tist), 2021, "Why Is It Called Unfolded Protein Response and Not Panic! At the ER," Twitter, November 18, 2021, 6:53 p.m.

30. Goudeau, J., Samaddar, M., Bohnert, K. A., and Kenyon, C., 2020, "Addendum: A Lysosomal Switch Triggers Proteostasis Renewal in the Immortal *C. elegans* Germ Lineage," *Nature* 580:E5, https://doi.org/10.1038/s41586-020-2108-0.

31. Zhang, Q., et al., 2018, "The Mitochondrial Unfolded Protein Response Is Mediated Cell-Non-autonomously by Retromer-Dependent Wnt Signaling," *Cell* 174:870–83.e17, https://doi.org/10.1016/j.cell.2018.06.029.

32. Kaushik, S., et al., 2021, "Autophagy and the Hallmarks of Aging," *Ageing Research Reviews* 72:101468, https://doi.org/10.1016/j.arr.2021.101468; Nieto-Torres, J. L., and Hansen, M., 2021, "Macroautophagy and Aging: The Impact of Cellular Recycling on Health and Longevity," *Molecular Aspects of Medicine* 82:101020, https://doi.org/10.1016/j.mam.2021.101020.

33. Aman, Y., et al., 2021, "Autophagy in Healthy Aging and Disease," *Nature Aging* 1:634–50, https://doi.org/10.1038/s43587-021-00098-4.

34. Aman et al. 2021.

35. Aman et al. 2021.

36. Montano, M., and Long, K., 2011, "RNA Surveillance: An Emerging Role for RNA Regulatory Networks in Aging," *Ageing Research Reviews* 10:216–24, https://doi.org/10.1016/j.arr.2010.02.002.

37. Rangaraju, S., et al., 2015, "Suppression of Transcriptional Drift Extends *C. elegans* Lifespan by Postponing the Onset of Mortality," *eLife* 4:e08833, https://doi.org/10.7554/eLife.08833.

38. Seo, M., et al., 2015, "RNA Helicase HEL-1 Promotes Longevity by Specifically Activating DAF-16/FOXO Transcription Factor Signaling in *Caenorhabditis elegans*," *Proceedings of the National Academy of Sciences of the USA* 112:E4246–55, https://doi.org/10.1073/pnas.1505451112.

39. Son, H. G., et al., 2017, "RNA Surveillance via Nonsense-Mediated mRNA Decay Is Crucial for Longevity in *daf-2*/Insulin/IGF-1 Mutant *C. elegans*," *Nature Communications* 8:14749, https://doi.org/10.1038/ncomms14749.

40. Kuroyanagi, H., Ohno, G., Sakane, H., Maruoka, H., and Hagiwara, M., 2010, "Visualization and Genetic Analysis of Alternative Splicing Regulation In Vivo Using Fluorescence Reporters in Transgenic *Caenorhabditis elegans*," *Nature Protocols* 5:1495–1517, https://doi.org/10.1038/nprot.2010.107; Heintz, C., et al., 2017, "Splicing Factor 1 Modulates Dietary Restriction and TORC1 Pathway Longevity in *C. elegans*," *Nature* 541:102–6, https://doi.org/10.1038/nature20789.

41. Kew, C., et al., 2020, "Evolutionarily Conserved Regulation of Immunity by the Splicing Factor RNP-6/PUF60," *eLife* 9:e57591, https://doi.org/10.7554/eLife.57591.

42. Rhoads, T. W., et al., 2018, "Caloric Restriction Engages Hepatic RNA Processing Mechanisms in Rhesus Monkeys," *Cell Metabolism* 27:677–88.e5, https://doi.org/10.1016/j.cmet.2018.01.014.

43. Ibanez-Ventoso, C., et al., 2006, "Modulated MicroRNA Expression during Adult Lifespan in *Caenorhabditis elegans*," *Aging Cell* 5:235–46, https://doi.org/10.1111/j.1474-9726

.2006.00210.x; Inukai, S., Pincus, Z., de Lencastre, A., and Slack, F. J., 2018, "A MicroRNA Feedback Loop Regulates Global MicroRNA Abundance during Aging," *RNA* 24:159–72, https://doi.org/10.1261/rna.062190.117; Smith-Vikos, T., et al., 2014, "MicroRNAs Mediate Dietary-Restriction-Induced Longevity through PHA-4/FOXA and SKN-1/Nrf Transcription Factors," *Current Biology* 24:2238–46, https://doi.org/10.1016/j.cub.2014.08.013; Inukai, S., de Lencastre, A., Turner, M., and Slack, F., 2012, "Novel MicroRNAs Differentially Expressed during Aging in the Mouse Brain," *PLOS One* 7:e40028, https://doi.org/10.1371/journal.pone .0040028; De Lencastre, A., et al., 2010, "MicroRNAs Both Promote and Antagonize Longevity in *C. elegans*," *Current Biology* 20:2159–68, https://doi.org/10.1016/j.cub.2010.11.015; Pincus, Z., Smith-Vikos, T., and Slack, F. J., 2011, "MicroRNA Predictors of Longevity in *Caenorhabditis elegans*," *PLOS Genetics* 7:e1002306, https://doi.org/10.1371/journal.pgen .1002306.

44. Pincus, Smith-Vikos, and Slack 2011; Ruediger, C., Karimzadegan, S., Lin, S., and Shapira, M., 2021, "miR-71 Mediates Age-Dependent Opposing Contributions of the Stress-Activated Kinase KGB-1 in *Caenorhabditis elegans*," *Genetics* 218: iya049, https://doi.org/10 .1093/genetics/iyab049; Hsieh, Y. W., Chang, C., and Chuang, C. F., 2012, "The MicroRNA mir-71 Inhibits Calcium Signaling by Targeting the TIR-1/Sarm1 Adaptor Protein to Control Stochastic L/R Neuronal Asymmetry in *C. elegans*," *PLOS Genetics* 8:e1002864, https://doi.org /10.1371/journal.pgen.1002864; Boulias, K., and Horvitz, H. R., 2012, "The *C. elegans* MicroRNA mir-71 Acts in Neurons to Promote Germline-Mediated Longevity through Regulation of DAF-16/FOXO," *Cell Metabolism* 15:439–50, https://doi.org/10.1016/j.cmet.2012.02 .014; Boehm, M., and Slack, F., 2005, "A Developmental Timing MicroRNA and Its Target Regulate Life Span in *C. elegans*," *Science* 310:1954–57, https://doi.org/10.1126/science .1115596.

45. Finger, F., et al., 2019, "Olfaction Regulates Organismal Proteostasis and Longevity via MicroRNA-Dependent Signaling," *Nature Metabolism* 1:350–59, https://doi.org/10.1038 /s42255-019-0033-z.

46. Vesely, C., and Jantsch, M. F., 2021, "An I for an A: Dynamic Regulation of Adenosine Deamination-Mediated RNA Editing," *Genes* 12:1026, https://doi.org/10.3390/genes 12071026.

47. Vesely and Jantsch 2021.

48. Sebastiani, P., et al., 2009, "RNA Editing Genes Associated with Extreme Old Age in Humans and with Lifespan in *C. elegans*," *PLOS One* 4:e8210, https://doi.org/10.1371/journal .pone.0008210.

49. Oguro, R., et al., 2012, "A Single Nucleotide Polymorphism of the Adenosine Deaminase, RNA-Specific Gene Is Associated with the Serum Triglyceride Level, Abdominal Circumference, and Serum Adiponectin Concentration," *Experimental Gerontology* 47:183–87, https://doi.org/10.1016/j.exger.2011.12.004.

50. Cortés-López, M., et al., 2018, "Global Accumulation of circRNAs during Aging in *Caenorhabditis elegans*," *BMC Genomics* 19:8, https://doi.org/10.1186/s12864-017-4386-y.

51. Weigelt, C. M., et al., 2020, "An Insulin-Sensitive Circular RNA that Regulates Lifespan in *Drosophila*," *Molecular Cell* 79:268–79.e5, https://doi.org/10.1016/j.molcel.2020.06.011.

52. Memczak, S., et al., 2013, "Circular RNAs Are a Large Class of Animal RNAs with Regulatory Potency," *Nature* 495:333–38, https://doi.org/10.1038/nature11928.

Chapter 9

1. Sachdev, C., 2018, "At Age 101, She's a World Champ Runner," NPR, Goats and Soda, January 1, 2018, https://www.npr.org/sections/goatsandsoda/2018/01/01/568665002/at-age-101-shes-a-world-champ-runner.

2. Lepers, R., Stapley, P. J., and Cattagni, T., 2016, "Centenarian Athletes: Examples of Ultimate Human Performance?," *Age and Ageing* 45:732–36, https://doi.org/10.1093/ageing/afw111.

3. Harman, D., 1972, "The Biologic Clock: The Mitochondria?," *Journal of the American Geriatrics Society* 20:145–47, https://doi.org/10.1111/j.1532-5415.1972.tb00787.x.

4. Edgar, D., and Trifunovic, A., 2009, "The mtDNA Mutator Mouse: Dissecting Mitochondrial Involvement in Aging," *Aging (Albany NY)* 1:1028–32, https://doi.org/10.18632/aging.100109; Wallace, D. C., 1999, "Mitochondrial Diseases in Man and Mouse," *Science* 283:1482–88, https://doi.org/10.1126/science.283.5407.1482.

5. Martin, G. M., Austad, S. N., and Johnson, T. E., 1996, "Genetic Analysis of Ageing: Role of Oxidative Damage and Environmental Stresses," *Nature Genetics* 13:25–34, https://doi.org/10.1038/ng0596-25.

6. Sun, J., and Tower, J., 1999, "FLP Recombinase-Mediated Induction of Cu/Zn-Superoxide Dismutase Transgene Expression Can Extend the Life Span of Adult *Drosophila melanogaster* Flies," *Molecular Cell Biology* 19:216–28; Parkes, T. L., et al., 1998, "Extension of *Drosophila* Lifespan by Overexpression of Human SOD1 in Motorneurons," *Nature Genetics* 19:171–74.

7. Schriner, S. E., et al., 2005, "Extension of Murine Life Span by Overexpression of Catalase Targeted to Mitochondria," *Science* 308:1909–11, https://doi.org/10.1126/science.1106653.

8. Yen, C. A., and Curran, S. P., 2021, "Incomplete Proline Catabolism Drives Premature Sperm Aging," *Aging Cell* 20:e13308, https://doi.org/10.1111/acel.13308.

9. Sun, Y., et al., 2020, "Lysosome Activity Is Modulated by Multiple Longevity Pathways and Is Important for Lifespan Extension in *C. elegans*," *eLife* 9:e55745, https://doi.org/10.7554/eLife.55745.

10. Salmon, A. B., et al., 2009, "The Long Lifespan of Two Bat Species Is Correlated with Resistance to Protein Oxidation and Enhanced Protein Homeostasis," *FASEB Journal* 23:2317–26, https://doi.org/10.1096/fj.08-122523.

11. Murphy, C. T., et al., 2003, "Genes that Act Downstream of DAF-16 to Influence the Lifespan of *Caenorhabditis elegans*," *Nature* 424:277–83.

12. Van Raamsdonk, J. M., and Hekimi, S., 2012, "Superoxide Dismutase Is Dispensable for Normal Animal Lifespan," *Proceedings of the National Academy of Sciences of the USA* 109:5785–90, https://doi.org/10.1073/pnas.1116158109; Honda, Y., Tanaka, M., and Honda, S., 2010, "Redox Regulation, Gene Expression and Longevity," *Geriatrics and Gerontology International* 10 (suppl 1):S59–69, https://doi.org/10.1111/j.1447-0594.2010.00591.x; Doonan, R., et al., 2008, "Against the Oxidative Damage Theory of Aging: Superoxide Dismutases Protect against Oxidative Stress but Have Little or No Effect on Life Span in *Caenorhabditis elegans*," *Genes and Development* 22:3236–41, https://doi.org/10.1101/gad.504808.

13. Bayne, A.-C.V., Mockett, R. J., Orr, W. C., and Sohal, R. S., 2005, "Enhanced Catabolism of Mitochondrial Superoxide/Hydrogen Peroxide and Aging in Transgenic *Drosophila*," *Biochemical Journal* 391:277–84, https://doi.org/10.1042/bj20041872.

14. De Waal, E. M., et al., 2013, "Elevated Protein Carbonylation and Oxidative Stress Do Not Affect Protein Structure and Function in the Long-Living Naked-Mole Rat: A Proteomic

Approach," *Biochemical and Biophysical Research Communications* 434:815–19, https://doi.org/10 .1016/j.bbrc.2013.04.019; Pérez, V. I., et al., 2009, "Is the Oxidative Stress Theory of Aging Dead?," *Biochimica et Biophysica Acta* 1790:1005–14, https://doi.org/10.1016/j.bbagen.2009.06.003.

15. Fraser, A. G., et al., 2000, "Functional Genomic Analysis of *C. elegans* Chromosome I by Systematic RNA Interference," *Nature* 408:325–30, https://doi.org/10.1038/35042517.

16. Lee, S. S., et al., 2003, "A Systematic RNAi Screen Identifies a Critical Role for Mito-chondria in *C. elegans* Longevity," *Nature Genetics* 33:40–48, https://doi.org/10.1038/ng1056; Dillin, A., et al., 2002, "Rates of Behavior and Aging Specified by Mitochondrial Function during Development," *Science* 298, 2398–2401, https://doi.org/10.1126/science.1077780.

17. Copeland, J. M., et al., 2009, "Extension of *Drosophila* Life Span by RNAi of the Mito-chondrial Respiratory Chain," *Current Biology* 19:1591–98, https://doi.org/10.1016/j.cub.2009 .08.016.

18. Dell'agnello, C., et al., 2007, "Increased Longevity and Refractoriness to Ca(2+)-Dependent Neurodegeneration in Surf1 Knockout Mice," *Human Molecular Genetics* 16:431–44, https://doi.org/10.1093/hmg/ddl477; Orsini, F., et al., 2004, "The Life Span Determinant p66Shc Localizes to Mitochondria where It Associates with Mitochondrial Heat Shock Protein 70 and Regulates Trans-membrane Potential," *Journal of Biological Chemistry* 279:25689–95, https://doi.org/10.1074/jbc.M401844200.

19. She would later win the prestigious Lasker Award for this discovery.

20. Durieux, J., Wolff, S., and Dillin, A., 2011, "The Cell-Non-autonomous Nature of Elec-tron Transport Chain-Mediated Longevity," *Cell* 144:79–91, https://doi.org/10.1016/j.cell .2010.12.016.

21. Walter, L., Baruah, A., Chang, H. W., Pace, H. M., and Lee, S. S., 2011, "The Homeobox Protein CEH-23 Mediates Prolonged Longevity in Response to Impaired Mitochondrial Elec-tron Transport Chain in *C. elegans*," *PLOS Biology* 9:e1001084, https://doi.org/10.1371/journal .pbio.1001084.

22. Maglioni, S., Schiavi, A., Runci, A., Shaik, A., and Ventura, N., 2014, "Mitochondrial Stress Extends Lifespan in *C. elegans* through Neuronal Hormesis," *Experimental Gerontology* 56:89–98, https://doi.org/10.1016/j.exger.2014.03.026.

23. Haynes, C. M., Fiorese, C. J., and Lin, Y. F., 2013, "Evaluating and Responding to Mito-chondrial Dysfunction: The Mitochondrial Unfolded-Protein Response and Beyond," *Trends in Cell Biology* 23:311–18, https://doi.org/10.1016/j.tcb.2013.02.002.

24. Nargund, A. M., Pellegrino, M. W., Fiorese, C. J., Baker, B. M., and Haynes, C. M., 2012, "Mitochondrial Import Efficiency of ATFS-1 Regulates Mitochondrial UPR Activation," *Science* 337:587–90, https://doi.org/10.1126/science.1223560.

25. Durieux, Wolff, and Dillin 2011.

26. Merkwirth, C., et al., 2016, "Two Conserved Histone Demethylases Regulate Mitochon-drial Stress-Induced Longevity," *Cell* 165:1209–23, https://doi.org/10.1016/j.cell.2016.04.012.

27. One major problem that mitochondria must continually solve is stoichiometry, or the ratios of proteins. Since the OXPHOS complexes require both nuclearly and mitochondrially encoded subunits, there must be a "mitonuclear balance" to coordinate the levels; if the compo-nents made in the nucleus and in the mitochondria are no longer coordinated to make functional complexes, as in the case of defective mtDNA or defective aminoacyl-tRNAs in the mitochondria, mitochondrial function also declines, and the UPR^{mt} is activated to alleviate the stress. This was shown in both worms and mice (Houtkooper, R. H., Mouchiroud, L., Ryu, D., Moullan, N.,

Katsyuba, E., Knott, G., Williams, R. W., and Auwerx, J., 2013, "Mitonuclear Protein Imbalance as a Conserved Longevity Mechanism," *Nature* 497:451–57, https://doi.org/10.1038 /nature12188).

28. Durieux, Wolff, and Dillin 2011.

29. Shao, L. W., Niu, R., and Liu, Y., 2016, "Neuropeptide Signals Cell Non-autonomous Mitochondrial Unfolded Protein Response," *Cell Research* 26:1182–96, https://doi.org/10.1038 /cr.2016.118.

30. Yin, J.-A., et al., 2017, "Genetic Variation in Glia–Neuron Signalling Modulates Ageing Rate," *Nature* 551:198–203, https://doi.org/10.1038/nature24463.

31. Owusu-Ansah, E., Song, W., and Perrimon, N., 2013, "Muscle Mitohormesis Promotes Longevity via Systemic Repression of Insulin Signaling," *Cell* 155:699–712, https://doi.org/10 .1016/j.cell.2013.09.021.

32. Song, W., et al., 2017, "Activin Signaling Mediates Muscle-to-Adipose Communication in a Mitochondria Dysfunction-Associated Obesity Model," *Proceedings of the National Academy of Sciences of the USA* 114:8596–8601, https://doi.org/10.1073/pnas.1708037114.

33. Kim, K. H., et al., 2013, "Autophagy Deficiency Leads to Protection from Obesity and Insulin Resistance by Inducing Fgf21 as a Mitokine," *Nature Medicine* 19:83–92, https://doi.org /10.1038/nm.3014.

34. Kim et al. 2013.

35. Zhang, Y., et al., 2012, "The Starvation Hormone, Fibroblast Growth Factor-21, Extends Lifespan in Mice," *eLife* 1:e00065, https://doi.org/10.7554/eLife.00065.

36. Ikonen, M., et al., 2003, "Interaction between the Alzheimer's Survival Peptide Humanin and Insulin-like Growth Factor-Binding Protein 3 Regulates Cell Survival and Apoptosis," *Proceedings of the National Academy of Sciences of the USA* 100:13042–47, https://doi.org/10.1073 /pnas.2135111100.

37. Lee, C., et al., 2015, "The Mitochondrial-Derived Peptide MOTS-c Promotes Metabolic Homeostasis and Reduces Obesity and Insulin Resistance," *Cell Metabolism* 21:443–54, https:// doi.org/10.1016/j.cmet.2015.02.009.

38. Zhang, Q., et al., 2010, "Circulating Mitochondrial DAMPs Cause Inflammatory Responses to Injury," *Nature* 464:104–7, https://doi.org/10.1038/nature08780.

39. Rea, S. L., Ventura, N., and Johnson, T. E., 2007, "Relationship between Mitochondrial Electron Transport Chain Dysfunction, Development, and Life Extension in *Caenorhabditis elegans*," *PLOS Biology* 5: e259, https://doi.org/10.1371/journal.pbio.0050259.

40. Yun, J., and Finkel, T., 2014, "Mitohormesis," *Cell Metabolism* 19:757–66, https://doi .org/10.1016/j.cmet.2014.01.011; Hughes, B. G., and Hekimi, S., 2011, "A Mild Impairment of Mitochondrial Electron Transport Has Sex-Specific Effects on Lifespan and Aging in Mice," *PLOS One* 6:e26116, https://doi.org/10.1371/journal.pone.0026116.

41. Wei, Y., and Kenyon, C., 2016, "Roles for ROS and Hydrogen Sulfide in the Longevity Response to Germline Loss in *Caenorhabditis elegans*," *Proceedings of the National Academy of Sciences of the USA* 113:E2832–41, https://doi.org/10.1073/pnas.1524727113.

42. Ristow, M., et al., 2009, "Antioxidants Prevent Health-Promoting Effects of Physical Exercise in Humans," *Proceedings of the National Academy of Sciences of the USA* 106:8665–70, https://doi.org/10.1073/pnas.0903485106.

43. Latorre-Pellicer, A., et al., 2016, "Mitochondrial and Nuclear DNA Matching Shapes Metabolism and Healthy Ageing," *Nature* 535:561–65, https://doi.org/10.1038/nature18618.

44. Sahin, E., et al., 2011, "Telomere Dysfunction Induces Metabolic and Mitochondrial Compromise," *Nature* 470:359–65, https://doi.org/10.1038/nature09787.

45. Tanaka, A., et al., 2010, "Proteasome and p97 Mediate Mitophagy and Degradation of Mitofusins Induced by Parkin," *Journal of Cell Biology* 191:1367–80, https://doi.org/10.1083/jcb.201007013.

46. Aguilaniu, H., Gustafsson, L., Rigoulet, M., and Nyström, T., 2003, "Asymmetric Inheritance of Oxidatively Damaged Proteins during Cytokinesis," *Science* 299:1751–53, https://doi.org/10.1126/science.1080418.

47. McFaline-Figueroa, J. R., et al., 2011, "Mitochondrial Quality Control during Inheritance Is Associated with Lifespan and Mother-Daughter Age Asymmetry in Budding Yeast," *Aging Cell* 10:885–95, https://doi.org/10.1111/j.1474-9726.2011.00731.x.

48. Henderson, K. A., Hughes, A. L., and Gottschling, D. E., 2014, "Mother-Daughter Asymmetry of pH Underlies Aging and Rejuvenation in Yeast," *eLife* 3:e03504, https://doi.org/10.7554/eLife.03504.

49. Katajisto, P., et al., 2015, "Asymmetric Apportioning of Aged Mitochondria between Daughter Cells Is Required for Stemness," *Science* 348:340–43, https://doi.org/10.1126/science.1260384.

50. Choi, C. S., et al., 2008, "Paradoxical Effects of Increased Expression of PGC-1alpha on Muscle Mitochondrial Function and Insulin-Stimulated Muscle Glucose Metabolism," *Proceedings of the National Academy of Sciences of the USA* 105:19926–31, https://doi.org/10.1073/pnas.0810339105; Arany, Z., et al., 2005, "Transcriptional Coactivator PGC-1 Alpha Controls the Energy State and Contractile Function of Cardiac Muscle," *Cell Metabolism* 1:259–71, https://doi.org/10.1016/j.cmet.2005.03.002; Ruas, J. L., et al., 2012, "A PGC-1α Isoform Induced by Resistance Training Regulates Skeletal Muscle Hypertrophy," *Cell* 151:1319–31, https://doi.org/10.1016/j.cell.2012.10.050; Austin, S., and St-Pierre, J., 2012, "PGC1α and Mitochondrial Metabolism: Emerging Concepts and Relevance in Ageing and Neurodegenerative Disorders," *Journal of Cell Science* 125:4963–71, https://doi.org/10.1242/jcs.113662.

51. Walker, D. W., and Benzer, S., 2004, "Mitochondrial 'Swirls' Induced by Oxygen Stress and in the *Drosophila* Mutant Hyperswirl," *Proceedings of the National Academy of Sciences of the USA* 101:10290–95, https://doi.org/10.1073/pnas.0403767101.

52. Palikaras, K., Lionaki, E., and Tavernarakis, N., 2015, "Coordination of Mitophagy and Mitochondrial Biogenesis during Ageing in *C. elegans*," *Nature* 521:525–28, https://doi.org/10.1038/nature14300.

53. Byrne, J. J., et al., 2019, "Disruption of Mitochondrial Dynamics Affects Behaviour and Lifespan in *Caenorhabditis elegans*," *Cellular and Molecular Life Sciences* 76:1967–85, https://doi.org/10.1007/s00018-019-03024-5.

54. Rana, A., et al., 2017, "Promoting Drp1-Mediated Mitochondrial Fission in Midlife Prolongs Healthy Lifespan of *Drosophila melanogaster*," *Nature Communications* 8:448, https://doi.org/10.1038/s41467-017-00525-4.

55. Lee et al. 2003.

56. Chaudhari, S. N., and Kipreos, E. T., 2017, "Increased Mitochondrial Fusion Allows the Survival of Older Animals in Diverse *C. elegans* Longevity Pathways," *Nature Communications* 8:182, https://doi.org/10.1038/s41467-017-00274-4.

57. For worms: Weir, H. J., et al., 2017, "Dietary Restriction and AMPK Increase Lifespan via Mitochondrial Network and Peroxisome Remodeling," *Cell Metabolism* 26:884–96.e5,

https://doi.org/10.1016/j.cmet.2017.09.024. For yeast: Bernhardt, D., Müller, M., Reichert, A. S., and Osiewacz, H. D., 2015, "Simultaneous Impairment of Mitochondrial Fission and Fusion Reduces Mitophagy and Shortens Replicative Lifespan," *Scientific Reports* 5:7885, https://doi.org/10.1038/srep07885.

58. Weir et al. 2017.

59. Tsang, W. Y., and Lemire, B. D., 2002, "Mitochondrial Genome Content Is Regulated during Nematode Development," *Biochemical and Biophysical Research Communications* 291:8–16, https://doi.org/10.1006/bbrc.2002.6394; Bratic, I., et al., 2009, "Mitochondrial DNA Level, but Not Active Replicase, Is Essential for *Caenorhabditis elegans* Development," *Nucleic Acids Research* 37:1817–28, https://doi.org/10.1093/nar/gkp018.

60. Cota, V., and Murphy, C. T., 2021, "Fission and PINK-1-Mediated Mitophagy Are Required for Insulin/IGF-1 Signaling Mutant Reproductive Longevity," *bioRxiv* 2021.08.16.456566, https://doi.org/10.1101/2021.08.16.456566.

61. Weir et al. 2017.

62. Yao, V., et al., 2018, "An Integrative Tissue-Network Approach to Identify and Test Human Disease Genes. *Nature Biotechnology* 36: 1091–99, https://doi.org/10.1038/nbt.4246.

63. Sohrabi, S., Mor, D. E., Kaletsky, R., Keyes, W., and Murphy, C. T., 2021, "High-Throughput Behavioral Screen in *C. elegans* Reveals Parkinson's Disease Drug Candidates," *Communications Biology* 4:203, https://doi.org/10.1038/s42003-021-01731-z.

64. Mor, D. E., et al., 2020, "Metformin Rescues Parkinson's Disease Phenotypes Caused by Hyperactive Mitochondria," *Proceedings of the National Academy of Sciences of the USA* 117:26438–47, https://doi.org/10.1073/pnas.2009838117.

65. Egan, B., and Zierath, J. R., 2013, "Exercise Metabolism and the Molecular Regulation of Skeletal Muscle Adaptation," *Cell Metabolism* 17:162–84, https://doi.org/10.1016/j.cmet .2012.12.012; Rowe, G. C., El-Khoury, R., Patten, I. S., Rustin, P., and Arany, Z., 2012, "PGC-1α Is Dispensable for Exercise-Induced Mitochondrial Biogenesis in Skeletal Muscle," *PLOS One* 7:e41817, https://doi.org/10.1371/journal.pone.0041817; Rowe, G. C., Safdar, A., and Arany, Z., 2014, "Running Forward: New Frontiers in Endurance Exercise Biology," *Circulation* 129:798–810, https://doi.org/10.1161/circulationaha.113.001590.

66. Tiku, V., et al., 2017, "Small Nucleoli Are a Cellular Hallmark of Longevity," *Nature Communications* 8:16083, https://doi.org/10.1038/ncomms16083.

67. Laranjeiro, R., et al., 2019, "Swim Exercise in *Caenorhabditis elegans* Extends Neuromuscular and Gut Healthspan, Enhances Learning Ability, and Protects against Neurodegeneration," *Proceedings of the National Academy of Sciences of the USA* 116:23829–39, https://doi.org /10.1073/pnas.1909210116.

68. Twilley, N., 2017, "A Pill to Make Exercise Obsolete," *New Yorker*, October 30, 2017, https://www.newyorker.com/magazine/2017/11/06/a-pill-to-make-exercise-obsolete.

Chapter 10

1. Alvarez, J., 2013, "'Wear Sunscreen': The Story behind the Commencement Speech that Kurt Vonnegut Never Gave," Open Culture, November 11, 2013, https://www.openculture .com/2013/11/wear-sunscreen-the-story-behind-the-commencement-speech-that-kurt -vonnegut-never-wrote.html.

2. Hoeijmakers, J.H.J., 2009, "DNA Damage, Aging, and Cancer," *New England Journal of Medicine* 361:1475–85, https://doi.org/10.1056/NEJMra0804615.

3. Maynard, S., et al., 2015, "DNA Damage, DNA repair, Aging, and Neurodegeneration," in *Aging: The Longevity Dividend*, edited by S. J. Olshansky, G. M. Martin, and J. L. Kirkland (New York: Cold Spring Harbor Laboratory Press).

4. Hisama, F. M., et al., 2015, "How Research on Human Progeroid and Antigeroid Syndromes Can Contribute to the Longevity Dividend Initiative," in *Aging: The Longevity Dividend*.

5. Wang, H., Lautrup, S., Caponio, D., Zhang, J., and Fang, E. F., 2021, "DNA Damage-Induced Neurodegeneration in Accelerated Ageing and Alzheimer's Disease," *International Journal of Molecular Sciences* 22:6748, https://doi.org/10.3390/ijms22136748.

6. Wang et al. 2021.

7. Sandre-Giovannoli, A. D., et al., 2003, "Lamin A Truncation in Hutchinson-Gilford Progeria," *Science* 300:2055, https://doi.org/10.1126/science.1084125.

8. McClintock, D., et al., 2007, "The Mutant Form of Lamin A that Causes Hutchinson-Gilford Progeria Is a Biomarker of Cellular Aging in Human Skin," *PLOS One* 2:e1269, https://doi.org/10.1371/journal.pone.0001269.

9. Koblan, L. W., et al., 2021, "In Vivo Base Editing Rescues Hutchinson-Gilford Progeria Syndrome in Mice," *Nature* 589:608–14, https://doi.org/10.1038/s41586-020-03086-7.

10. Sánchez-López, A., et al., 2021, "Cardiovascular Progerin Suppression and Lamin A Restoration Rescues Hutchinson-Gilford Progeria Syndrome," *Circulation* 144:1777–94, https://doi.org/10.1161/CIRCULATIONAHA.121.055313.

11. Schaible, R., Sussman, M., and Kramer, B. H., 2014, "Aging and Potential for Self-Renewal: *Hydra* Living in the Age of Aging—a Mini-Review," *Gerontology* 60:548–56, https://doi.org/10.1159/000360397.

12. In contrast to the non-senescence strategy used by *Hydra*, a recent analysis of sea urchins with different lifespans (*Mesocentrotus franciscanus*, 100 years vs. *Lytechinus variegatus*, 4 years) showed no obvious relationship between regeneration rates with age and lifespan, so it is not the only way to live long: Bodnar, A. G., 2015, "Cellular and Molecular Mechanisms of Negligible Senescence: Insight from the Sea Urchin," *Invertebrate Reproduction and Development* 59:23–27, https://doi.org/10.1080/07924259.2014.938195.

13. Takahashi, K., and Yamanaka, S., 2006, "Induction of Pluripotent Stem Cells from Mouse Embryonic and Adult Fibroblast Cultures by Defined Factors," *Cell* 126:663–76, https://doi.org/10.1016/j.cell.2006.07.024; Takahashi, K., et al., 2007, "Induction of Pluripotent Stem Cells from Adult Human Fibroblasts by Defined Factors," *Cell* 131:861–72, https://doi.org/10.1016/j.cell.2007.11.019.

14. Xiao, B., Ng, H. H., Takahashi, R., and Tan, E.-K., 2016, "Induced Pluripotent Stem Cells in Parkinson's Disease: Scientific and Clinical Challenges," *Journal of Neurology, Neurosurgery and Psychiatry* 87:697, https://doi.org/10.1136/jnnp-2015-312036.

15. Beyret, E., Martinez Redondo, P., Platero Luengo, A., and Izpisua Belmonte, J. C., 2018, "Elixir of Life: Thwarting Aging with Regenerative Reprogramming," *Circulation Research* 122:128–41, https://doi.org/10.1161/CIRCRESAHA.117.311866.

16. Ocampo, A., et al., 2016, "In Vivo Amelioration of Age-Associated Hallmarks by Partial Reprogramming," *Cell* 167:1719–33.e12, https://doi.org/10.1016/j.cell.2016.11.052.

17. Lu, Y., et al., 2020, "Reprogramming to Recover Youthful Epigenetic Information and Restore Vision," *Nature* 588:124–29, https://doi.org/10.1038/s41586-020-2975-4.

18. Rodríguez-Matellán, A., Alcazar, N., Hernández, F., Serrano, M., and Ávila, J., 2020, "Reprogramming Ameliorates Aging Features in Dentate Gyrus Cells and Improves Memory in Mice," *Stem Cell Reports* 15:1056–66, https://doi.org/10.1016/j.stemcr.2020.09.010.

19. Holstege, H., et al., 2014, "Somatic Mutations Found in the Healthy Blood Compartment of a 115-Yr-Old Woman Demonstrate Oligoclonal Hematopoiesis," *Genome Research* 24:733–42, https://doi.org/10.1101/gr.162131.113.

20. Hernandez-Segura, A., Nehme, J., and Demaria, M., 2018, "Hallmarks of Cellular Senescence," *Trends in Cell Biology* 28:436–53, https://doi.org/10.1016/j.tcb.2018.02.001.

21. Greider, C. W., and Blackburn, E. H., 1987, "The Telomere Terminal Transferase of *Tetrahymena* is a Ribonucleoprotein Enzyme with Two Kinds of Primer Specificity," *Cell* 51:887–98, https://doi.org/10.1016/0092-8674(87)90576-9.

22. Raices, M., Maruyama, H., Dillin, A., and Karlseder, J., 2005, "Uncoupling of Longevity and Telomere Length in *C. elegans*," *PLOS Genetics* 1: e30, https://doi.org/10.1371/journal.pgen.0010030.

23. Lim, C. S., Mian, I. S., Dernburg, A. F., and Campisi, J., 2001, "*C. elegans* clk-2, a Gene that Limits Life Span, Encodes a Telomere Length Regulator Similar to Yeast Telomere Binding Protein Tel2p," *Current Biology* 11:1706–10, https://doi.org/10.1016/s0960-9822(01)00526-7; Ermolaeva, M. A., et al., 2013, "DNA Damage in Germ Cells Induces an Innate Immune Response that Triggers Systemic Stress Resistance," *Nature* 501:416–20, https://doi.org/10.1038/nature12452.

24. Geronimus, A. T., et al., 2015, "Race-Ethnicity, Poverty, Urban Stressors, and Telomere Length in a Detroit Community-Based Sample," *Journal of Health and Social Behavior* 56:199–224, https://doi.org/10.1177/0022146515582100; Needham, B. L., et al., 2013, "Socioeconomic Status, Health Behavior, and Leukocyte Telomere Length in the National Health and Nutrition Examination Survey, 1999–2002," *Social Science and Medicine* 85:1–8, https://doi.org/10.1016/j.socscimed.2013.02.023.

25. Rehkopf, D. H., et al., 2013, "Longer Leukocyte Telomere Length in Costa Rica's Nicoya Peninsula: A Population-Based Study," *Experimental Gerontology* 48:1266–73, https://doi.org/10.1016/j.exger.2013.08.005.

26. Tiku, V., et al., 2017, "Small Nucleoli Are a Cellular Hallmark of Longevity," *Nature Communications* 8:16083, https://doi.org/10.1038/ncomms16083.

27. Neurohr, G. E., et al., 2019, "Excessive Cell Growth Causes Cytoplasm Dilution and Contributes to Senescence," *Cell* 176:1083–97.e18, https://doi.org/10.1016/j.cell.2019.01.018.

28. Neurohr et al. 2019.

29. Tiku et al. 2017.

30. Lengefeld, J., et al., 2021, "Cell Size Is a Determinant of Stem Cell Potential during Aging," *Science Advances* 7:eabk0271, https://doi.org/10.1126/sciadv.abk0271.

31. Coppé, J. P., et al., 2008, "Senescence-Associated Secretory Phenotypes Reveal Cell-Nonautonomous Functions of Oncogenic RAS and the p53 Tumor Suppressor," *PLOS Biology* 6:2853–68, https://doi.org/10.1371/journal.pbio.0060301.

32. Glück, S., et al., 2017, "Innate Immune Sensing of Cytosolic Chromatin Fragments through cGAS Promotes Senescence," *Nature Cell Biology* 19:1061–70, https://doi.org/10.1038

/ncb3586; Dou, Z., et al., 2017, "Cytoplasmic Chromatin Triggers Inflammation in Senescence and Cancer," *Nature* 550:402–6, https://doi.org/10.1038/nature24050.

33. De Cecco, M., et al., 2019, "L1 Drives IFN in Senescent Cells and Promotes Age-Associated Inflammation," *Nature* 566:73–78, https://doi.org/10.1038/s41586-018-0784-9.

34. Ermolaeva et al. 2013.

35. Coppé, J. P., Desprez, P. Y., Krtolica, A., and Campisi, J., 2010, "The Senescence-Associated Secretory Phenotype: The Dark Side of Tumor Suppression," *Annual Review of Pathology* 5:99–118, https://doi.org/10.1146/annurev-pathol-121808-102144.

36. Terlecki-Zaniewicz, L., et al., 2018, "Small Extracellular Vesicles and Their miRNA Cargo Are Anti-apoptotic Members of the Senescence-Associated Secretory Phenotype," *Aging (Albany NY)* 10:1103–32, https://doi.org/10.18632/aging.101452.

37. Wiley, C. D., et al., 2017, "Analysis of Individual Cells Identifies Cell-to-Cell Variability following Induction of Cellular Senescence," *Aging Cell* 16:1043–50, https://doi.org/10.1111/acel.12632.

38. Zhu, Y., et al., 2015, "The Achilles' Heel of Senescent Cells: From Transcriptome to Seno-lytic Drugs," *Aging Cell* 14:644–58, https://doi.org/10.1111/acel.12344; Martel, J., et al., 2020, "Emerging Use of Senolytics and Senomorphics against Aging and Chronic Diseases," *Medicinal Research Reviews* 40:2114–31, https://doi.org/10.1002/med.21702; Schafer, M. J., Haak, A. J., Tschumperlin, D. J., and LeBrasseur, N. K., 2018, "Targeting Senescent Cells in Fibrosis: Pathology, Paradox, and Practical Considerations," *Current Rheumatology Reports* 20:3, https://doi.org/10.1007/s11926-018-0712-x.

39. Jones, M. W., Brett, K., Han, N., and Wyatt, H. A., 2021, "Hyperbaric Physics," in *StatPearls* (Treasure Island, FL: StatPearls).

40. Kamran, P., et al., 2013, "Parabiosis in Mice: A Detailed Protocol," *Journal of Visualized Experiments*, no. 80, e50556, https://doi.org/10.3791/50556.

41. Kosoff, M., 2016, "Peter Thiel Wants to Inject Himself with Young People's Blood," *Vanity Fair*, August 1, 2016, https://www.vanityfair.com/news/2016/08/peter-thiel-wants-to-inject-himself-with-young-peoples-blood; Buhr, S., 2017, "No, Peter Thiel Is Not Harvesting the Blood of the Young," *TechCrunch*, June 14, 2017, https://techcrunch.com/2017/06/14/no-peter-thiel-is-not-harvesting-the-blood-of-the-young/; Haynes, G., 2017, "Ambrosia: The Startup Harvesting the Blood of the Young," *Guardian*, August 21, 2017, https://www.theguardian.com/society/shortcuts/2017/aug/21/ambrosia-the-startup-harvesting-the-blood-of-the-young; "The Blood Boy," *Silicon Valley*, season 4, episode 5 (HBO, 2017).

42. Ludwig, F. C., and Elashoff, R. M., 1972, "Mortality in Syngeneic Rat Parabionts of Different Chronological Age," *Transactions of the New York Academy of Sciences* 34:582–87, https://doi.org/10.1111/j.2164-0947.1972.tb02712.x.

43. Wagers, A. J., Sherwood, R. I., Christensen, J. L., and Weissman, I. L., 2002, "Little Evidence for Developmental Plasticity of Adult Hematopoietic Stem Cells," *Science* 297:2256–59, https://doi.org/10.1126/science.1074807.

44. Conboy, I. M., et al., 2005, "Rejuvenation of Aged Progenitor Cells by Exposure to a Young Systemic Environment," *Nature* 433:760–64, https://doi.org/10.1038/nature03260.

45. Zhang, B., et al., 2021, "Multi-omic Rejuvenation and Lifespan Extension upon Exposure to Youthful Circulation," *bioRxiv*, 2021.11.11.468258, https://doi.org/10.1101/2021.11.11.468258.

46. Loffredo, F. S., et al., 2013, "Growth Differentiation Factor 11 Is a Circulating Factor that Reverses Age-Related Cardiac Hypertrophy," *Cell* 153:828–39, https://doi.org/10.1016/j.cell .2013.04.015.

47. Elabd, C., et al., 2014, "Oxytocin Is an Age-Specific Circulating Hormone that Is Necessary for Muscle Maintenance and Regeneration," *Nature Communications* 5:4082, https://doi .org/10.1038/ncomms5082; Saito, Y., et al., 1998, "Klotho Protein Protects against Endothelial Dysfunction," *Biochemical and Biophysical Research Communications* 248:324–29, https://doi .org/10.1006/bbrc.1998.8943; Zhou, Y., et al., 2021, "Integration of FGF21 Signaling and Metabolomics in High-Fat Diet-Induced Obesity," *Journal of Proteome Research* 20:3900–3912, https://doi.org/10.1021/acs.jproteome.1c00197.

48. Villeda, S. A., et al., 2011, "The Ageing Systemic Milieu Negatively Regulates Neurogenesis and Cognitive Function," *Nature* 477:90–94, https://doi.org/10.1038/nature10357.

49. Villeda, S. A., et al., 2014, "Young Blood Reverses Age-Related Impairments in Cognitive Function and Synaptic Plasticity in Mice," *Nature Medicine* 20:659–63, https://doi.org/10.1038 /nm.3569.

50. Castellano, J. M., et al., 2017, "Human Umbilical Cord Plasma Proteins Revitalize Hippocampal Function in Aged Mice," *Nature* 544:488–92, https://doi.org/10.1038/nature22067.

51. Horowitz, A. M., et al., 2020, "Blood Factors Transfer Beneficial Effects of Exercise on Neurogenesis and Cognition to the Aged Brain," *Science* 369:167–73, https://doi.org/10.1126 /science.aaw2622.

Notes to text box: "Extremophiles: Nature's Superheroes" (Chapter 10, pages 179–180)

a. Innis, M. A., Myambo, K. B., Gelfand, D. H., and Brow, M. A., 1988, "DNA Sequencing with *Thermus aquaticus* DNA Polymerase and Direct Sequencing of Polymerase Chain Reaction-Amplified DNA," *Proceedings of the National Academy of Sciences of the USA* 85:9436–40, https://doi.org/10.1073/pnas.85.24.9436.

b. Moseley, B. E., and Setlow, J. K., 1968, "Transformation in *Micrococcus radiodurans* and the Ultraviolet Sensitivity of Its Transforming DNA," *Proceedings of the National Academy of Sciences of the USA* 61:176–83, https://doi.org/10.1073/pnas.61.1.176.

c. Gaudin, M., et al., 2013, "Hyperthermophilic Archaea Produce Membrane Vesicles That Can Transfer DNA," *Environmental Microbiology Reports* 5:109–16, https://doi.org/10.1111/j .1758-2229.2012.00348.x.

d. Krupovic, M., Gonnet, M., Hania, W. B., Forterre, P., and Erauso, G., 2013, "Insights into Dynamics of Mobile Genetic Elements in Hyperthermophilic Environments from Five New *Thermococcus* Plasmids," *PLOS One* 8:e49044, https://doi.org/10.1371/journal.pone.0049044.

e. Van Wolferen, M., Ajon, M., Driessen, A. J., and Albers, S. V., 2013, "How Hyperthermophiles Adapt to Change Their Lives: DNA Exchange in Extreme Conditions," *Extremophiles* 17:545–63, https://doi.org/10.1007/s00792-013-0552-6.

f. Vreeland, R. H., Rosenzweig, W. D., and Powers, D. W., 2000, "Isolation of a 250 Million-Year-Old Halotolerant Bacterium from a Primary Salt Crystal," *Nature* 407:897–900, https:// doi.org/10.1038/35038060.

g. Madhusoodanan, J., 2014, "Microbial Stowaways to Mars Identified," *Nature*, May 19, 2014, https://doi.org/10.1038/nature.2014.15249.

h. Mastascusa, V., et al., 2014, "Extremophiles Survival to Simulated Space Conditions: An Astrobiology Model Study," *Origins of Life and Evolution of Biospheres* 44:231–37, https://doi .org/10.1007/s11084-014-9397-y.

i. Sloan, D., Alves Batista, R., and Loeb, A., 2017, "The Resilience of Life to Astrophysical Events," *Scientific Reports* 7:5419, https://doi.org/10.1038/s41598-017-05796-x.

j. Vecchi, M., et al., 2021, "The Toughest Animals of the Earth versus Global Warming: Effects of Long-Term Experimental Warming on Tardigrade Community Structure of a Temperate Deciduous Forest," *Ecology and Evolution* 11:9856–63, https://doi.org/10.1002/ece3 .7816.

Chapter 11

1. Kluger, J., 2013, "Too Old to Be a Dad?," *Time*, April 11, 2013, https://healthland.time .com/2013/04/11/too-old-to-be-a-dad/.

2. Shadyab, A. H., et al., 2017, "Maternal Age at Childbirth and Parity as Predictors of Longevity among Women in the United States: The Women's Health Initiative," *American Journal of Public Health* 107:113–19, https://doi.org/10.2105/AJPH.2016.303503.

3. Sun, F., et al., 2015, "Extended Maternal Age at Birth of Last Child and Women's Longevity in the Long Life Family Study," *Menopause* 22:26–31, https://doi.org/10.1097/GME .0000000000000276.

4. Perls, T. T., Alpert, L., and Fretts, R. C., 1997, "Middle-Aged Mothers Live Longer," *Nature* 389:133, https://doi.org/10.1038/38148.

5. Fagan, E., et al., 2017, "Telomere Length Is Longer in Women with Late Maternal Age," *Menopause* 24:497–501, https://doi.org/10.1097/GME.0000000000000795.

6. Steiner, A. Z., et al., 2017, "Association between Biomarkers of Ovarian Reserve and Infertility Among Older Women of Reproductive Age," *JAMA* 318:1367–76, https://doi.org/10 .1001/jama.2017.14588.

7. Liss, J., et al., 2017, "Clinical Utility of Different Anti-Mullerian Hormone: AMH Assays for the Purpose of Pregnancy Prediction," *Gynecological Endocrinology* 33:791–96, https://doi .org/10.1080/09513590.2017.1318370; Depmann, M., et al., 2017, "Anti-Mullerian Hormone Does Not Predict Time to Pregnancy: Results of a Prospective Cohort Study," *Gynecological Endocrinology* 33:644–48, https://doi.org/10.1080/09513590.2017.1306848; Warzecha, D., Szymusik, I., Pietrzak, B., and Wielgos, M., 2017, "Anti-Mullerian Hormone: A Marker of Upcoming Menopause or a Questionable Guesswork?," *Neuro Endocrinology Letters* 38:75–82.

8. Stolk, L., et al., 2012, "Meta-analyses Identify 13 Loci Associated with Age at Menopause and Highlight DNA Repair and Immune Pathways," *Nature Genetics* 44:260–68, https://doi .org/10.1038/ng.1051; He, C., et al., 2009, "Genome-wide Association Studies Identify Loci Associated with Age at Menarche and Age at Natural Menopause," *Nature Genetics* 41:724–28, https://doi.org/10.1038/ng.385; Stolk, L., et al., 2009, "Loci at Chromosomes 13, 19 and 20 Influence Age at Natural Menopause," *Nature Genetics* 41:645–47, https://doi.org/10.1038 /ng.387.

9. Ossewaarde, M. E., et al., 2005, "Age at Menopause, Cause-Specific Mortality and Total Life Expectancy," *Epidemiology* 16:556–62, https://doi.org/10.1097/01.ede.0000165392 .35273.d4.

10. Mills, P. K., Beeson, W. L., Phillips, R. L., and Fraser, G. E., 1989, "Prospective Study of Exogenous Hormone Use and Breast Cancer in Seventh-Day Adventists," *Cancer* 64:591–97, https://doi.org/10.1002/1097-0142(19890801)64:3<591::aid-cncr2820640305>3.0.co;2-u.

11. Levine, M. E., et al., 2016, "Menopause Accelerates Biological Aging," *Proceedings of the National Academy of Sciences of the USA* 113:9327–32, https://doi.org/10.1073/pnas.1604558113.

12. Horvath, S., 2013, "DNA Methylation Age of Human Tissues and Cell Types," *Genome Biology* 14:R115, https://doi.org/10.1186/gb-2013-14-10-r115.

13. Levine et al. 2016.

14. Levine et al. 2016.

15. Hsin, H., and Kenyon, C., 1999, "Signals from the Reproductive System Regulate the Lifespan of *C. elegans*," *Nature* 399:362–66.

16. Arantes-Oliveira, N., Apfeld, J., Dillin, A., and Kenyon, C., 2002, "Regulation of Life-Span by Germ-Line Stem Cells in *Caenorhabditis elegans*," *Science* 295:502–5.

17. Ghazi, A., Henis-Korenblit, S., and Kenyon, C., 2009, "A Transcription Elongation Factor that Links Signals from the Reproductive System to Lifespan Extension in *Caenorhabditis elegans*," *PLOS Genetics* 5:e1000639, https://doi.org/10.1371/journal.pgen.1000639; Amrit, F.R.G., et al., 2019, "The Longevity-Promoting Factor, TCER-1, Widely Represses Stress Resistance and Innate Immunity," *Nature Communications* 10:3042, https://doi.org/10.1038/s41467-019-10759-z; Amrit, F.R., et al., 2016, "DAF-16 and TCER-1 Facilitate Adaptation to Germline Loss by Restoring Lipid Homeostasis and Repressing Reproductive Physiology in *C. elegans*," *PLOS Genetics* 12:e1005788, https://doi.org/10.1371/journal.pgen.1005788.

18. Shi, C., Booth, L. N., and Murphy, C. T., 2019, "Insulin-like Peptides and the mTOR-TFEB Pathway Protect *Caenorhabditis elegans* Hermaphrodites from Mating-Induced Death," *eLife* 8: e46413, https://doi.org/10.7554/eLife.46413; Lee, H. J., et al., 2019, "Prostaglandin Signals from Adult Germ Stem Cells Delay Somatic Aging of *Caenorhabditis elegans*," *Nature Metabolism* 1:790–810, https://doi.org/10.1038/s42255-019-0097-9.

19. Flatt, T., et al., 2008, "*Drosophila* Germ-Line Modulation of Insulin Signaling and Lifespan," *Proceedings of the National Academy of Sciences of the USA* 105:6368–73, https://doi.org/10.1073/pnas.0709128105.

20. Cargill, S. L., Carey, J. R., Muller, H. G., and Anderson, G., 2003, "Age of Ovary Determines Remaining Life Expectancy in Old Ovariectomized Mice," *Aging Cell* 2:185–90, https://doi.org/10.1046/j.1474-9728.2003.00049.x.

21. Min, K. J., Lee, C. K., and Park, H. N., 2012, "The Lifespan of Korean Eunuchs," *Current Biology* 22:R792–93, https://doi.org/10.1016/j.cub.2012.06.036; Le Bourg, E.. 2015, "No Ground for Advocating that Korean Eunuchs Lived Longer than Intact Men," *Gerontology* 62:69–70, https://doi.org/10.1159/000435854.

22. Jiang, Z., and Shen, H., 2021, "Mitochondria: Emerging Therapeutic Strategies for Oocyte Rescue," *Reproductive Sciences*, https://doi.org/10.1007/s43032-021-00523-4.

23. Cota, V., and Murphy, C. T., 2021, "Fission and PINK-1-Mediated Mitophagy Are Required for Insulin/IGF-1 Signaling Mutant Reproductive Longevity," *bioRxiv*, 2021.08.16.456566, https://doi.org/10.1101/2021.08.16.456566; Lesnik, C., Cota, V.,

Sohrabi, S., and Murphy, C. T., "BCAA Metabolism Is Critical for Oocyte Quality" (MS in preparation).

24. Hamatani, T., et al., 2004, "Age-Associated Alteration of Gene Expression Patterns in Mouse Oocytes," *Human Molecular Genetics* 13:2263–78; Steuerwald, N. M., Bermudez, M. G., Wells, D., Munne, S., and Cohen, J., 2007, "Maternal Age-Related Differential Global Expression Profiles Observed in Human Oocytes," *Reproductive Biomedicine Online* 14:700–708, https://doi.org/10.1016/s1472-6483(10)60671-2.

25. Hughes, S. E., Evason, K., Xiong, C., and Kornfeld, K., 2007, "Genetic and Pharmacological Factors that Influence Reproductive Aging in Nematodes," *PLOS Genetics* 3:e25.

26. Luo, S., Kleemann, G. A., Ashraf, J. M., Shaw, W. M., and Murphy, C. T., 2010, "TGF-Beta and Insulin Signaling Regulate Reproductive Aging via Oocyte and Germline Quality Maintenance," *Cell* 143:299–312, https://doi.org/10.1016/j.cell.2010.09.013.

27. Luo et al. 2010.

28. Hamatani et al. 2004.

29. Savage-Dunn, C., et al., 2000, "SMA-3 smad Has Specific and Critical Functions in DBL-1/SMA-6 TGFbeta-Related Signaling," *Developmental Biology* 223:70–76.

30. Luo, S., Shaw, W. M., Ashraf, J., and Murphy, C. T., 2009, "TGF-Beta Sma/Mab Signaling Mutations Uncouple Reproductive Aging from Somatic Aging," *PLOS Genetics* 5:e1000789, https://doi.org/10.1371/journal.pgen.1000789.

31. Shaw, W. M., Luo, S., Landis, J., Ashraf, J., and Murphy, C. T., 2007, "The *C. elegans* TGF-Beta Dauer Pathway Regulates Longevity via Insulin Signaling," *Current Biology* 17:1635–45, https://doi.org/10.1016/j.cub.2007.08.058.

32. Huang, C., Xiong, C., and Kornfeld, K., 2004, "Measurements of Age-Related Changes of Physiological Processes that Predict Lifespan of *Caenorhabditis elegans*," *Proceedings of the National Academy of Sciences of the USA* 101:8084–89.

33. Luo et al. 2009.

34. Luo et al. 2010; Templeman, N. M., et al., 2018, "Insulin Signaling Regulates Oocyte Quality Maintenance with Age via Cathepsin B Activity," *Current Biology* 28:753–60.e4, https://doi.org/10.1016/j.cub.2018.01.052.

35. Templeman et al. 2018.

36. Balboula, A. Z., et al., 2010, "Cathepsin B Activity Is Related to the Quality of Bovine Cumulus Oocyte Complexes and Its Inhibition Can Improve Their Developmental Competence," *Molecular Reproduction and Development* 77:439–48, https://doi.org/10.1002/mrd.21164.

37. Perls, Alpert, and Fretts 1997.

38. Hawkes, K., 2020, "The Centrality of Ancestral Grandmothering in Human Evolution," *Integrative and Comparative Biology* 60:765–81, https://doi.org/10.1093/icb/icaa029. Interestingly, the Sarah Treem play *The How and the Why* revolves around two female scientists studying the grandmother effect.

39. Maliha, G., and Murphy, C. T., 2016, "A Simple Offspring-to-Mother Size Ratio Predicts Post-reproductive Lifespan," *bioRxiv*, 048835, https://doi.org/10.1101/048835.

40. Bodnar, A. G., 2015, "Cellular and Molecular Mechanisms of Negligible Senescence: Insight from the Sea Urchin," *Invertebrate Reproduction and Development* 59:23–27, https://doi.org/10.1080/07924259.2014.938195.

41. Luo et al. 2009.

Chapter 12

1. "Interview with Elizabeth Arias: Decline in Life Expectancy in 2020," CDC, National Center for Health Statistics, February 19, 2021, https://www.cdc.gov/nchs/pressroom /podcasts/2021/20210219/20210219.htm.

2. Ostan, R., et al., 2016, "Gender, Aging and Longevity in Humans: An Update of an Intriguing/Neglected Scenario Paving the Way to a Gender-Specific Medicine," *Clinical Science (London)* 130:1711–25, https://doi.org/10.1042/cs20160004.

3. Ostan et al. 2016; Jack, C. R., Jr., et al., 2015, "Age, Sex and APOE ε4 Effects on Memory, Brain Structure and β-amyloid across the Adult Lifespan," *JAMA Neurology* 72:511–19, https:// doi.org/10.1001/jamaneurol.2014.4821; Davis, E. J., et al., 2020, "A Second X Chromosome Contributes to Resilience in a Mouse Model of Alzheimer's Disease," *Science Translational Medicine* 12:eaaz5677, https://doi.org/10.1126/scitranslmed.aaz5677.

4. Zeng, Y., et al., 2018, "Sex Differences in Genetic Associations with Longevity," *JAMA Network Open* 1:e181670, https://doi.org/10.1001/jamanetworkopen.2018.1670.

5. Anselmi, C. V., et al., 2009, "Association of theFOXO3A Locus with Extreme Longevity in a Southern Italian Centenarian Study," *Rejuvenation Research* 12:95–104, https://doi.org/10 .1089/rej.2008.0827.

6. Gems, D., 2014, "Evolution of Sexually Dimorphic Longevity in Humans," *Aging (Albany NY)* 6:84–91, https://doi.org/10.18632/aging.100640.

7. Hawkes, K., 2020, "The Centrality of Ancestral Grandmothering in Human Evolution," *Integrative and Comparative Biology* 60:765–81, https://doi.org/10.1093/icb/icaa029.

8. Dobson, R., 2006, "The Stress of Marriage Shortens Your Life by a Year (if You're the Wife)," *Independent*, February 26, 2006, https://www.independent.co.uk/life-style/health-and-families /health-news/the-stress-of-marriage-shortens-your-life-by-a-year-if-youre-the-wife-5335547.html.

9. Ostan et al. 2016.

10. Jasienska, G., Nenko, I., and Jasienski, M., 2006, "Daughters Increase Longevity of Fathers, but Daughters and Sons Equally Reduce Longevity of Mothers," *American Journal of Human Biology* 18:422–25, https://doi.org/10.1002/ajhb.20497.

11. Adebowale, A. S., 2018, "Spousal Age Difference and Associated Predictors of Intimate Partner Violence in Nigeria," *BMC Public Health* 18:212, https://doi.org/10.1186/s12889-018 -5118-1.

12. Metzler, S., Heinze, J., and Schrempf, A., 2016, "Mating and Longevity in Ant Males," *Ecology and Evolution* 6:8903–6, https://doi.org/10.1002/ece3.2474.

13. Becker, J. B., Prendergast, B. J., and Liang, J. W., 2016, "Female Rats Are Not More Variable than Male Rats: A Meta-analysis of Neuroscience Studies," *Biology of Sex Differences* 7:34, https://doi.org/10.1186/s13293-016-0087-5.

14. Becker, Prendergast, and Liang 2016.

15. Smarr, B., and Kriegsfeld, L. J., 2022, "Female Mice Exhibit Less Overall Variance, with a Higher Proportion of Structured Variance, than Males at Multiple Timescales of Continuous Body Temperature and Locomotive Activity Records," *Biology of Sex Differences* 13:41, https:// doi.org/10.1186/s13293-022-00451-1.

16. Harrison, D. E., et al., 2014, "Acarbose, 17-α-estradiol, and Nordihydroguaiaretic Acid Extend Mouse Lifespan Preferentially in Males," *Aging Cell* 13:273–82, https://doi.org/10.1111 /acel.12170.

17. Kanfi, Y., et al., 2012, "The Sirtuin SIRT6 Regulates Lifespan in Male Mice," *Nature* 483:218–21, https://doi.org/10.1038/nature10815.

18. Kanfi et al. 2012.

19. Kane, A. E., Sinclair, D. A., Mitchell, J. R., and Mitchell, S. J., 2018, "Sex Differences in the Response to Dietary Restriction in Rodents," *Current Opinion in Physiology* 6:28–34, https://doi.org/10.1016/j.cophys.2018.03.008.

20. Davis, E. J., Lobach, I., and Dubal, D. B., 2019, "Female XX Sex Chromosomes Increase Survival and Extend Lifespan in Aging Mice," *Aging Cell* 18:e12871, https://doi.org/10.1111/acel.12871.

21. Tower, J., 2006, "Sex-Specific Regulation of Aging and Apoptosis," *Mechanisms of Ageing and Development* 127:705–18, https://doi.org/10.1016/j.mad.2006.05.001.

22. Garratt, M., 2020, "Why Do Sexes Differ in Lifespan Extension? Sex-Specific Pathways of Aging and Underlying Mechanisms for Dimorphic Responses," *Nutrition and Healthy Aging* 5:247–59, https://doi.org/10.3233/NHA-190067; Clancy, D. J., et al., 2001, "Extension of Life-Span by Loss of CHICO, a *Drosophila* Insulin Receptor Substrate Protein," *Science* 292:104–6, https://doi.org/10.1126/science.1057991; Giannakou, M. E., et al., 2004, "Long-Lived *Drosophila* with Overexpressed dFOXO in Adult Fat Body," *Science* 305:361, https://doi.org/10.1126/science.1098219; Tatar, M., et al., 2001, "A Mutant *Drosophila* Insulin Receptor Homolog that Extends Life-Span and Impairs Neuroendocrine Function," *Science* 292:107–10, https://doi.org/10.1126/science.1057987.

23. Ryan, D. A., et al., 2014, "Sex, Age, and Hunger Regulate Behavioral Prioritization through Dynamic Modulation of Chemoreceptor Expression," *Current Biology* 24:2509–17, https://doi.org/10.1016/j.cub.2014.09.032.

24. Gems, D., and Riddle, D. L., 2000, "Genetic, Behavioral and Environmental Determinants of Male Longevity in *Caenorhabditis elegans*," *Genetics* 154:1597–1610.

25. Klass, M. R., 1983, "A Method for the Isolation of Longevity Mutants in the Nematode *Caenorhabditis elegans* and Initial Results," *Mechanisms of Ageing and Development* 22:279–86, https://doi.org/10.1016/0047-6374(83)90082-9.

26. Shi, C., Runnels, A. M., and Murphy, C. T., 2017, "Mating and Male Pheromone Kill *Caenorhabditis* Males through Distinct Mechanisms," *eLife* 6:e23493, https://doi.org/10.7554/eLife.23493.

27. Hsin, H., and Kenyon, C., 1999, "Signals from the Reproductive System Regulate the Lifespan of *C. elegans*. *Nature* 399:362–66.

28. McCulloch, D., and Gems, D., 2003, "Evolution of Male Longevity Bias in Nematodes," *Aging Cell* 2:165–73, https://doi.org/10.1046/j.1474-9728.2003.00047.x.

29. Lee, S. J., Murphy, C. T., and Kenyon, C., 2009, "Glucose Shortens the Life Span of *C. elegans* by Downregulating DAF-16/FOXO Activity and Aquaporin Gene Expression," *Cell Metabolism* 10:379–91, https://doi.org/10.1016/j.cmet.2009.10.003; Liggett, M. R., Hoy, M. J., Mastroianni, M,. and Mondoux, M. A., 2015, "High-Glucose Diets Have Sex-Specific Effects on Aging in *C. elegans*: Toxic to Hermaphrodites but Beneficial to Males," *Aging (Albany NY)* 7:383–88, https://doi.org/10.18632/aging.100759.

30. Honjoh, S., Ihara, A., Kajiwara, Y., Yamamoto, T., and Nishida, E., 2017, "The Sexual Dimorphism of Dietary Restriction Responsiveness in *Caenorhabditis elegans*," *Cell Reports* 21:3646–52, https://doi.org/10.1016/j.celrep.2017.11.108.

31. Honjoh et al. 2017.

32. Garrison, J. L., et al., 2012, "Oxytocin/Vasopressin-Related Peptides Have an Ancient Role in Reproductive Behavior," *Science* 338:540–43, https://doi.org/10.1126/science.1226201.

33. Fetter-Pruneda, I., et al., 2021, "An Oxytocin/Vasopressin-Related Neuropeptide Modulates Social Foraging Behavior in the Clonal Raider Ant," *PLOS Biology* 19:e3001305, https://doi.org/10.1371/journal.pbio.3001305.

34. Shi, C., Runnels, A. M., and Murphy, C. T., 2016, "Mating-Induced Male Death and Pheromone Toxin-Regulated Androstasis," *bioRxiv*, 034181, https://doi.org/10.1101/034181.

35. Headland, M., Clifton, P. M., Carter, S., and Keogh, J. B., 2016, "Weight-Loss Outcomes: A Systematic Review and Meta-analysis of Intermittent Energy Restriction Trials Lasting a Minimum of 6 Months," *Nutrients* 8:354, https://doi.org/10.3390/nu8060354.

36. Headland et al. 2016.

37. Chapman, T., Herndon, L. A., Heifetz, Y., Partridge, L., and Wolfner, M. F., 2001, "The Acp26Aa Seminal Fluid Protein Is a Modulator of Early Egg Hatchability in *Drosophila melanogaster*," *Proceedings of the Royal Society B: Biological Sciences* 268:1647–54, https://doi.org/10.1098/rspb.2001.1684; Chapman, T., Hutchings, J., and Partridge, L., 1993, "No Reduction in the Cost of Mating for *Drosophila melanogaster* Females Mating with Spermless Males," *Proceedings of the Royal Society B: Biological Sciences* 253:211–17, https://doi.org/10.1098/rspb.1993.0105.

38. Chapman et al. 2001.

39. Chapman, T., et al., 2003, "The Sex Peptide of *Drosophila melanogaster*: Female Post-Mating Responses Analyzed by Using RNA Interference," *Proceedings of the National Academy of Sciences of the USA* 100:9923–28, https://doi.org/10.1073/pnas.1631635100.

40. Shi, C., and Murphy, C. T., 2014, "Mating Induces Shrinking and Death in *Caenorhabditis* Mothers," *Science* 343:536–40, https://doi.org/10.1126/science.1242958.

41. Gems, D., and Riddle, D. L., 1996, "Longevity in *Caenorhabditis elegans* Reduced by Mating but Not Gamete Production," *Nature* 379:723–25.

42. Shi and Murphy 2014; Shi, C., Booth, L. N., and Murphy, C. T., 2019, "Insulin-like Peptides and the mTOR-TFEB Pathway Protect *Caenorhabditis elegans* Hermaphrodites from Mating-Induced Death," *eLife* 8:e46413, https://doi.org/10.7554/eLife.46413.

43. Morsci, N. S., Haas, L. A., and Barr, M. M., 2011, "Sperm Status Regulates Sexual Attraction in *Caenorhabditis elegans*," *Genetics* 189:1341–46, https://doi.org/10.1534/genetics.111.133603; Shi and Murphy 2014.

44. Shi and Murphy 2014.

45. Night Editor, 2013, "Sex Is 'Kiss of Death' for Female Worms because PATRIARCHY," *Jezebel*, December 23, 2013, https://jezebel.com/sex-is-kiss-of-death-for-female-worms-because-patriar-1488279886.

46. Maures, T. J., et al., 2014, "Males Shorten the Life Span of *C. elegans* Hermaphrodites via Secreted Compounds," *Science* 343:541–44, https://doi.org/10.1126/science.1244160.

47. Maures et al. 2014.

48. Gendron, C. M., et al., 2014, "*Drosophila* Life Span and Physiology Are Modulated by Sexual Perception and Reward," *Science* 343:544–48, https://doi.org/10.1126/science.1243339.

49. Booth, L. N., Maures, T. J., Yeo, R. W., Tantilert, C., and Brunet, A., 2019, "Self-Sperm Induce Resistance to the Detrimental Effects of Sexual Encounters with Males in Hermaphroditic Nematodes," *eLife* 8:e46418, https://doi.org/10.7554/eLife.46418.

50. Shi, Booth, and Murphy 2019.

51. Morsci, Haas, and Barr 2011.

52. Shi, Booth, and Murphy 2019.

53. Morsci, Haas, and Barr 2011.

54. Shi, Runnels, and Murphy 2016.

55. Maures et al. 2014.

56. Shi, Runnels, and Murphy 2016, 2017.

57. Shi, Runnels, and Murphy 2017.

58. Shi, Runnels, and Murphy 2017.

59. Seah, N. E., et al., 2016, "Autophagy-Mediated Longevity Is Modulated by Lipoprotein Biogenesis," *Autophagy* 12:261–72, https://doi.org/10.1080/15548627.2015.1127464; Garigan, D., et al., 2002, "Genetic Analysis of Tissue Aging in *Caenorhabditis elegans*: A Role for Heat-Shock Factor and Bacterial Proliferation," *Genetics* 161:1101–12.

60. Shi, Runnels, and Murphy 2016, 2017.

61. Shi and Murphy 2014.

62. Morsci, Haas, and Barr 2011.

63. Shi, Runnels, and Murphy 2017.

64. Morsci, Haas, and Barr 2011.

65. Shi and Murphy 2014.

66. Shi and Murphy 2014.

67. Grosser, B. I., Monti-Bloch, L., Jennings-White, C., and Berliner, D. L., 2000, "Behavioral and Electrophysiological Effects of Androstadienone, a Human Pheromone," *Psychoneuroendocrinology* 25:289299, https://doi.org/10.1016/S0306-4530(99)00056-6.

68. Moore, A. J., Gowaty, P. A., and Moore, P. J., 2003, "Females Avoid Manipulative Males and Live Longer," *Journal of Evolutionary Biology* 16:523–30, https://doi.org/10.1046/j.1420-9101.2003.00527.x.

69. Rantala, M. J., Jokinen, I., Kortet, R., Vainikka, A., and Suhonen, J., 2002, "Do Pheromones Reveal Male Immunocompetence?," *Proceedings of the Royal Society B: Biological Sciences* 269:1681–85, https://doi.org/10.1098/rspb.2002.2056.

70. Ober, C., et al., 1997, "HLA and Mate Choice in Humans," *American Journal of Human Genetics* 61:497–504, https://doi.org/10.1086/515511.

71. Savic, I., Berglund, H., Gulyas, B., and Roland, P., 2001, "Smelling of Odorous Sex Hormone-like Compounds Causes Sex-Differentiated Hypothalamic Activations in Humans," *Neuron* 31:661–68, https://doi.org/10.1016/S0896-6273(01)00390-7.

72. Strassmann, B. I., 1999, "Menstrual Synchrony Pheromones: Cause for Doubt," *Human Reproduction* 14:579–80, https://doi.org/10.1093/humrep/14.3.579.

73. Grosser et al. 2000; Gustavson, A. R., Dawson, M. E., and Bonett, D. G., 1987, "Androstenol, a Putative Human Pheromone, Affects Human (*Homo sapiens*) Male Choice Performance," *Journal of Comparative Psychology* 101:210–12, https://doi.org/10.1037/0735-7036.101.2.210; Mast, T. G., and Samuelsen, C. L., 2009, "Human Pheromone Detection by the Vomeronasal Organ: Unnecessary for Mate Selection?," *Chemical Senses* 34:529–31, https://doi.org/10.1093/chemse/bjp030.

74. Savic et al. 2001.

75. Mast and Samuelsen 2009.

76. Kohl, J. V., 2012, "Human Pheromones and Food Odors: Epigenetic Influences on the Socioaffective Nature of Evolved Behaviors," *Socioaffective Neuroscience and Psychology* 2:17338, https://doi.org/10.3402/snp.v2i0.17338.

77. Doctor Ben, 2017, "Emperor Wuzong (武宗)—Zhengde (正德) Reign Period of the Ming Dynasty (1506 AD–1521 AD)," drben.net, last updated May 22, 2017, http://www.drben.net/ChinaReport/Sources/History/Ming/Ming_Dynasty-Reign_Wuzong-1505AD-1521AD.html.

78. McMahon, K., 2016, *Celestial Women: Imperial Wives and Concubines in China from Song to Qing* (Lanham, MD: Rowman and Littlefield). Regarding the Wuzong emperor, subtitles of the contents are: "An Emperor Who Liked to Roam," "No More Keeping Track of Visitations," "Men Who 'Slept and Rose with the Emperor,'" and "Stolen Women and Muslim Dancers."

79. Shi, Runnels, and Murphy 2016.

80. Shi, Runnels, and Murphy 2016.

Chapter 13

1. Apfeld, J., and Kenyon, C., 1999, "Regulation of Lifespan by Sensory Perception in *Caenorhabditis elegans*," *Nature* 402:804–9.

2. Schroeder, N. E., et al., 2013, "Dauer-Specific Dendrite Arborization in *C. elegans* Is Regulated by KPC-1/Furin," *Current Biology* 23:1527–35, https://doi.org/10.1016/j.cub.2013.06.058; Golden, J. W., and Riddle, D. L., 1984, "The *Caenorhabditis elegans* Dauer Larva: Developmental Effects of Pheromone, Food, and Temperature," *Developmental Biology* 102:368–78; Albert, P. S., Brown, S. J., and Riddle, D. L., 1981, "Sensory Control of Dauer Larva Formation in *Caenorhabditis elegans*," *Journal of Comparative Neurology* 198:435–51.

3. Lee, H., et al., 2011, "Nictation, a Dispersal Behavior of the Nematode *Caenorhabditis elegans*, Is Regulated by IL2 Neurons," *Nature Neuroscience* 15:107–12, https://doi.org/10.1038/nn.2975.

4. Bargmann, C. I., and Horvitz, H. R., 1991, "Control of Larval Development by Chemosensory Neurons in *Caenorhabditis elegans*," *Science* 251:1243–46.

5. Alcedo, J., and Kenyon, C., 2004, "Regulation of *C. elegans* Longevity by Specific Gustatory and Olfactory Neurons," *Neuron* 41:45–55.

6. Libina, N., Berman, J. R., and Kenyon, C., 2003, "Tissue-Specific Activities of *C. elegans* DAF-16 in the Regulation of Lifespan," *Cell* 115:489–502.

7. Hsu, A. L., Murphy, C. T., and Kenyon, C., 2003, "Regulation of Aging and Age-Related Disease by DAF-16 and Heat-Shock Factor," *Science* 300:1142–45; Higuchi-Sanabria, R., et al., 2018, "Spatial Regulation of the Actin Cytoskeleton by HSF-1 during Aging," *Molecular Biology of the Cell* 29:2522–27, https://doi.org/10.1091/mbc.E18-06-0362.

8. Bishop, N. A., and Guarente, L., 2007, "Two Neurons Mediate Diet-Restriction-Induced Longevity in *C. elegans*," *Nature* 447:545–49, https://doi.org/10.1038/nature05904; An, J. H., and Blackwell, T. K., 2003, "SKN-1 Links *C. elegans* Mesendodermal Specification to a Conserved Oxidative Stress Response," *Genes and Development* 17:1882–93, https://doi.org/10.1101/gad.1107803.

9. Cornils, A., Gloeck, M., Chen, Z., Zhang, Y., and Alcedo, J., 2011, "Specific Insulin-like Peptides Encode Sensory Information to Regulate Distinct Developmental Processes,"

Development 138:1183–93, https://doi.org/10.1242/dev.060905; Fernandes de Abreu, D. A., et al., 2014, "An Insulin-to-Insulin Regulatory Network Orchestrates Phenotypic Specificity in Development and Physiology," *PLOS Genetics* 10:e1004225, https://doi.org/10.1371/journal.pgen.1004225; Zhang, B., et al., 2018, "Brain-Gut Communications via Distinct Neuroendocrine Signals Bidirectionally Regulate Longevity in *C. elegans*," *Genes and Development* 32:258–70, https://doi.org/10.1101/gad.309625.117; Burkewitz, K., et al., 2015, "Neuronal CRTC-1 Governs Systemic Mitochondrial Metabolism and Lifespan via a Catecholamine Signal," *Cell* 160:842–55, https://doi.org/10.1016/j.cell.2015.02.004; Miller, H. A., Dean, E. S., Pletcher, S. D., and Leiser, S. F., 2020, "Cell Non-autonomous Regulation of Health and Longevity," *eLife* 9:e62659, https://doi.org/10.7554/eLife.62659.

10. Finger, F., and Hoppe, T., 2014, "MicroRNAs Meet Calcium: Joint Venture in ER Proteostasis," *Science Signaling* 7:re11, https://doi.org/10.1126/scisignal.2005671.

11. Boulias, K., and Horvitz, H. R., 2012, "The *C. elegans* MicroRNA mir-71 Acts in Neurons to Promote Germline-Mediated Longevity through Regulation of DAF-16/FOXO," *Cell Metabolism* 15:439–50, https://doi.org/10.1016/j.cmet.2012.02.014.

12. Maier, W., Adilov, B., Regenass, M., and Alcedo, J., 2010, "A Neuromedin U Receptor Acts with the Sensory System to Modulate Food Type-Dependent Effects on *C. elegans* Lifespan," *PLOS Biology* 8:e1000376, https://doi.org/10.1371/journal.pbio.1000376.

13. Frakes, A. E., et al., 2020, "Four Glial Cells Regulate ER Stress Resistance and Longevity via Neuropeptide Signaling in *C. elegans*," *Science* 367:436–40, https://doi.org/10.1126/science.aaz6896.

14. Nässel, D. R., Kubrak, O. I., Liu, Y., Luo, J., and Lushchak, O. V., 2013, "Factors that Regulate Insulin Producing Cells and Their Output in *Drosophila*," *Frontiers in Physiology* 4:252, https://doi.org/10.3389/fphys.2013.00252; Pierce, S. B., et al., 2001, "Regulation of DAF-2 Receptor Signaling by Human Insulin and *ins-1*, a Member of the Unusually Large and Diverse *C. elegans* Insulin Gene Family," *Genes and Development* 15:672–86.

15. Broughton, S., et al., 2008, "Reduction of DILP2 in *Drosophila* Triages a Metabolic Phenotype from Lifespan Revealing Redundancy and Compensation among DILPs," *PLOS One* 3:e3721, https://doi.org/10.1371/journal.pone.0003721; Flatt, T., et al., 2008, "*Drosophila* Germ-Line Modulation of Insulin Signaling and Lifespan," *Proceedings of the National Academy of Sciences of the USA* 105:6368–73, https://doi.org/10.1073/pnas.0709128105.

16. Ostojic, I., et al., 2014, "Positive and Negative Gustatory Inputs Affect *Drosophila* Lifespan Partly in Parallel to dFOXO Signaling," *Proceedings of the National Academy of Sciences of the USA* 111:8143–48, https://doi.org/10.1073/pnas.1315466111; Libert, S., et al., 2007, "Regulation of *Drosophila* Life Span by Olfaction and Food-Derived Odors," *Science* 315:1133–37, https://doi.org/10.1126/science.1136610.

17. Libert et al. 2007.

18. Ro, J., et al., 2016, "Serotonin Signaling Mediates Protein Valuation and Aging," *eLife* 5:e16843, https://doi.org/10.7554/eLife.16843; Ro, J., Harvanek, Z. M., and Pletcher, S. D., 2014, "FLIC: High-Throughput, Continuous Analysis of Feeding Behaviors in *Drosophila*," *PLOS One* 9:e101107, https://doi.org/10.1371/journal.pone.0101107.

19. Gendron, C. M., et al., 2014, "*Drosophila* Life Span and Physiology Are Modulated by Sexual Perception and Reward," *Science* 343:544–48, https://doi.org/10.1126/science.1243339.

20. Riera, C. E., et al., 2017, "The Sense of Smell Impacts Metabolic Health and Obesity," *Cell Metabolism* 26:198–211.e5, https://doi.org/10.1016/j.cmet.2017.06.015.

21. Riera et al. 2017.

22. Lee, S. J., and Kenyon, C., 2009, "Regulation of the Longevity Response to Temperature by Thermosensory Neurons in *Caenorhabditis elegans*," *Current Biology* 19:715–22, https://doi.org/10.1016/j.cub.2009.03.041.

23. Chen, Y. C., et al., 2016, "A *C. elegans* Thermosensory Circuit Regulates Longevity through *crh-1*/CREB-Dependent *flp-6* Neuropeptide Signaling," *Developmental Cell* 39:209–23, https://doi.org/10.1016/j.devcel.2016.08.021.

24. Xiao, R., et al., 2013, "A Genetic Program Promotes *C. elegans* Longevity at Cold Temperatures via a Thermosensitive TRP Channel," *Cell* 152:806–17, https://doi.org/10.1016/j.cell.2013.01.020; Zhang, B., et al., 2015, "Environmental Temperature Differentially Modulates *C. elegans* Longevity through a Thermosensitive TRP Channel," *Cell Reports* 11:1414–24, https://doi.org/10.1016/j.celrep.2015.04.066.

25. Mair, W., et al., 2011, "Lifespan Extension Induced by AMPK and Calcineurin Is Mediated by CRTC-1 and CREB," *Nature* 470:404–8, https://doi.org/10.1038/nature09706.

26. Riera, C. E., et al., 2014, "TRPV1 Pain Receptors Regulate Longevity and Metabolism by Neuropeptide Signaling," *Cell* 157:1023–36, https://doi.org/10.1016/j.cell.2014.03.051.

27. NobelPrize.org, 2022, press release, Nobel Prize Outreach AB 2022, accessed December 20, 2022, https://www.nobelprize.org/prizes/medicine/2021/press-release/.

28. Riera et al. 2014.

29. Gray, J. M., et al., 2004, "Oxygen Sensation and Social Feeding Mediated by a *C. elegans* Guanylate Cyclase Homologue," *Nature* 430:317–22.

30. Shen, C., Nettleton, D., Jiang, M., Kim, S. K., and Powell-Coffman, J. A., 2005, "Roles of the HIF-1 Hypoxia-Inducible Factor during Hypoxia Response in *Caenorhabditis elegans*," *Journal of Biological Chemistry* 280:20580–88, https://doi.org/10.1074/jbc.M501894200.

31. Hwang, W., et al., 2015, "Inhibition of Elongin C Promotes Longevity and Protein Homeostasis via HIF-1 in *C. elegans*," *Aging Cell* 14:995–1002, https://doi.org/10.1111/acel.12390.

32. Leiser, S. F., et al., 2015, "Cell Nonautonomous Activation of Flavin-Containing Monooxygenase Promotes Longevity and Health Span," *Science* 350:1375–78, https://doi.org/10.1126/science.aac9257.

33. Tepper, R. G., Ashraf, J., Kaletsky, R., Kleemann, G., Murphy, C. T., and Bussemaker, H. J., 2013, "PQM-1 Complements DAF-16 as a Key Transcriptional Regulator of DAF-2-Mediated Development and Longevity," *Cell* 154:676–90, https://doi.org/10.1016/j.cell.2013.07.006.

34. Heimbucher, T., Hog, J., Gupta, P., and Murphy, C. T., 2020, "PQM-1 Controls Hypoxic Survival via Regulation of Lipid Metabolism," *Nature Communications* 11:4627, https://doi.org/10.1038/s41467-020-18369-w.

35. Poon, P. C., Kuo, T.-H., Linford, N. J., Roman, G., and Pletcher, S. D., 2010, "Carbon Dioxide Sensing Modulates Lifespan and Physiology in *Drosophila*," *PLOS Biology* 8:e1000356, https://doi.org/10.1371/journal.pbio.1000356.

36. Lin, C.-T., He, C.-W., Huang, T.-T., and Pan, C.-L., 2017, "Longevity Control by the Nervous System: Sensory Perception, Stress Response and Beyond," *Translational Medicine of Aging* 1:41–51, https://doi.org/10.1016/j.tma.2017.07.001.

37. Murphy, C. T., Lee, S. J., and Kenyon, C., 2007, "Tissue Entrainment by Feedback Regulation of Insulin Gene Expression in the Endoderm of *Caenorhabditis elegans*," *Proceedings of the National Academy of Sciences of the USA* 104:19046–50.

38. Chakraborty, T. S., et al., 2019, "Sensory Perception of Dead Conspecifics Induces Aversive Cues and Modulates Lifespan through Serotonin in *Drosophila*," *Nature Communications* 10:2365, https://doi.org/10.1038/s41467-019-10285-y.

39. Iliff, A. J., et al., 2021, "The Nematode C. *elegans* Senses Airborne Sound," *Neuron* 109:3633–46.e7, https://doi.org/10.1016/j.neuron.2021.08.035.

Chapter 14

1. Deak, F., and Sonntag, W. E., 2012, "Aging, Synaptic Dysfunction, and Insulin-like Growth Factor (IGF)-1," *Journals of Gerontology, Series A, Biological Sciences and Medical Sciences* 67A:611–25, https://doi.org/10.1093/gerona/gls118.

2. VanGuilder, H. D., Yan, H., Farley, J. A., Sonntag, W. E., and Freeman, W. M., 2010, "Aging Alters the Expression of Neurotransmission-Regulating Proteins in the Hippocampal Synaptoproteome," *Journal of Neurochemistry* 113:1577–88, https://doi.org/10.1111/j.1471-4159.2010 .06719.x.

3. Kauffman, A. L., Ashraf, J. M., Corces-Zimmerman, M. R., Landis, J. N., and Murphy, C. T., 2010, "Insulin Signaling and Dietary Restriction Differentially Influence the Decline of Learning and Memory with Age," *PLOS Biology* 8:e1000372, https://doi.org/10.1371/journal .pbio.1000372; Li, L. B., et al., 2016, "The Neuronal Kinesin UNC-104/KIF1A Is a Key Regulator of Synaptic Aging and Insulin Signaling-Regulated Memory," *Current Biology* 26:605–15, https://doi.org/10.1016/j.cub.2015.12.068.

4. Kaletsky, R., et al., 2016, "The C. *elegans* Adult Neuronal IIS/FOXO Transcriptome Reveals Adult Phenotype Regulators," *Nature* 529:92–96, https://doi.org/10.1038/nature16483; Kaletsky, R., et al., 2018, "Transcriptome Analysis of Adult *Caenorhabditis elegans* Cells Reveals Tissue-Specific Gene and Isoform Expression," *PLOS Genetics* 14:e1007559, https://doi.org /10.1371/journal.pgen.1007559.

5. Arey, R. N., and Murphy, C. T., 2017, "Conserved Regulators of Cognitive Aging: From Worms to Humans," *Behavioural Brain Research* 322:299–310, https://doi.org/10.1016/j.bbr .2016.06.035.

6. Simons Foundation, n.d., "Plasticity and the Aging Brain: Projects," accessed December 20, 2022, https://www.simonsfoundation.org/collaborations/plasticity-and-the-aging -brain/projects.

7. Ghosh, H. S., 2019, "Adult Neurogenesis and the Promise of Adult Neural Stem Cells," *Journal of Experimental Neuroscience* 13:1179069519856876, https://doi.org/10.1177 /1179069519856876. This is actually the topic of some debate; some studies conclude that there is little hippocampal neurogenesis after age seven.

8. Eriksson, P. S., et al., 1998, "Neurogenesis in the Adult Human Hippocampus," *Nature Medicine* 4:1313–17, https://doi.org/10.1038/3305; Cope, E. C., and Gould, E., 2019, "Adult Neurogenesis, Glia, and the Extracellular Matrix," *Cell Stem Cell* 24:690–705, https://doi.org /10.1016/j.stem.2019.03.023.

9. Sorrells, S. F., et al., 2018, "Human Hippocampal Neurogenesis Drops Sharply in Children to Undetectable Levels in Adults," *Nature* 555:377–81, https://doi.org/10.1038/nature25975; Franjic, D., et al., 2021, "Transcriptomic Taxonomy and Neurogenic Trajectories of Adult Human, Macaque, and Pig Hippocampal and Entorhinal Cells," *Neuron* 110:452–69.e14, https://doi.org/10.1016/j.neuron.2021.10.036.

10. Mertens, J., et al., 2021, "Age-Dependent Instability of Mature Neuronal Fate in Induced Neurons from Alzheimer's Patients," *Cell Stem Cell* 28:1533–48.e6, https://doi.org/10.1016/j .stem.2021.04.004; Vadodaria, K. C., Jones, J. R., Linker, S., and Gage, F. H., 2020, "Modeling Brain Disorders Using Induced Pluripotent Stem Cells," *Cold Spring Harbor Perspectives in Biology* 12:a035659, https://doi.org/10.1101/cshperspect.a035659.

11. Lu, Y., et al., 2020, "Reprogramming to Recover Youthful Epigenetic Information and Restore Vision," *Nature* 588:124–29, https://doi.org/10.1038/s41586-020-2975-4.

12. Rodríguez-Matellán, A., Alcazar, N., Hernández, F., Serrano, M., and Ávila, J., 2020, "Reprogramming Ameliorates Aging Features in Dentate Gyrus Cells and Improves Memory in Mice," *Stem Cell Reports* 15:1056–66, https://doi.org/10.1016/j.stemcr.2020.09.010.

13. Zimmerman, B., Rypma, B., Gratton, G., and Fabiani, M., 2021, "Age-Related Changes in Cerebrovascular Health and Their Effects on Neural Function and Cognition: A Comprehensive Review," *Psychophysiology* 58:e13796, https://doi.org/10.1111/psyp.13796.

14. Martínez-Salazar, B., et al., 2022, "COVID-19 and the Vasculature: Current Aspects and Long-Term Consequences," *Frontiers in Cell and Developmental Biology* 10:824851, https://doi .org/10.3389/fcell.2022.824851.

15. Everson, C. A., Henchen, C. J., Szabo, A., and Hogg, N., 2014, "Cell Injury and Repair Resulting from Sleep Loss and Sleep Recovery in Laboratory Rats," *Sleep* 37:1929–40, https:// doi.org/10.5665/sleep.4244; Chen, X., Redline, S., Shields, A. E., Williams, D. R., and Williams, M. A., 2014, "Associations of Allostatic Load with Sleep Apnea, Insomnia, Short Sleep Duration, and Other Sleep Disturbances: Findings from the National Health and Nutrition Examination Survey 2005 to 2008," *Annals of Epidemiology* 24:612–19, https://doi.org/10.1016/j.annepidem .2014.05.014; McEwen, B. S., 2006, "Sleep Deprivation as a Neurobiologic and Physiologic Stressor: Allostasis and Allostatic Load," *Metabolism* 55:S20–23, https://doi.org/10.1016/j .metabol.2006.07.008.

16. Gordleeva, S., Kanakov, O., Ivanchenko, M., Zaikin, A., and Franceschi, C., 2020, "Brain Aging and Garbage Cleaning: Modelling the Role of Sleep, Glymphatic System, and Microglia Senescence in the Propagation of Inflammaging," *Seminars in Immunopathology* 42:647–65, https://doi.org/10.1007/s00281-020-00816-x; Carroll, J. E., et al., 2016, "Partial Sleep Deprivation Activates the DNA Damage Response (DDR) and the Senescence-Associated Secretory Phenotype (SASP) in Aged Adult Humans," *Brain, Behavior, and Immunity* 51:223–29, https:// doi.org/10.1016/j.bbi.2015.08.024.

17. Roselli, C., Ramaswami, M., Boto, T., and Cervantes-Sandoval, I., 2021, "The Making of Long-Lasting Memories: A Fruit Fly Perspective," *Frontiers in Behavioral Neuroscience* 15:662129, https://doi.org/10.3389/fnbeh.2021.662129; Yin, J. C., et al., 1994, "Induction of a Dominant Negative CREB Transgene Specifically Blocks Long-Term Memory in *Drosophila*," *Cell* 79:49–58, https://doi.org/10.1016/0092-8674(94)90399-9; Dubnau, J., et al., 2003, "The Staufen/Pumilio Pathway Is Involved in *Drosophila* Long-Term Memory," *Current Biology* 13:286–96, https://doi.org/10.1016/s0960-9822(03)00064-2.

18. Dash, P. K., Hochner, B., and Kandel, E. R., 1990, "Injection of the cAMP-Responsive Element into the Nucleus of *Aplysia* Sensory Neurons Blocks Long-Term Facilitation," *Nature* 345:718–21, https://doi.org/10.1038/345718a0.

19. Arey and Murphy 2017.

20. Kauffman et al. 2010; Kauffman, A., et al., 2011, "*C. elegans* Positive Butanone Learning, Short-Term, and Long-Term Associative Memory Assays," *Journal of Visualized Experiments*, no. 49, 2490, https://doi.org/10.3791/2490.

21. Kauffman et al. 2010, 2011.

22. Kauffman et al. 2010; Li et al. 2016; Lakhina, V., et al., 2015, "Genome-wide Functional Analysis of CREB/Long-Term Memory-Dependent Transcription Reveals Distinct Basal and Memory Gene Expression Programs," *Neuron* 85:330–45, https://doi.org/10.1016/j.neuron .2014.12.029; Arey, R. N., Stein, G. M., Kaletsky, R., Kauffman, A., and Murphy, C. T., 2018, "Activation of Galphaq Signaling Enhances Memory Consolidation and Slows Cognitive Decline," *Neuron* 98:562–74.e5, https://doi.org/10.1016/j.neuron.2018.03.039; Arey, R. N., Kaletsky, R., and Murphy, C. T., 2019, "Nervous System-wide Profiling of Presynaptic mRNAs Reveals Regulators of Associative Memory," *Scientific Reports* 9:20314, https://doi.org/10.1038/s41598 -019-56908-8; Stein, G. M., and Murphy, C. T., 2012, "The Intersection of Aging, Longevity Pathways, and Learning and Memory in *C. elegans*," *Frontiers in Genetics* 3:259, https://doi.org /10.3389/fgene.2012.00259; Stein, G. M., and Murphy, C. T., 2014, "*C. elegans* Positive Olfactory Associative Memory Is a Molecularly Conserved Behavioral Paradigm," *Neurobiology of Learning and Memory* 115:86–94, https://doi.org/10.1016/j.nlm.2014.07.011.

23. Arey, Kaletsky, and Murphy 2019.

24. Kauffman et al. 2010.

25. Making more CREB: Kauffman et al. 2010. CREB activators: Arey et al. 2018.

26. Arey et al. 2018.

27. Villeda, S. A., et al., 2014, "Young Blood Reverses Age-Related Impairments in Cognitive Function and Synaptic Plasticity in Mice," *Nature Medicine* 20:659–63, https://doi.org/10 .1038/nm.3569; Villeda, S. A., et al., 2011, "The Ageing Systemic Milieu Negatively Regulates Neurogenesis and Cognitive Function," *Nature* 477:90–94, https://doi.org/10.1038 /nature10357.

28. White, C. W., 3rd, et al., 2020, "Age-Related Loss of Neural Stem Cell O-GlcNAc Promotes a Glial Fate Switch through STAT3 Activation," *Proceedings of the National Academy of Sciences of the USA* 117:22214–24, https://doi.org/10.1073/pnas.2007439117; Gontier, G., et al., 2018, "Tet2 Rescues Age-Related Regenerative Decline and Enhances Cognitive Function in the Adult Mouse Brain," *Cell Reports* 22:1974–81, https://doi.org/10.1016/j.celrep.2018.02 .001; Smith, L. K., et al., 2015, "Beta2-Microglobulin Is a Systemic Pro-aging Factor that Impairs Cognitive Function and Neurogenesis," *Nature Medicine* 21:932–37, https://doi.org/10.1038 /nm.3898; Bouchard, J., and Villeda, S. A., 2015, "Aging and Brain Rejuvenation as Systemic Events," *Journal of Neurochemistry* 132:5–19, https://doi.org/10.1111/jnc.12969.

29. Lin, T,. et al., 2018, "Systemic Inflammation Mediates Age-Related Cognitive Deficits," *Frontiers in Aging Neuroscience* 10:00236, https://doi.org/10.3389/fnagi.2018.00236.

30. Reddy, O. C., and van der Werf, Y. D., 2020, "The Sleeping Brain: Harnessing the Power of the Glymphatic System through Lifestyle Choices," *Brain Sciences* 10:868, https://doi.org /10.3390/brainsci10110868.

31. Horowitz, A. M., et al., 2020, "Blood Factors Transfer Beneficial Effects of Exercise on Neurogenesis and Cognition to the Aged Brain," *Science* 369:167–73, https://doi.org/10.1126/science.aaw2622.

32. De Miguel, Z., et al., 2021, "Exercise Plasma Boosts Memory and Dampens Brain Inflammation via Clusterin," *Nature* 600:494–99, https://doi.org/10.1038/s41586-021-04183-x.

33. Yao, V., et al., 2018, "An Integrative Tissue-Network Approach to Identify and Test Human Disease Genes," *Nature Biotechnology* 36:1091–99, https://doi.org/10.1038/nbt.4246.

34. Cohen, E., et al., 2009, "Reduced IGF-1 Signaling Delays Age-Associated Proteotoxicity in Mice," *Cell* 139:1157–69, https://doi.org/10.1016/j.cell.2009.11.014; Cohen, E., and Dillin, A., 2008, "The Insulin Paradox: Aging, Proteotoxicity and Neurodegeneration," *Nature Reviews Neuroscience* 9:759–67, https://doi.org/10.1038/nrn2474.

35. Gabriele, R.M.C., Abel, E., Fox, N. C., Wray, S., and Arber, C., 2022, "Knockdown of Amyloid Precursor Protein: Biological Consequences and Clinical Opportunities," *Frontiers in Neuroscience* 16:835645, https://doi.org/10.3389/fnins.2022.835645.

36. Kessissoglou, I. A., et al., 2020, "The *Drosophila* Amyloid Precursor Protein Homologue Mediates Neuronal Survival and Neuroglial Interactions," *PLOS Biology* 18:e3000703, https://doi.org/10.1371/journal.pbio.3000703.

37. Gabriele et al. 2022.

38. Lee, W. J., et al., 2022, "Regional Aβ-Tau Interactions Promote Onset and Acceleration of Alzheimer's Disease Tau Spreading," *Neuron* 110:1932–43.e5, https://doi.org/10.1016/j.neuron.2022.03.034.

39. Li, T., Cao, H. X., and Ke, D., 2021, "Type 2 Diabetes Mellitus Easily Develops into Alzheimer's Disease via Hyperglycemia and Insulin Resistance," *Current Medical Science* 41:1165–71, https://doi.org/10.1007/s11596-021-2467-2.

40. Lopez-Rodriguez, A. B., et al., 2021, "Acute Systemic Inflammation Exacerbates Neuroinflammation in Alzheimer's Disease: IL-1β Drives Amplified Responses in Primed Astrocytes and Neuronal Network Dysfunction," *Alzheimer's and Dementia* 17:1735–55, https://doi.org/10.1002/alz.12341.

41. Grundke-Iqbal, I., et al., 1986, "Microtubule-Associated Protein Tau: A Component of Alzheimer Paired Helical Filaments," *Journal of Biological Chemistry* 261:6084–89.

42. Arriagada, P. V., Growdon, J. H., Hedley-Whyte, E. T., and Hyman, B. T., 1992, "Neurofibrillary Tangles but Not Senile Plaques Parallel Duration and Severity of Alzheimer's Disease," *Neurology* 42:631–39, https://doi.org/10.1212/wnl.42.3.631.

43. Novak, P., et al., 2021, "ADAMANT: A Placebo-Controlled Randomized Phase 2 Study of AADvac1, an Active Immunotherapy against Pathological Tau in Alzheimer's Disease," *Nature Aging* 1:521–34, https://doi.org/10.1038/s43587-021-00070-2.

44. Spires-Jones, T. L., and Hyman, B. T., 2014, "The Intersection of Amyloid Beta and Tau at Synapses in Alzheimer's Disease," *Neuron* 82:756–71, https://doi.org/10.1016/j.neuron.2014.05.004.

45. Goedert, M., 2018, "Tau Filaments in Neurodegenerative Diseases," *FEBS Letters* 592:2383–91, https://doi.org/10.1002/1873-3468.13108; Yoshiyama, Y., et al., 2007, "Synapse Loss and Microglial Activation Precede Tangles in a P301S Tauopathy Mouse Model," *Neuron* 53:337–51, https://doi.org/10.1016/j.neuron.2007.01.010.

46. Spires-Jones and Hyman 2014.

47. Largo-Barrientos, P., et al., 2021, "Lowering Synaptogyrin-3 Expression Rescues Tau-Induced Memory Defects and Synaptic Loss in the Presence of Microglial Activation," *Neuron* 109:767–77.e5, https://doi.org/10.1016/j.neuron.2020.12.016.

48. Gorbunova, V., et al., 2021, "The Role of Retrotransposable Elements in Ageing and Age-Associated Diseases," *Nature* 596:43–53, https://doi.org/10.1038/s41586-021 -03542-y.

49. Chang, Y. H., and Dubnau, J., 2019, "The Gypsy Endogenous Retrovirus Drives Non-cell-autonomous Propagation in a *Drosophila* TDP-43 Model of Neurodegeneration," *Current Biology* 29:3135–52.e4, https://doi.org/10.1016/j.cub.2019.07.071; Guo, C., et al., 2018, "Tau Activates Transposable Elements in Alzheimer's Disease," *Cell Reports* 23:2874–80, https://doi .org/10.1016/j.celrep.2018.05.004.

50. Ramirez, P., et al., 2021, "Pathogenic Tau Accelerates Aging-Associated Activation of Transposable Elements in the Mouse Central Nervous System," *Progress in Neurobiology* 208:102181, https://doi.org/10.1016/j.pneurobio.2021.102181; Guo, C., et al., 2018, "Tau Activates Transposable Elements in Alzheimer's Disease," *Cell Reports* 23:2874–80, https://doi .org/10.1016/j.celrep.2018.05.004.

51. Tam, O. H., Ostrow, L. W., and Gale Hammell, M., 2019, "Diseases of the nERVous System: Retrotransposon Activity in Neurodegenerative Disease," *Mobile DNA* 10:32, https:// doi.org/10.1186/s13100-019-0176-1.

52. Guengerich, P., 2020, "Repurposing an HIV Drug to Treat Alzheimer's," *Discover*, January 29, 2020, https://discover.vumc.org/2020/01/repurposing-an-hiv-drug-to-treat-alzheimers /; Stulpin, C., 2020, "ART Reduces Alzheimer's Prevalence among Patients with HIV," *Healio News*, October 8, 2020, https://www.healio.com/news/infectious-disease/20201008/art -reduces-alzheimers-prevalence-among-patients-with-hiv; Mast, N., Verwilst, P., Wilkey, C. J., Guengerich, F. P., and Pikuleva, I. A., 2020, "In Vitro Activation of Cytochrome P450 46A1 (CYP46A1) by Efavirenz-Related Compounds," *Journal of Medicinal Chemistry* 63:6477–88, https://doi.org/10.1021/acs.jmedchem.9b01383.

53. Gold, J., et al., 2019, "Safety and Tolerability of Triumeq in Amyotrophic Lateral Sclerosis: The Lighthouse Trial," *Amyotrophic Lateral Sclerosis and Frontotemporal Degeneration* 20:595–604, https://doi.org/10.1080/21678421.2019.1632899.

54. Yamazaki, Y., Zhao, N., Caulfield, T. R., Liu, C.-C., and Bu, G., 2019, "Apolipoprotein E and Alzheimer Disease: Pathobiology and Targeting Strategies," *Nature Reviews Neurology* 15:501–18, https://doi.org/10.1038/s41582-019-0228-7.

55. Park, J. E., et al., 2022, "Diagnostic Blood Biomarkers in Alzheimer's Disease," *Biomedicines* 10:169, https://doi.org/10.3390/biomedicines10010169; Hardy-Sosa, A., et al., 2022, "Diagnostic Accuracy of Blood-Based Biomarker Panels: A Systematic Review," *Frontiers in Aging Neuroscience* 14:683689, https://doi.org/10.3389/fnagi.2022.683689.

56. National Institute on Aging, n.d., "How Is Alzheimer's Disease Treated?," accessed December 20, 2022, https://www.nia.nih.gov/health/how-alzheimers-disease-treated.

57. Hashimoto, M., et al., 2005, "Does Donepezil Treatment Slow the Progression of Hippocampal Atrophy in Patients with Alzheimer's Disease?," *American Journal of Psychiatry* 162:676–82, https://doi.org/10.1176/appi.ajp.162.4.676.

58. Cacabelos, R.. 2007, "Donepezil in Alzheimer's Disease: From Conventional Trials to Pharmacogenetics," *Neuropsychiatric Disease and Treatment* 3:303–33.

59. Makin, S., 2018, "The Amyloid Hypothesis on Trial," *Nature* 559:S4–7, https://doi.org/10.1038/d41586-018-05719-4.

60. Knopman, D. S., Jones, D. T., and Greicius, M. D., 2021, "Failure to Demonstrate Efficacy of Aducanumab: An Analysis of the EMERGE and ENGAGE Trials as Reported by Biogen, December 2019," *Alzheimer's and Dementia* 17:696–701, https://doi.org/10.1002/alz.12213.

61. Novak et al. 2021.

62. Haghani, A., et al., 2020, "Mouse Brain Transcriptome Responses to Inhaled Nanoparticulate Matter Differed by Sex and APOE in Nrf2-Nfkb Interactions," *eLife* 9:e54822, https://doi.org/10.7554/eLife.54822.

63. Alzheimer's Impact Movement and Alzheimer's Association, 2020, "Race, Ethnicity and Alzheimer's," factsheet, March 2020, https://www.alz.org/aaic/downloads2020/2020_Race_and_Ethnicity_Fact_Sheet.pdf.

64. Bruney, G. 2020, "Very Abbreviated History of the Destruction of Black Neighborhoods," *Esquire*, May 30, 2020, https://www.esquire.com/news-politics/a32719786/george-floyd-protests-riots-black-comminity-destruction-history/.

65. Alzheimer's Association International Conference, 2017, "Stressful Life Experiences Age the Brain by Four Years, African Americans Most at Risk," press release, https://www.alz.org/aaic/releases_2017/AAIC17-Sun-briefing-racial-disparities.asp.

66. Alexander, D., and Currie, J., 2017, "Is It Who You Are or Where You Live? Residential Segregation and Racial Gaps in Childhood Asthma," National Bureau of Economic Research Working Paper 23622, July 2017, https://www.nber.org/papers/w23622.

67. Alzheimer's Association International Conference 2017.

68. Alzheimer's Impact Movement and Alzheimer's Association 2020.

69. Kauffman et al. 2011.

70. Kaletsky, R., et al., 2016, "The *C. elegans* Adult Neuronal IIS/FOXO Transcriptome Reveals Adult Phenotype Regulators," *Nature* 529:92–96, https://doi.org/10.1038/nature16483.

71. Kauffman, A. L., Ashraf, J. M., Corces-Zimmerman, M. R., Landis, J. N., and Murphy, C. T., 2010, "Insulin Signaling and Dietary Restriction Differentially Influence the Decline of Learning and Memory with Age," *PLOS Biology* 8:e1000372, https://doi.org/10.1371/journal.pbio.1000372.

72. Arey et al. 2018.

73. Sonntag, W., et al., 2013, "Insulin-like Growth Factor-1 in CNS and Cerebrovascular Aging," *Frontiers in Aging Neuroscience* 5:27, https://doi.org/10.3389/fnagi.2013.00027.

74. Dillin, A., et al., 2002, "Rates of Behavior and Aging Specified by Mitochondrial Function during Development," *Science* 298:2398–2401, https://doi.org/10.1126/science.1077780.

75. Kauffman et al. 2010.

76. Kauffman et al. 2010.

77. Willette, A. A., et al., 2012, "Calorie Restriction Reduces the Influence of Glucoregulatory Dysfunction on Regional Brain Volume in Aged Rhesus Monkeys," *Diabetes* 61:1036–42, https://doi.org/10.2337/db11-1187.

78. Chou, A., et al., 2017, "Inhibition of the Integrated Stress Response Reverses Cognitive Deficits after Traumatic Brain Injury," *Proceedings of the National Academy of Sciences of the USA* 114:E6420–26, https://doi.org/10.1073/pnas.1707661114.

79. Krukowski, K., et al., 2020, "Small Molecule Cognitive Enhancer Reverses Age-Related Memory Decline in Mice," *eLife* 9:e62048, https://doi.org/10.7554/eLife.62048.

80. Minhas, P. S., et al., 2021, "Restoring Metabolism of Myeloid Cells Reverses Cognitive Decline in Ageing," *Nature* 590:122–28, https://doi.org/10.1038/s41586-020-03160-0.

81. Vadini, F., et al., 2020, "Liraglutide Improves Memory in Obese Patients with Prediabetes or Early Type 2 Diabetes: A Randomized, Controlled Study," *International Journal of Obesity* 44:1254–63, https://doi.org/10.1038/s41366-020-0535-5.

82. Campbell, J. M., et al., 2018, "Metformin Use Associated with Reduced Risk of Dementia in Patients with Diabetes: A Systematic Review and Meta-analysis," *Journal of Alzheimer's Disease* 65:1225–36, https://doi.org/10.3233/JAD-180263; Justice, J. N., et al., 2018, "A Framework for Selection of Blood-Based Biomarkers for Geroscience-Guided Clinical Trials: Report from the TAME Biomarkers Workgroup," *Geroscience* 40:419–36, https://doi.org/10.1007/s11357 -018-0042-y.

Notes to text box (Chapter 14, page 270)

a. Belluck, P., and Robbins, R., 2021, "F.D.A. Approves Alzheimer's Drug Despite Fierce Debate over Whether It Works," *New York Times*, last updated July 20, 2021, https://www .nytimes.com/2021/06/07/health/aduhelm-fda-alzheimers-drug.html; Chappell, B., 2021, "3 Experts Have Resigned from an FDA Committee over Alzheimer's Drug Approval," NPR, last updated June 11, 2021, https://www.npr.org/2021/06/11/1005567149/3-experts-have -resigned-from-an-fda-committee-over-alzheimers-drug-approval?utm_campaign=nprandutm _source=twitter.comandutm_term=nprnewsandutm_medium=social.

b. Toobin, J., 2021, "The Road to Aduhelm: What One Ex-FDA Adviser Called 'Probably the Worst Drug Approval Decision in Recent US History' for an Alzheimer's Treatment," CNN Politics, last updated September 27, 2021, https://www.cnn.com/2021/09/26/politics /alzheimers-drug-aduhelm-fda-approval/index.html.

Note to text box (Chapter 14, page 272)

a. Stamps, J. J., Bartoshuk, L. M., and Heilman, K. M., 2013, "A Brief Olfactory Test for Alzheimer's Disease," *Journal of the Neurological Sciences* 333:19–24, https://doi.org/10.1016/j.jns .2013.06.033.

Chapter 15

1. Barlow, D., 2015, "Denise Barlow: A Career in Epigenetics," interview, *RNA Biology* 12:105–8, https://doi.org/10.1080/15476286.2015.1018711.

2. Roseboom, T., de Rooij, S., and Painter, R., 2006, "The Dutch Famine and Its Long-Term Consequences for Adult Health," *Early Human Development* 82:485–91, https://doi.org/10 .1016/j.earlhumdev.2006.07.001.

3. Roseboom, de Rooij, and Painter 2006.

4. De Rooij, S. R., Wouters, H., Yonker, J. E., Painter, R. C., and Roseboom, T. J., 2010, "Prenatal Undernutrition and Cognitive Function in Late Adulthood," *Proceedings of the National Academy of Sciences of the USA* 107:16881–86, https://doi.org/10.1073/pnas.1009459107.

5. De Rooij et al. 2010.

6. Cheng, Q., et al., 2020, "Prenatal and Early-Life Exposure to the Great Chinese Famine Increased the Risk of Tuberculosis in Adulthood across Two Generations," *Proceedings of the National Academy of Sciences of the USA* 117:27549–55, https://doi.org/10.1073/pnas.2008336117.

7. Yellow Horse Brave Heart, M., 2000, "Wakiksuyapi: Carrying the Historical Trauma of the Lakota," *Tulane Studies in Social Welfare* 21:245–66

8. Shulevitz, J., 2014, "The Science of Suffering," *New Republic*, November 16, 2014, https://newrepublic.com/article/120144/trauma-genetic-scientists-say-parents-are-passing-ptsd-kids.

9. Stern, L., and Hulko, W., 2016, "Historical Trauma, PTSD, and Dementia: Implications for Trauma-Informed Social Work," *Gerontologist* 56:311, https://doi.org/10.1093/geront/gnw162.1270.

10. Waehrer, G. M., Miller, T. R., Silverio Marques, S. C., Oh, D. L., and Burke Harris, N., 2020, "Disease Burden of Adverse Childhood Experiences across 14 States," *PLOS One* 15:e0226134, https://doi.org/10.1371/journal.pone.0226134.

11. Van Meel, E. R,. et al., 2020, "Parental Psychological Distress during Pregnancy and the Risk of Childhood Lower Lung Function and Asthma: A Population-Based Prospective Cohort Study," *Thorax* 75:1074–81, https://doi.org/10.1136/thoraxjnl-2019-214099.

12. Marlin, B. J., Mitre, M., D'Amour J. A., Chao, M. V., and Froemke, R. C., 2015, "Oxytocin Enables Maternal Behaviour by Balancing Cortical Inhibition," *Nature* 520:499–504, https://doi.org/10.1038/nature14402.

13. Zheng, X., et al., 2021, "Sperm Epigenetic Alterations Contribute to Inter- and Transgenerational Effects of Paternal Exposure to Long-Term Psychological Stress via Evading Offspring Embryonic Reprogramming," *Cell Discovery* 7:101, https://doi.org/10.1038/s41421-021 -00343-5; Bošković, A., and Rando, O. J., 2018, "Transgenerational Epigenetic Inheritance," *Annual Review of Genetics* 52:21–41, https://doi.org/10.1146/annurev-genet-120417-031404.

14. Dias, B. G., and Ressler, K. J., 2014, "Parental Olfactory Experience Influences Behavior and Neural Structure in Subsequent Generations," *Nature Neuroscience* 17:89–96, https://doi.org/10.1038/nn.3594.

15. Posner, R., et al., 2019, "Neuronal Small RNAs Control Behavior Transgenerationally," *Cell* 177:1814–26.e15, https://doi.org/10.1016/j.cell.2019.04.029.

16. Davison, E. H., et al., 2006, "Late-Life Emergence of Early-Life Trauma:The Phenomenon of Late-Onset Stress Symptomatology among Aging Combat Veterans," *Research on Aging* 28:84–114, https://doi.org/10.1177/0164027505281560.

17. *Encyclopedia.com*, 2022, "The Disastrous Effects of Lysenkoism on Soviet Agriculture," Science and Its Times: Understanding the Social Significance of Scientific Discovery, November 29, 2022, https://www.encyclopedia.com/science/encyclopedias-almanacs-transcripts-and -maps/disastrous-effects-lysenkoism-soviet-agriculture; https://www.readex.com/blog/inherit -problem-how-lysenkoism-ruined-soviet-plant-genetics-and-perpetuated-famine-under.

18. Bateson, W., and Pellew, C., 1915, "On the Genetics of 'Rogues' among Culinary Peas (*Pisum sativum*)," *Journal of Genetics* 5:13–36, https://doi.org/10.1007/BF02982150; Just, E. E., *The Biology of the Cell Surface* (Philadelphia: P. Blakiston's Son, 1939).

19. Beadle, G. W., and McClintock, B., 1928, "A Genic Disturbance of Meiosis in *Zea mays*," *Science* 68:433, https://doi.org/10.1126/science.68.1766.433.

20. McClintock, B., 1984, "The Significance of Responses of the Genome to Challenge," *Science* 226:792–801, https://doi.org/10.1126/science.15739260.

21. Lynch, V. J., Leclerc, R. D., May, G., and Wagner, G. P., 2011, "Transposon-Mediated Rewiring of Gene Regulatory Networks Contributed to the Evolution of Pregnancy in Mammals," *Nature Genetics* 43:1154–59, https://doi.org/10.1038/ng.917.

22. De Cecco, M., et al., 2013, "Transposable Elements Become Active and Mobile in the Genomes of Aging Mammalian Somatic Tissues," *Aging (Albany NY)* 5:867–83, https://doi.org/10.18632/aging.100621; De Cecco, M., et al., 2013, "Genomes of Replicatively Senescent Cells Undergo Global Epigenetic Changes Leading to Gene Silencing and Activation of Transposable Elements," *Aging Cell* 12:247–56, https://doi.org/10.1111/acel.12047.

23. Chang, Y.-H., and Dubnau, J., 2019, "The Gypsy Endogenous Retrovirus Drives Noncell-autonomous Propagation in a *Drosophila* TDP-43 Model of Neurodegeneration," *Current Biology* 29:3135–52.e4, https://doi.org/10.1016/j.cub.2019.07.071; Krug, L., et al., 2017, "Retrotransposon Activation Contributes to Neurodegeneration in a *Drosophila* TDP-43 model of ALS," *PLOS Genetics* 13:e1006635, https://doi.org/10.1371/journal.pgen.1006635.

24. De Melo, A., et al., 2021, "The Role of Microglia in Prion Diseases and Possible Therapeutic Targets: A Literature Review," *Prion* 15:191–206, https://doi.org/10.1080/19336896.2021.1991771.

25. Harvey, Z. H., Chakravarty, A. K., Futia, R. A., and Jarosz, D. F., 2020, "A Prion Epigenetic Switch Establishes an Active Chromatin State," *Cell* 180:928–40.e14, https://doi.org/10.1016/j.cell.2020.02.014.

26. Harvey et al. 2020.

27. Si, K., Choi, Y. B., White-Grindley, E., Majumdar, A., and Kandel, E. R., 2010, "*Aplysia* CPEB Can Form Prion-like Multimers in Sensory Neurons that Contribute to Long-Term Facilitation," *Cell* 140:421–35, https://doi.org/10.1016/j.cell.2010.01.008; Si, K., and Kandel, E. R., 2016, "The Role of Functional Prion-like Proteins in the Persistence of Memory," *Cold Spring Harbor Perspectives in Biology* 8:a021774, https://doi.org/10.1101/cshperspect.a021774; Mastushita-Sakai, T., White-Grindley, E., Samuelson, J., Seidel, C., and Si, K., 2010, "*Drosophila* Orb2 Targets Genes Involved in Neuronal Growth, Synapse Formation, and Protein Turnover," *Proceedings of the National Academy of Sciences of the USA* 107:11987–92, https://doi.org/10.1073/pnas.1004433107; Shorter, J., and Lindquist, S., 2005, "Prions as Adaptive Conduits of Memory and Inheritance," *Nature Reviews Genetics* 6:435–50, https://doi.org/10.1038/nrg1616.

28. Moore, L. D., Le, T., and Fan, G., 2013, "DNA Methylation and Its Basic Function," *Neuropsychopharmacology* 38:23–38, https://doi.org/10.1038/npp.2012.112.

29. Herb, B. R., et al., 2012, "Reversible Switching between Epigenetic States in Honeybee Behavioral Subcastes," *Nature Neuroscience* 15:1371–73, https://doi.org/10.1038/nn.3218.

30. Horvath, S., 2013, "DNA Methylation Age of Human Tissues and Cell Types," *Genome Biology* 14:R115, https://doi.org/10.1186/gb-2013-14-10-r115.

31. Hannum, G., et al., 2013, "Genome-wide Methylation Profiles Reveal Quantitative Views of Human Aging Rates," *Molecular Cell* 49:359–67, https://doi.org/10.1016/j.molcel.2012.10.016.

32. Liu, Z., et al., 2020, "Underlying Features of Epigenetic Aging Clocks In Vivo and In Vitro," *Aging Cell* 19:e13229, https://doi.org/10.1111/acel.13229; Levine, M. E., 2020, "Assessment of Epigenetic Clocks as Biomarkers of Aging in Basic and Population Research," *Journals of gerontology, Series A, Biological Sciences and Medical Sciences* 75:463–65, https://doi.org/10.1093/gerona/glaa021; Schmitz, L. L., et al., 2021, "The Socioeconomic Gradient in Epigenetic

Ageing Clocks: Evidence from the Multi-ethnic Study of Atherosclerosis and the Health and Retirement Study," *Epigenetics* 17:589–611, published online ahead of print Juy 6, 2021, https://doi.org/10.1080/15592294.2021.1939479.

33. Levine, M. E., et al., 2016, "Menopause Accelerates Biological Aging," *Proceedings of the National Academy of Sciences of the USA* 113:9327–32, https://doi.org/10.1073/pnas.1604558113.

34. Buckley, M. T., et al., 2022, "Cell Type-Specific Aging Clocks to Quantify Aging and Rejuvenation in Regenerative Regions of the Brain," *bioRxiv*, 2022.01.10.475747, https://doi.org/10.1101/2022.01.10.475747.

35. Olins, D. E., and Olins, A. L., 1978, "Nucleosomes: The Structural Quantum in Chromosomes," *American Scientist* 66:704–11.

36. Farrelly, L. A., et al., 2019, "Histone Serotonylation Is a Permissive Modification that Enhances TFIID Binding to H3K4me3," *Nature* 567:535–39, https://doi.org/10.1038/s41586-019-1024-7.

37. Ma, Z., et al., 2018, "Epigenetic Drift of H3K27me3 in Aging Links Glycolysis to Healthy Longevity in *Drosophila*," *eLife* 7:e35368, https://doi.org/10.7554/eLife.35368.

38. Cheung, P., et al., 2018, "Single-Cell Chromatin Modification Profiling Reveals Increased Epigenetic Variations with Aging," *Cell* 173:1385–97.e14, https://doi.org/10.1016/j.cell.2018.03.079.

39. Benayoun, B. A., et al., 2019, "Remodeling of Epigenome and Transcriptome Landscapes with Aging in Mice Reveals Widespread Induction of Inflammatory Responses," *Genome Research* 29:697–709, https://doi.org/10.1101/gr.240093.118.

40. Greer, E. L., et al., 2010, "Members of the H3K4 Trimethylation Complex Regulate Lifespan in a Germline-Dependent Manner in *C. elegans*," *Nature* 466:383–87, https://doi.org/10.1038/nature09195.

41. Maures, T. J., Greer, E. L., Hauswirth, A. G., and Brunet, A., 2011, "The H3K27 Demethylase UTX-1 Regulates *C. elegans* Lifespan in a Germline-Independent, Insulin-Dependent Manner," *Aging Cell* 10:980–90, https://doi.org/10.1111/j.1474-9726.2011.00738.x; Jin, C., et al., 2011, "Histone Demethylase UTX-1 Regulates *C. elegans* Life Span by Targeting the Insulin/IGF-1 Signaling Pathway," *Cell Metabolism* 14:161–72, https://doi.org/10.1016/j.cmet.2011.07.001.

42. Guillermo, A.R.R., et al., 2021, "H3K27 Modifiers Regulate Lifespan in *C. elegans* in a Context-Dependent Manner," *BMC Biology* 19:59, https://doi.org/10.1186/s12915-021-00984-8; Merkwirth, C., et al., 2016, "Two Conserved Histone Demethylases Regulate Mitochondrial Stress-Induced Longevity," *Cell* 165:1209–23, https://doi.org/10.1016/j.cell.2016.04.012.

43. Greer, E. L., et al., 2011, "Transgenerational Epigenetic Inheritance of Longevity in *Caenorhabditis elegans*," *Nature* 479:365–71, https://doi.org/10.1038/nature10572.

44. Han, S., et al., 2017, "Mono-unsaturated Fatty Acids Link H3K4me3 Modifiers to *C. elegans* Lifespan," *Nature* 544:185–90, https://doi.org/10.1038/nature21686.

45. Lee, T. W., David, H. S., Engstrom, A. K., Carpenter, B. S., and Katz, D. J., 2019, "Repressive H3K9me2 Protects Lifespan against the Transgenerational Burden of COMPASS Activity in *C. elegans*," *eLife* 8:e48498, https://doi.org/10.7554/eLife.48498.

46. Liu, Z. C., and Ambros, V., 1989, "Heterochronic Genes Control the Stage-Specific Initiation and Expression of the Dauer Larva Developmental Program in *Caenorhabditis elegans*," *Genes and Development* 3:2039–49, https://doi.org/10.1101/gad.3.12b.2039; Lee, R. C., and Ambros, V., 2001, "An Extensive Class of Small RNAs in *Caenorhabditis elegans*," *Science* 294:862–64, https://doi.org/10.1126/science.1065329.

47. Zia, A., Farkhondeh, T., Sahebdel, F., Pourbagher-Shahri, A. M., and Samarghandian, S., 2021, "Key miRNAs in Modulating Aging and Longevity: A Focus on Signaling Pathways and Cellular Targets," *Current Molecular Pharmacology* 15:736–62, https://doi.org/10.2174/1874467214666210917141541; Kinser, H. E., and Pincus, Z., 2020, "MicroRNAs as Modulators of Longevity and the Aging Process," *Human Genetics* 139:291–308, https://doi.org/10.1007/s00439-019-02046-0; Boehm, M., and Slack, F., 2005, "A Developmental Timing MicroRNA and Its Target Regulate Life Span in *C. elegans*," *Science* 310:1954–57, https://doi.org/10.1126/science.1115596.

48. Zia et al. 2021; Kinser and Pincus 2020; Boehm and Slack 2005.

49. Baugh, L. R., and Hu, P. J., 2020, "Starvation Responses throughout the *Caenorhabditis elegans* Life Cycle," *Genetics* 216:837–78, https://doi.org/10.1534/genetics.120.303565.

50. Angelo, G., and Van Gilst, M. R., 2009, "Starvation Protects Germline Stem Cells and Extends Reproductive Longevity in *C. elegans*," *Science* 326:954–58, https://doi.org/10.1126/science.1178343.

51. Jobson, M. A., et al., 2011, "Transgenerational Effects of Early Life Starvation on Growth, Reproduction, and Stress Resistance in *Caenorhabditis elegans*," *Genetics* 201:201–12, https://doi.org/10.1534/genetics.115.178699; Greer et al. 2011.

52. Rechavi, O., et al., 2014, "Starvation-Induced Transgenerational Inheritance of Small RNAs in *C. elegans*," *Cell* 158:277–87, https://doi.org/10.1016/j.cell.2014.06.020.

53. Zhang, Y., Lu, H., and Bargmann, C. I., 2005, "Pathogenic Bacteria Induce Aversive Olfactory Learning in *Caenorhabditis elegans*," *Nature* 438:179–84, https://doi.org/10.1038/nature04216.

54. Jin, X., Pokala, N., and Bargmann, C. I., 2016, "Distinct Circuits for the Formation and Retrieval of an Imprinted Olfactory Memory," *Cell* 164:632–43, https://doi.org/10.1016/j.cell.2016.01.007.

55. Moore, R. S., Kaletsky, R., and Murphy, C. T., 2019, "Piwi/PRG-1 Argonaute and TGF-β Mediate Transgenerational Learned Pathogenic Avoidance," *Cell* 177:1827–41.e12, https://doi.org/10.1016/j.cell.2019.05.024.

56. Moore, Kaletsky, and Murphy 2019.

57. Kaletsky, R., et al., 2020, "*C. elegans* 'Reads' Bacterial Non-coding RNAs to Learn Pathogenic Avoidance," *bioRxiv*, 2020.01.26.920322, https://doi.org/10.1101/2020.01.26.920322.

58. Kaletsky, R., et al., 2020, "*C. elegans* Interprets Bacterial Non-coding RNAs to Learn Pathogenic Avoidance," *Nature* 586:445–51, https://doi.org/10.1038/s41586-020-2699-5.

59. Moore, R. S., et al., 2021, "The Role of the Cer1 Transposon in Horizontal Transfer of Transgenerational Memory," *Cell* 184:4697–4712.e18, https://doi.org/10.1016/j.cell.2021.07.022.

60. Lim, A. I., et al., 2021, "Prenatal Maternal Infection Promotes Tissue-Specific Immunity and Inflammation in Offspring," *Science* 373:eabf3002, doi:https://doi.org/10.1126/science.abf3002.

Note to text box: "Caveats to Reports of Human Transgenerational Phenomena" (Chapter 15, pages 285–286)

a. Birney, E., 2015, "Why I'm Sceptical about the Idea of Genetically Inherited Trauma," *Guardian*, September 11, 2015, https://www.theguardian.com/science/blog/2015/sep/11/why-im-sceptical-about-the-idea-of-genetically-inherited-trauma-epigenetics.

Chapter 16

1. Bana, B., and Cabreiro, F., 2019, "The Microbiome and Aging," *Annual Review of Genetics* 53:239–61, https://doi.org/10.1146/annurev-genet-112618–043650.

2. Yap, C. X., et al., 2021, "Autism-Related Dietary Preferences Mediate Autism-Gut Microbiome Associations," *Cell* 184:5916–31.e17, https://doi.org/10.1016/j.cell.2021.10.015.

3. Bárcena, C., et al., 2019, "Healthspan and Lifespan Extension by Fecal Microbiota Transplantation into Progeroid Mice," *Nature Medicine* 25:1234–42, https://doi.org/10.1038/s41591-019-0504-5.

4. Samuel, B. S., Rowedder, H., Braendle, C., Felix, M. A., and Ruvkun, G., 2016, "*Caenorhabditis elegans* Responses to Bacteria from Its Natural Habitats," *Proceedings of the National Academy of Sciences of the USA* 113:E3941–49, https://doi.org/10.1073/pnas.1607183113.

5. Rera, M., Clark, R. I., and Walker, D. W., 2012, "Intestinal Barrier Dysfunction Links Metabolic and Inflammatory Markers of Aging to Death in *Drosophila*," *Proceedings of the National Academy of Sciences of the USA* 109:21528–33, https://doi.org/10.1073/pnas.1215849110.

6. Clark, R. I., et al., 2015, "Distinct Shifts in Microbiota Composition during *Drosophila* Aging Impair Intestinal Function and Drive Mortality," *Cell Reports* 12:1656–67, https://doi.org/10.1016/j.celrep.2015.08.004.

7. Anderson, K. E., et al., 2018, "The Queen's Gut Refines with Age: Longevity Phenotypes in a Social Insect Model," *Microbiome* 6:108, https://doi.org/10.1186/s40168-018-0489-1; Kunugi, H., and Mohammed Ali, A., 2019, "Royal Jelly and Its Components Promote Healthy Aging and Longevity: From Animal Models to Humans," *International Journal of Molecular Sciences* 20:4662, https://doi.org/10.3390/ijms20194662; Liu, P., et al., 2021, "Overwintering Honeybees Maintained Dynamic and Stable Intestinal Bacteria," *Scientific Reports* 11:22233, https://doi.org/10.1038/s41598-021-01204-7.

8. Pascale, A., et al., 2018, "Microbiota and Metabolic Diseases," *Endocrine* 61:357–71, https://doi.org/10.1007/s12020-018-1605-5.

9. Brice, K. L., et al., 2019, "The Koala (*Phascolarctos cinereus*) Faecal Microbiome Differs with Diet in a Wild Population," *PeerJ* 7:e6534, https://doi.org/10.7717/peerj.6534.

10. Rothschild, D., et al., 2018, "Environment Dominates over Host Genetics in Shaping Human Gut Microbiota," *Nature* 555:210–15, https://doi.org/10.1038/nature25973.

11. Holscher, H. D., 2017, "Dietary Fiber and Prebiotics and the Gastrointestinal Microbiota," *Gut Microbes* 8:172–84, https://doi.org/10.1080/19490976.2017.1290756; Suez, J., et al., 2014, "Artificial Sweeteners Induce Glucose Intolerance by Altering the Gut Microbiota," *Nature* 514:181–86, https://doi.org/10.1038/nature13793.

12. Reveles, K. R., Patel, S., Forney, L., and Ross, C. N., 2019, "Age-Related Changes in the Marmoset Gut Microbiome," *American Journal of Primatology* 81:e22960, https://doi.org/10.1002/ajp.22960; Scott, K. A., et al., 2017, "Revisiting Metchnikoff: Age-Related Alterations in Microbiota-Gut-Brain Axis in the Mouse," *Brain, Behavior, and Immunity* 65:20–32, https://doi.org/10.1016/j.bbi.2017.02.004; Langille, M. G., et al., 2014, "Microbial Shifts in the Aging Mouse Gut," *Microbiome* 2:50, https://doi.org/10.1186/s40168-014-0050-9.

13. Bárcena et al. 2019.

14. Luan, Z., et al., 2020, "Metagenomics Study Reveals Changes in Gut Microbiota in Centenarians: A Cohort Study of Hainan Centenarians," *Frontiers in Microbiology* 11:1474, https://doi.org/10.3389/fmicb.2020.01474.

15. Hughes, G. M., Leech, J., Puechmaille, S. J., Lopez, J. V., and Teeling, E. C., 2018, "Is There a Link between Aging and Microbiome Diversity in Exceptional Mammalian Longevity?," *PeerJ* 6:e4174, https://doi.org/10.7717/peerj.4174.

16. Anderson et al. 2018.

17. Wu, L., et al., 2019, "A Cross-Sectional Study of Compositional and Functional Profiles of Gut Microbiota in Sardinian Centenarians," *mSystems* 4:e00325-19, https://doi.org/10.1128/mSystems.00325-19.

18. Biagi, E., et al., 2010, "Through Ageing, and Beyond: Gut Microbiota and Inflammatory Status in Seniors and Centenarians," *PLOS One* 5:e10667, https://doi.org/10.1371/journal.pone.0010667.

19. Luan, Z., et al., 2020, "Metagenomics Study Reveals Changes in Gut Microbiota in Centenarians: A Cohort Study of Hainan Centenarians," *Frontiers in Microbiology* 11:1474, https://doi.org/10.3389/fmicb.2020.01474.

20. Ridaura, V. K., et al., 2013, "Gut Microbiota from Twins Discordant for Obesity Modulate Metabolism in Mice," *Science* 341:1241214, https://doi.org/10.1126/science.1241214.

21. D'Amato, A., et al., 2020, "Faecal Microbiota Transplant from Aged Donor Mice Affects Spatial Learning and Memory via Modulating Hippocampal Synaptic Plasticity- and Neurotransmission-Related Proteins in Young Recipients," *Microbiome* 8:140, https://doi.org/10.1186/s40168-020-00914-w.

22. Kundu, P., et al., 2019, "Neurogenesis and Prolongevity Signaling in Young Germ-Free Mice Transplanted with the Gut Microbiota of Old Mice," *Science Translational Medicine* 11:eaau4760, https://doi.org/10.1126/scitranslmed.aau4760.

23. Sugimoto, Y., et al., 2019, "A Metagenomic Strategy for Harnessing the Chemical Repertoire of the Human Microbiome," *Science* 366:eaax9176, https://doi.org/10.1126/science.aax9176.

24. Bárcena et al 2019.

25. Smith, P., et al., 2017, "Regulation of Life Span by the Gut Microbiota in the Short-Lived African Turquoise Killifish," *eLife* 6:e27014, https://doi.org/10.7554/eLife.27014.

26. Boehme, M., et al., 2021, "Microbiota from Young Mice Counteracts Selective Age-Associated Behavioral Deficits," *Nature Aging* 1:666–76, https://doi.org/10.1038/s43587-021-00093-9.

27. Bonaz, B., Bazin, T., and Pellissier, S., 2018, "The Vagus Nerve at the Interface of the Microbiota-Gut-Brain Axis," *Frontiers in Neuroscience* 12:49, https://doi.org/10.3389/fnins.2018.00049; Cryan, J. F., et al., 2019, "The Microbiota-Gut-Brain Axis," *Physiological Reviews* 99:1877–2013, https://doi.org/10.1152/physrev.00018.2018.

28. Tran, T.T.T., et al., 2019, "APOE Genotype Influences the Gut Microbiome Structure and Function in Humans and Mice: Relevance for Alzheimer's Disease Pathophysiology," *FASEB Journal* 33:8221–31, https://doi.org/10.1096/fj.201900071R.

29. Braak, H., Rüb, U., Gai, W. P., and Del Tredici, K., 2003, "Idiopathic Parkinson's Disease: Possible Routes by which Vulnerable Neuronal Types May Be Subject to Neuroinvasion by an Unknown Pathogen," *Journal of Neural Transmission* 110:517–36, https://doi.org/10.1007/s00702-002-0808-2.

30. Sampson, T. R., et al., 2016, "Gut Microbiota Regulate Motor Deficits and Neuroinflammation in a Model of Parkinson's Disease," *Cell* 167:1469–80.e12, https://doi.org/10.1016/j.cell.2016.11.018.

31. Yap et al. 2021.

32. Pellanda, P., Ghosh, T. S., and O'Toole, P. W., 2021, "Understanding the Impact of Age-Related Changes in the Gut Microbiome on Chronic Diseases and the Prospect of Elderly-Specific Dietary Interventions," *Current Opinion in Biotechnology* 70:48–55, https://doi.org/10.1016/j.copbio.2020.11.001.

33. Van de Wouw, M., et al., 2021, "Kefir Ameliorates Specific Microbiota-Gut-Brain Axis Impairments in a Mouse Model Relevant to Autism Spectrum Disorder," *Brain, Behavior, and Immunity* 97:119–34, https://doi.org/10.1016/j.bbi.2021.07.004.

34. Buettner, D., and Skemp, S., 2016, "Blue Zones: Lessons From the World's Longest Lived," *American Journal of Lifestyle Medicine* 10:318–21, https://doi.org/10.1177/1559827616637066.

35. Pellanda, Ghosh, and O'Toole 2021.

36. Ghosh, T. S., et al., 2020, "Mediterranean Diet Intervention Alters the Gut Microbiome in Older People Reducing Frailty and Improving Health Status: The NU-AGE 1-Year Dietary Intervention across Five European Countries," *Gut* 69:1218–28, https://doi.org/10.1136/gutjnl-2019-319654.

37. Rangan, P., et al., 2019, "Fasting-Mimicking Diet Modulates Microbiota and Promotes Intestinal Regeneration to Reduce Inflammatory Bowel Disease Pathology," *Cell Reports* 26:2704–19.e6, https://doi.org/10.1016/j.celrep.2019.02.019.

Notes to text box (Chapter 16, page 306)

a. Garigan, D., et al., 2002, "Genetic Analysis of Tissue Aging in *Caenorhabditis elegans*: A Role for Heat-Shock Factor and Bacterial Proliferation," *Genetics* 161:1101–12.

b. Kauffman, A. L., Ashraf, J. M., Corces-Zimmerman, M. R., Landis, J. N., and Murphy, C. T., 2010, "Insulin Signaling and Dietary Restriction Differentially Influence the Decline of Learning and Memory with Age," *PLOS Biology* 8:e1000372, https://doi.org/10.1371/journal.pbio.1000372; MacNeil, L. T., Watson, E., Arda, H. E., Zhu, L. J., and Walhout, A. J., 2013, "Diet-Induced Developmental Acceleration Independent of TOR and Insulin in *C. elegans*," *Cell* 153:240–52, https://doi.org/10.1016/j.cell.2013.02.049.

c. Donato, V., et al., 2017, "*Bacillus subtilis* Biofilm Extends *Caenorhabditis elegans* Longevity through Downregulation of the Insulin-Like Signalling Pathway," *Nature Communications* 8:14332, https://doi.org/10.1038/ncomms14332.

d. Hartsough, L. A., et al., 2020, "Optogenetic Control of Gut Bacterial Metabolism to Promote Longevity," *eLife* 9, https://doi.org/10.7554/eLife.56849; Han, B., et al., 2017, "Microbial Genetic Composition Tunes Host Longevity," *Cell* 169:1249–62.e13, https://doi.org/10.1016/j.cell.2017.05.036.

e. Cabreiro, F., et al., 2013, "Metformin Retards Aging in *C. elegans* by Altering Microbial Folate and Methionine Metabolism," *Cell* 153:228–39, https://doi.org/10.1016/j.cell.2013.02.035.

f. Montalvo-Katz, S., Huang, H., Appel, M. D., Berg, M., and Shapira, M., 2013, "Association with Soil Bacteria Enhances p38-Dependent Infection Resistance in Caenorhabditis elegans," *Infection and Immunity* 81:514–20, https://doi.org/10.1128/iai.00653–12.

Chapter 17

1. Isaak, A., 2020, "The Ultra-rich Are Investing in Companies Trying to Reverse Aging. Is It Going to Work?," CNBC, February 23, 2020, https://www.cnbc.com/2020/02/19/the-ultra-rich-are-investing-in-companies-trying-to-reverse-aging.html.

2. Nanalyze, 2019, "The Top 10 Companies Working to Increase Longevity," August 17, 2019, https://www.nanalyze.com/2019/08/top-10-companies-longevity/.

3. Zimmermann, A., Madreiter-Sokolowski, C., Stryeck, S., and Abdellatif, M., 2021, "Targeting the Mitochondria-Proteostasis Axis to Delay Aging," *Frontiers in Cell and Developmental Biology* 9:656201, https://doi.org/10.3389/fcell.2021.656201.

4. Ristow, M., et al., 2009, "Antioxidants Prevent Health-Promoting Effects of Physical Exercise in Humans," *Proceedings of the National Academy of Sciences of the USA* 106:8665–70, https://doi.org/10.1073/pnas.0903485106.

5. Cross, R., 2019, "Autophagy: Drugging the Yin and Yang of the Cell," *C&EN*, June 2, 2019, https://cen.acs.org/business/start-ups/Autophagy-Drugging-yin-yang-cell/97/i22.

6. Rabouw, H. H., et al., 2019, "Small Molecule ISRIB Suppresses the Integrated Stress Response within a Defined Window of Activation," *Proceedings of the National Academy of Sciences of the USA* 116:2097–2102, https://doi.org/10.1073/pnas.1815767116.

7. Chou, A., et al., 2017, "Inhibition of the Integrated Stress Response Reverses Cognitive Deficits after Traumatic Brain Injury," *Proceedings of the National Academy of Sciences of the USA* 114:E6420–26, https://doi.org/10.1073/pnas.1707661114; Krukowski, K., et al., 2020, "Small Molecule Cognitive Enhancer Reverses Age-Related Memory Decline in Mice," *eLife* 9:e62048, https://doi.org/10.7554/eLife.62048.

8. Krukowski et al. 2020.

9. Onken, B., et al., 2020, "Metformin Treatment of Diverse *Caenorhabditis* Species Reveals the Importance of Genetic Background in Longevity and Healthspan Extension Outcomes. *Aging Cell* 21:e13488, https://doi.org/10.1111/acel.13488; Seliger, S. L., et al., 2020, "Baseline Characteristics and Patient-Reported Outcomes of ADPKD Patients in the Multicenter TAME-PKD Clinical Trial," *Kidney360* 1:1363–72, https://doi.org/10.34067/KID.0004002020.

10. Aliper, A., et al., 2017, "Towards Natural Mimetics of Metformin and Rapamycin. *Aging (Albany NY)* 9:2245–68, https://doi.org/10.18632/aging.101319.

11. US National Library of Medicine, 2022, "Participatory Evaluation (of) Aging (with) Rapamycin (for) Longevity Study (PEARL)," ClinicalTrials.gov, id: NCT04488601, last updated October 4, 2022, https://clinicaltrials.gov/ct2/show/NCT04488601.

12. Curran, S. P., and Ruvkun, G., 2007, "Lifespan Regulation by Evolutionarily Conserved Genes Essential for Viability," *PLOS Genetics* 3:e56, https://doi.org/10.1371/journal.pgen.0030056; Tacutu, R., et al., 2012, "Prediction of *C. elegans* Longevity Genes by Human and Worm Longevity Networks," *PLOS One* 7:e48282, https://doi.org/10.1371/journal.pone.0048282.

13. Harrison, D. E., et al., 2009, "Rapamycin Fed Late in Life Extends Lifespan in Genetically Heterogeneous Mice," *Nature* 460:392–95, https://doi.org/10.1038/nature08221.

14. Dillin, A., Crawford, D. K., and Kenyon, C., 2002, "Timing Requirements for Insulin/IGF-1 Signaling in *C. elegans*," *Science* 298:830–34.

15. Venz, R., Pekec, T., Katic, I., Ciosk, R., and Ewald, C. Y., 2021, "End-of-Life Targeted Auxin-Mediated Degradation of DAF-2 Insulin/IGF-1 Receptor Promotes Longevity Free from Growth-Related Pathologies," *eLife* 10:e71335, https://doi.org/10.7554/eLife.71335.

16. Hedges, C. P., Shetty, B., Broome, S. C., MacRae, C., Koutsifeli, P., Buckels, E. J., et al., 2023, "Dietary Supplementation of Clinically Utilized PI3K p110α Inhibitor Extends the Lifespan of Male and Female Mice," *Nature Aging*, https://doi.org/10.1038/s43587-022-00349-y.

17. Fan, W., and Evans, R. M., 2017, "Exercise Mimetics: Impact on Health and Performance," *Cell Metabolism* 25:242–47, https://doi.org/10.1016/j.cmet.2016.10.022.

18. Guerrieri, D., Moon, H. Y., and van Praag, H., 2017, "Exercise in a Pill: The Latest on Exercise-Mimetics," *Brain Plasticity* 2:153–69, https://doi.org/10.3233/BPL-160043.

19. Merrill, G. F., Kurth, E. J., Hardie, D. G., and Winder, W. W., 1997, "AICA Riboside Increases AMP-Activated Protein Kinase, Fatty Acid Oxidation, and Glucose Uptake in Rat Muscle," *American Journal of Physiology* 273:E1107–12, https://doi.org/10.1152/ajpendo.1997.273.6.E1107; Asby, D. J., et al., 2015, "AMPK Activation via Modulation of De Novo Purine Biosynthesis with an Inhibitor of ATIC Homodimerization," *Chemistry and Biology* 22:838–48, https://doi.org/10.1016/j.chembiol.2015.06.008.

20. Macmichael, S. 2012, "Spanish Police Arrest Ten as They Break up 'Next Generation Superdrug' Doping Ring in Operacion Skype," Road.cc, March 20, 2012, https://road.cc/content/news/55184-spanish-police-arrest-ten-they-break-next-generation-superdrug-doping-ring.

21. Asby et al. 2015; Associated Press, 2014, "Doping Probe Launched into Russian Walkers," *ESPN*, July 11, 2014, http://www.espn.com/espn/wire?section=trackandfield&id=11201932; *Canadian Cycling Magazine*, 2012, "New Doping Scandal Hits Cycling and Athletics," March 22, 2012, https://cyclingmagazine.ca/sections/news/new-doping-scandal-hits-cycling-and-athletics/.

22. Ryu, D., et al., 2016, "Urolithin A Induces Mitophagy and Prolongs Lifespan in *C. elegans* and Increases Muscle Function in Rodents," *Nature Medicine* 22:879–88, https://doi.org/10.1038/nm.4132.

23. Germain, D. J., Leavey, D. C., Van Hout, P.M.C., and McVeigh, P. J., 2021, "2,4 Dinitrophenol: It's Not Just for Men," *International Journal of Drug Policy* 95:102987, https://doi.org/10.1016/j.drugpo.2020.102987; Petróczi, A., et al., 2015, "Russian Roulette with Unlicensed Fat-Burner Drug 2,4-dinitrophenol (DNP): Evidence from a Multidisciplinary Study of the Internet, Bodybuilding Supplements and DNP Users," *Substance Abuse Treatment, Prevention, and Policy* 10:39, https://doi.org/10.1186/s13011-015-0034-1.

24. Hall, S. S., 2004, "Kenyon's Ageless Quest," *Smithsonian Magazine*, March 2004, https://www.smithsonianmag.com/science-nature/kenyons-ageless-quest-103525532/; Staats, S., et al., 2018, "Dietary Resveratrol Does Not Affect Life Span, Body Composition, Stress Response, and Longevity-Related Gene Expression in *Drosophila melanogaster*," *International Journal of Molecular Sciences* 19:223, https://doi.org/10.3390/ijms19010223.

25. Wood, J. G., et al., 2004, "Sirtuin Activators Mimic Caloric Restriction and Delay Ageing in Metazoans," *Nature* 430:686–89, https://doi.org/10.1038/nature02789; Valenzano, D. R., et al., 2006, "Resveratrol Prolongs Lifespan and Retards the Onset of Age-Related Markers in a Short-Lived Vertebrate," *Current Biology* 16:296–300, https://doi.org/10.1016/j.cub.2005.12.038; Baur, J. A., et al., 2006, "Resveratrol Improves Health and Survival of Mice on a

High-Calorie Diet," *Nature* 444:337–42, https://doi.org/10.1038/nature05354; Fischer, N., et al., 2017, "The Resveratrol Derivatives Trans-3,5-dimethoxy-4-fluoro-4'-hydroxystilbene and Trans-2,4',5-trihydroxystilbene Decrease Oxidative Stress and Prolong Lifespan in *Caenorhabditis elegans*," *Journal of Pharmacy and Pharmacology* 69:73–81, https://doi.org/10.1111/jphp .12657.

26. Hubbard, B. P., et al., 2013, "Evidence for a Common Mechanism of SIRT1 Regulation by Allosteric Activators," *Science* 339:1216–19, https://doi.org/10.1126/science.1231097.

27. Bass, T. M., Weinkove, D., Houthoofd, K., Gems, D., and Partridge, L., 2007, "Effects of Resveratrol on Lifespan in *Drosophila melanogaster* and *Caenorhabditis elegans*," *Mechanisms of Ageing and Development* 128:546–52, https://doi.org/10.1016/j.mad.2007.07.007; Burnett, C., et al., 2011, "Absence of Effects of Sir2 Overexpression on Lifespan in *C. elegans* and *Drosophila*," *Nature* 477:482–85, https://doi.org/10.1038/nature10296.

28. Bieganowski, P., and Brenner, C., 2004, "Discoveries of Nicotinamide Riboside as a Nutrient and Conserved *NRK* Genes Establish a Preiss-Handler Independent Route to NAD$^+$ in Fungi and Humans," *Cell* 117:495–502, https://doi.org/10.1016/S0092-8674(04)00416-7.

29. Timmerman, L., 2013, "GlaxoSmithKline Shuts Down Sirtris, Five Years after $720M Buyout," *Xconomy*, March 12, 2013, https://xconomy.com/boston/2013/03/12/glaxosmithkline -shuts-down-sirtris-five-years-after-720m-buyout/.

30. Extra Credits, 2018, "Pellagra—a Medical Mystery—Extra History," YouTube, July 26, 2018, https://www.youtube.com/watch?v=reYKBgdrZsM.

31. Taber, S (@SarahTaber_bww), 2021, "The pellagra epidemic came down to the basic fabric of society," Twitter, July 10, 2021, 2:28 a.m., https://twitter.com/SarahTaber_bww/status /1413671334853423107.

32. Imai, S., Armstrong, C. M., Kaeberlein, M., and Guarente, L., 2000, "Transcriptional Silencing and Longevity Protein Sir2 is an NAD-Dependent Histone Deacetylase," *Nature* 403:795–800, https://doi.org/10.1038/35001622.

33. Palmer, R. D., Elnashar, M. M., and Vaccarezza, M., 2021, "Precursor Comparisons for the Upregulation of Nicotinamide Adenine Dinucleotide: Novel Approaches for Better Aging," *Aging Medicine* 4:214–20, https://doi.org/10.1002/agm2.12170; Life Biosciences, n.d., "Jumpstart Fertility—Reversing Reproductive Decline," Lifespan.io: Crowdsourcing the Cure for Aging, accessed December 21, 2022, https://www.lifespan.io/road-maps/the-rejuvenation -roadmap/jumpstart-fertility-dna-repair.

34. Horowitz, A. M., et al., 2020, "Blood Factors Transfer Beneficial Effects of Exercise on Neurogenesis and Cognition to the Aged Brain," *Science* 369:167–73, https://doi.org/10.1126 /science.aaw2622.

35. Coppé, J. P., Desprez, P. Y., Krtolica, A., and Campisi, J., 2010, "The Senescence-Associated Secretory Phenotype: The Dark Side of Tumor Suppression," *Annual Review of Pathology* 5:99–118, https://doi.org/10.1146/annurev-pathol-121808-102144.

36. Kim, E.-C., and Kim, J.-R., 2019, "Senotherapeutics: Emerging Strategy for Healthy Aging and Age-Related Disease," *BMB Reports* 52:47–55, https://doi.org/10.5483/BMBRep .2019.52.1.293.

37. Atzmon, G., et al., 2010, "Genetic Variation in Human Telomerase Is Associated with Telomere Length in Ashkenazi Centenarians," *Proceedings of the National Academy of Sciences of the USA* 107:1710–17, https://doi.org/10.1073/pnas.0906191106.

38. Zhang, C., et al., 2018, "KMT2A Promotes Melanoma Cell Growth by Targeting hTERT Signaling Pathway," *Cell Death and Disease* 8:e2940, https://doi.org/10.1038/cddis .2017.285; Reyes-Uribe, P., et al., 2018, "Exploiting TERT Dependency as a Therapeutic Strategy for NRAS-Mutant Melanoma," *Oncogene* 37:4058–72, https://doi.org/10.1038/s41388-018 -0247-7.

39. Swain, F., 2019, "Buyer Beware of This $1 Million Gene Therapy for Aging," *MIT Technology Review*, December 6, 2019, https://www.technologyreview.com/2019/12/06/131657 /buyer-beware-of-this-1-million-gene-therapy-for-aging/; Mullin, E., 2019, "Scientists Dodge FDA to Offer a $1 Million Anti-aging Treatment in Colombia," *OneZero*, December 5, 2019, https://onezero.medium.com/scientists-dodge-fda-to-offer-a-1-million-anti-aging-treatment -in-colombia-38756dfb3ad1.

40. Lewis, R., 2019, "Gene Therapy Update: Remembering Jesse Gelsinger," *DNA Science* (blog), September 26, 2019, https://dnascience.plos.org/2019/09/26/gene-therapy-update -remembering-jesse-gelsinger/.

41. *Journal of Clinical Investigation*, 2008, "Why Gene Therapy Caused Leukemia in Some 'Boy in the Bubble Syndrome' Patients," *ScienceDaily*, August 10, 2008, https://www .sciencedaily.com/releases/2008/08/080807175438.htm.

42. US National Library of Medicine, 2022, "Gene Therapy for APOE4 Homozygote of Alzheimer's Disease," ClinicalTrials.gov, id: NCT03634007, last updated December 2, 2022, https://clinicaltrials.gov/ct2/show/NCT03634007.

43. US National Library of Medicine, 2022, "Double-Blind, Multicenter, Sham Surgery Controlled Study of CERE-120 in Subjects with Idiopathic Parkinson's Disease," ClinicalTrials .gov, id: NCT00400634, last updated November 10, 2022, https://clinicaltrials.gov/ct2/show /NCT00400634.

44. Jiang, T., et al., 2020, "Chemical Modifications of Adenine Base Editor mRNA and Guide RNA Expand Its Application Scope," *Nature Communications* 11:1979, https://doi.org /10.1038/s41467-020-15892-8.

45. Koblan, L. W., et al., 2021, "In Vivo Base Editing Rescues Hutchinson-Gilford Progeria Syndrome in Mice," *Nature* 589:608–14, https://doi.org/10.1038/s41586-020-03086-7.

46. Kim, J., et al., 2019, "Patient-Customized Oligonucleotide Therapy for a Rare Genetic Disease," *New England Journal of Medicine* 381:1644–52, https://doi.org/10.1056/NEJMoa 1813279.

47. Lu, Y., et al., 2020, "Reprogramming to Recover Youthful Epigenetic Information and Restore Vision," *Nature* 588:124–29, https://doi.org/10.1038/s41586-020-2975-4.

48. Barardo, D., et al., 2017, "The DrugAge Database of Aging-Related Drugs," *Aging Cell* 16:594–97, https://doi.org/10.1111/acel.12585; "DrugAge: The Database of Ageing-Related Drugs," Human Ageing Genomic Resources, genomics.senescence.info/drugs/.

49. Miller, R. A., et al., 2007, "An Aging Interventions Testing Program: Study Design and Interim Report," *Aging Cell* 6:565–75, https://doi.org/10.1111/j.1474-9726.2007.00311.x; Banse, S. A., et al., 2019, "Automated Lifespan Determination across *Caenorhabditis* Strains and Species Reveals Assay-Specific Effects of Chemical Interventions," *Geroscience* 41:945–60, https://doi.org/10.1007/s11357-019-00108-9.

50. Strong, R., et al., 2016, "Longer Lifespan in Male Mice Treated with a Weakly Estrogenic Agonist, an Antioxidant, an α-Glucosidase Inhibitor or a Nrf2-Inducer," *Aging Cell* 15:872–84, https://doi.org/10.1111/acel.12496.

51. Janssens, G. E., and Houtkooper, R. H., 2020, "Identification of Longevity Compounds with Minimized Probabilities of Side Effects," *Biogerontology* 21:709–19, https://doi.org/10.1007/s10522-020-09887-7; Wang, Z., Clark, N. R., and Ma'ayan, A., 2016, "Drug Induced Adverse Events Prediction with the LINCS L1000 Data Bioinformatics," https://doi.org/10.1093/bioinformatics/btw168, https://maayanlab.net/SEP-L1000/; Wang, Z., et al., 2016, "Extraction and Analysis of Signatures from the Gene Expression Omnibus by the Crowd," *Nature Communications* 7:12846, https://doi.org/10.1038/ncomms12846.

52. Bioage, n.d., "We Are Mapping Human Aging to Develop a Pipeline of Therapies that Treat Disease and Extend Healthy Lifespan," accessed December 21, 2022, https://bioagelabs.com/?fbclid=IwAR32Tjgt2bgJ-uphqvNtPPjd-7QmQ-tzn0sJctTttNphuFTzuQ0gkd9bONM#pipeline.

53. Reynolds, J. C., et al., 2021, "MOTS-c Is an Exercise-Induced Mitochondrial-Encoded Regulator of Age-Dependent Physical Decline and Muscle Homeostasis," *Nature Communications* 12:470, https://doi.org/10.1038/s41467-020-20790-0.

54. Bobrov, E., et al., 2018, "PhotoAgeClock: Deep Learning Algorithms for Development of Non-invasive Visual Biomarkers of Aging," *Aging (Albany NY)* 10:3249–59, https://doi.org/10.18632/aging.101629; Chen, W., et al., 2015, "Three-Dimensional Human Facial Morphologies as Robust Aging Markers," *Cell Research* 25:574–87, https://doi.org/10.1038/cr.2015.36; Xia, X., et al., 2020, "Three-Dimensional Facial-Image Analysis to Predict Heterogeneity of the Human Ageing Rate and the Impact of Lifestyle," *Nature Metabolism* 2:946–57, https://doi.org/10.1038/s42255-020-00270-x.

55. Bobrov et al. 2018.

56. Buhr, S., 2017, "A New Lawsuit Alleges Anti-aging Startup Elysium Health Hasn't Paid Its Sole Supplier," *TechCrunch*, January 6, 2017, https://techcrunch.com/2017/01/16/a-new-lawsuit-alleges-anti-aging-startup-elysium-health-hasnt-paid-its-supplier-and-is-in-breach-of-agreement/.

57. Wilfond, B. S., Porter, K. M., Creevy, K. E., Kaeberlein, M., and Promislow, D., 2018, "Research to Promote Longevity and Health Span in Companion Dogs: A Pediatric Perspective," *American Journal of Bioethics* 18:64–65, https://doi.org/10.1080/15265161.2018.1513591.

58. Kulkarni, A. S., et al., 2018, "Metformin Regulates Metabolic and Nonmetabolic Pathways in Skeletal Muscle and Subcutaneous Adipose Tissues of Older Adults," *Aging Cell* 17:e12723, https://doi.org/10.1111/acel.12723.

59. Lee, C. G., et al., 2021, "Effect of Metformin and Lifestyle Interventions on Mortality in the Diabetes Prevention Program and Diabetes Prevention Program Outcomes Study," *Diabetes Care* 44:2775–82, https://doi.org/10.2337/dc21-1046.

60. Konopka, A. R., et al., 2018, "Metformin Inhibits Mitochondrial Adaptations to Aerobic Exercise Training in Older Adults," *Aging Cell* 18:e12880, https://doi.org/10.1111/acel.12880; Cai, T., et al., 2021, "Effect of Metformin on Testosterone Levels in Male Patients with Type 2 Diabetes Mellitus Treated with Insulin," *Frontiers in Endocrinology* 12:813067, https://doi.org/10.3389/fendo.2021.813067.

61. Wilfond et al. 2018; Kaeberlein, M., Creevy, K. E., and Promislow, D. E., 2016, "The Dog Aging Project: Translational Geroscience in Companion Animals," *Mammalian Genome* 27:279–88, https://doi.org/10.1007/s00335-016-9638-7.

62. Cotman, C. W., and Head, E., 2008, "The Canine (Dog) Model of Human Aging and Disease: Dietary, Environmental and Immunotherapy Approaches," *Journal of Alzheimer's*

Disease 15:685–707, https://doi.org/10.3233/jad-2008-15413; Dewey, C. W., Davies, E. S., Xie, H., and Wakshlag, J. J., 2019, "Canine Cognitive Dysfunction: Pathophysiology, Diagnosis, and Treatment," *Veterinary Clinics of North America: Small Animal Practice* 49:477–99, https://doi.org/10.1016/j.cvsm.2019.01.013.

63. Bodenham, D., 2022, "The Pill that Could Extend the Lifespan of Pet Dogs by Years," BoughtByMany [now ManyPets], April 5, 2022, https://boughtbymany.com/news/article/pill-could-extend-life-of-pet-dogs/.

64. Weintraub, A., 2019, "Gene Therapy to Fend off Aging? Buzzy Harvard Startup Rejuvenate Bio Says It Works in Mice," Fierce Biotech, November 4, 2019, https://www.fiercebiotech.com/research/gene-therapy-to-fend-off-aging-buzzy-harvard-startup-rejuvenate-bio-says-it-works-mice.

65. Warner, H., et al., 2005, "Science Fact and the SENS Agenda," *EMBO Reports* 6:1006–8, https://doi.org/ 10.1038/sj.embor.7400555.

66. Cox, H., 2017, "Aubrey de Grey: Scientist Who Says Humans Can Live for 1,000 Years," *Financial Times*, February 8, 2017, https://www.ft.com/content/238cc916-e935-11e6-967b-c88452263daf.

67. Estep, P. W., Kaeberlein, M., Kapahi, P., Kennedy, B. K., Lithgow, G. J., Martin, G. M., Melov, S., Powers, R. W., and Tissenbaum, H. A., 2006, "Life Extension Pseudoscience and the SENS Plan," *MIT Technology Review*, https://www2.technologyreview.com/sens/docs/estepetal.pdf.

Notes to text box (Chapter 17, page 330)

a. Rizki, G., et al., 2011, "The Evolutionarily Conserved Longevity Determinants HCF-1 and SIR-2.1/SIRT1 Collaborate to Regulate DAF-16/FOXO," *PLOS Genetics* 7:e1002235, https://doi.org/10.1371/journal.pgen.1002235.

b. Burnett, C., et al., 2011, "Absence of Effects of Sir2 Overexpression on Lifespan in *C. elegans* and *Drosophila*," *Nature* 477:482–85, https://doi.org/10.1038/nature10296.

Note to text box (Chapter 17, page 335)

a. Medeiros, J., 2017, "Ageing Is a Disease. Gene Therapy Could Be the 'Cure,'" *Wired*, March 23, 2017, https://www.wired.co.uk/article/elizabeth-parrish-bioviva-gene-therapy.

INDEX

Page number in *italics* refer to figures.

A NOTE ON THE TYPE

This book has been composed in Arno, an Old-style serif typeface in the classic Venetian tradition, designed by Robert Slimbach at Adobe.